Humanity and Environment:

A Cultural Ecology

I. G. Simmons

University of Durham

LONGMAN

Addison Wesley Longman Limited
Edinburgh Gate, Harlow
Essex CM20 2JE
England

and Associated Companies throughout the World

First published 1997

ISBN 0 582 22547 7

British Library Cataloguing-in-Publication Data
A catalogue record for this book is available from the British Library.

Library of Congress Cataloging-in-Publication Data
A catalog entry for this title is available from the Library of Congress.

Set by 32 in Palatino 10/12pt

Produced through Longman Malaysia, TCP

Contents

Preface

These days, most books from mainstream publishers are texts aimed at students on particular courses. I am very glad that Addison Wesley Longman (initially the idea came from Vanessa Lawrence) has allowed me to write this book, which might, we hope, provoke some courses into being but is unlikely to cater to many already in existence. It is, however, written at a level of knowledge that is intended to place it within the grasp of undergraduates at an early stage of their courses. If its breadth (a risky enterprise, this) enables some of my colleagues to say, 'yes, this is the kind of thing geographers should be aiming at', then I will be very pleased indeed. However, I will wait just a little longer for the cheering to break out: in the case of my earlier books, most have been reviewed much more warmly outside the subject than within it.

It does represent what I think geography can now see as a medium-term ambition, namely that a great breadth of subject-matter and a variety of approaches (frameworks, discourses) are valid and that none is by definition outside the boundaries of our subject. To bring them into one focus is difficult and I have no doubt that a later generation will do it much better than this attempt; however, I wanted to show that developments in geography since the 1970s have had some effect. There is then another agenda here: to encourage my professional colleagues not to give up on the idea of geography as an integrating discipline. Moreover, writing about many of the topics is a great pleasure: science-based environmental history, environmental impact studies and environment-related poetry do not seem very far apart to me. That has, among other things, been one of the ways in which geography has become immeasurably more exciting during my time in it and why I regret none of the epistemological developments that have taken place.

So as retirement approaches, it seemed appropriate to put together material from various parts of my intellectual life in what seems to me to be largely a body of knowledge without any meaningful divisions. For everyday purposes, the modules rule (if only the students knew just how short-changed they are by this compartmentalization), but in the field or looking from the aircraft window, the ecology of the landscape is mostly one and indivisible. But books have to be written in sentences, one after the other, and not even the most gung-ho IT staff development seminar has got round that yet. Even CD-ROM and the WWW are, basically, laid out that way even though skipping is easy. We need a James Joyce perhaps as much as another dozen program compilers. However, one result is that a particular topic sometimes appears more than once: this is deliberate, for I want some things to be seen from more than one angle or in more than one context. I hope, then, that this book will be useful in contexts that might be labelled environmental studies or cultural studies as well as geography and its variants.

As always, thanks are due. My colleagues at Durham have contributed enormously to this book simply by being themselves: progressive, unselfish, concerned for their subject, and good fun. I do not think I have ever heard one of them disparage another because of the kind of work they did, and I take that to be a matter for celebration. So thank you, #academic. We have a wonderful group of support staff, too: the tea is adequately strong, the maps clear and the photographs much enhanced from our amateur negatives. Even large amounts of photocopying get done without complaint and our rooms are clean first thing in the morning. For this book, I have had a great deal of help in the later stages from Vicki Innes, who gophered the linework illustrations when I was getting very badly behind due to an obstructive neck injury. Her rapid appreciation of what was needed and could be quickly got has been very valuable indeed.

I have rather moved away from formal dedications of books but the spirit of this one brings my late Aunt Grace (Davis) very much to mind. She was an enthusiast for nature, being a long-standing member of the then Lincolnshire Naturalists' Trust, for example. We sang 'All Things Bright and Beautiful' at her funeral without any trace of irony. She was also an enthusiast for culture: she bought me an Arena season ticket to the Proms in the summer of 1955, for which I am still grateful on pretty well a daily basis. So this book keeps her memory alive.

I. G. Simmons
Durham, December 1995

Acknowledgements

We are grateful to the following for permission to reproduce copyright material:

Figure 1.3 from Intermediate Technology Publications, *Women and the Transport of Water* (Curtis, 1986); Figure 1.4 courtesy of British Airways Holidays, *British Airways Holidays 1996*, 1/e; Figures 1.5, 1.8, 1.9, 1.12 and 3.4 from Routledge, *Global Ecology, Environmental Change and Social Flexibility* (Smil, 1993); Figures 1.6 and 6.4 from Durham County Council, *County Durham Waste Disposal Plan 1984*, 1/e; Figure 1.10 from Oxford University Press, *World without End: Economics, Environment and Sustainable Development*, 1/e (Pearce and Warford, 1993); Figure 1.11 from *Man, Energy, Society* by Cook. Copyright © 1976 by W. H. Freeman and Company. Used with permission; Figure 2.2 reproduced by permission of Oxford University Press, *Eye and Brain: The Psychology of Seeing*, 2/e (Gregory, 1972); Figure 2.4 adapted by P. Williamson for IGBP, *Global Change: Reducing Uncertainties*; Figure 2.5 from Edward Arnold, *The Gobal Casino. An Introduction to Environmental Issues* (Middleton, 1995); Figures 2.6 and 2.9 from Edward Arnold, *Quaternary Environments* (Williams *et al.*, 1993); Figure 2.8 from Springer-Verlag GmbH & Co. KG, *Sea Levels, Land Levels and Tide Gauges* (Emergy and Aubrey, 1991); Figures 2.10, 3.5 and 3.6 from Routledge, *An Introduction to Global Environmental Issues* (Pickering and Owen, 1994); Figure 2.11 from Edward Arnold, *Biogeography: Natural and Cultural* (Simmons, 1979); Figures 2.12–2.14 and Tables 2.5–2.7 from Academic Press, *Global Biogeochemical Cycles* (Butcher *et al.*, 1992); Figure 3.1 from Addison Wesley Longman, *The Food Resource* (Pierce, 1990); Figure 3.2 from Simon and Schuster, Myriad Editions Limited, *The New State of the World* (M. Kidron and R. Segal, 1991); Figure 3.3 from Food and Agriculture organization of the United Nations, *Wood for Energy Forest Topics*, Report No. 1, Rome; Figure 3.7 and Table 1.2 from World Resources Institute, *World Resources 1994–95*; Figure 3.8 from Edward Arnold, *Earth, Air and Water* (Simmons, 1991); Figures 4.1, 4.2, 4.4, 4.11 and 4.12 from Blackwell Publishers, *Changing the Face of the Earth* (Simmons, 1989); Figure 4.6 from The White Horse Press, 'Man against the sea', *Environmental and History* **1**, 7–8 (Elvin and Ninghu, 1995); Figure 4.7 from Professor M. J. Tooley, *The Gardens of Gertrude Jekyll in the North of England* (Tooley and Tooley, 1982); Figure 4.8 reprinted with permission of John Wiley &

Sons, Inc., *General Energetics: Energy in the Biosphere and Civilization*, Smil, © 1991 John Wiley & Sons, Inc.; Figure 4.10 from Chatto & Windus, *A Hundred Years of Ceylon Tea 1867–1967* (Forrest, 1967); Figure 4.13 from Blackwell Publishers, *The Human Impact*, 3/e (Goudie, 1981); Figures 4.14 and 5.9 from Blackwell Publishers, *Land Degradation: Creation and Destruction* (Johnson and Lewis, 1995); Figure 5.2 reprinted by permission of John Wiley & Sons, Inc., *Environmental Issues in the 1990s* (Mannion and Bowlby, 1992), © 1992 John Wiley & Sons, Inc.; Figure 5.4 from *The Conversion of Energy* (Summers, 1971), artist: Dan Todd, © 1971 by Scientific American, Inc. All rights reserved; Figure 5.5 from MIT Press Ltd, *Scientists on Gaia* (Schneider and Boston, 1991); Figure 5.6 from Addison Wesley Longman, *Why Economists Disagree* (Cole *et al.*, 1983); Figure 5.7 from Energy and Resource Quality: the ecology of the economic process p. 112, John Wiley and Sons Inc. (Charles Hall, 1986); Figure 5.8 from the Environment, Politics and the Future p. 14, John Wiley and Sons Inc.; Figure 6.1 from Chapman & Hall, *Planning and Ecology* (Roberts and Roberts, 1984); Figure 6.2 from Earthscan Publications, *Vital Signs 1995–1996* (Brown *et al.*, 1995); Figure 6.3 from Port of Tyne Authority, *Tyne Landscape*, published by the Joint Committee as the Improvement of the River Tyne in 1969; Figure 6.5 from Croon Helm, *The Spatial Organisation of Multinational Corporations* (Clarke, 1985); Table 2.2 from Cambridge University Press, originally printed in Warrick and Oerlemans (1990), Sea level rise, in Houghton, Jenkins and Ephraums (eds), *Climate Change, IPCC Scientific Assessment* p 257–81; Tables 3.1–3.3 from Oxford University Press, *World Resources 1992–93* (World Resources Institute, 1993); Table 3.4 reprinted with permission from 'Photosynthesis and fish production in the sea', *Science* **166**, 72–76 (Ryther, 1969), © 1969 American Association for the Advancement of Science; Table 4.1 from Edward Arnold, *Earth, Air and Water* (Simmons, 1991); Table 4.2 from Cambridge University Press, *The Earth as Transformed by Human Action* (Turner *et al.*, 1990); Table 5.1 from Columbia University Press, *Ecological Economics: The Science and Management of Sustainability* (Costanza, 1991); Table 5.2 from Routledge, *The Environment in Question* (Cooper and Palmer, 1992).

Faber & Faber Ltd/Random House Inc for extracts from the poems 'Ode to Gaea', 'Woods' in "Bucolics" & 'In Praise of Limestone' by W.H. Auden in *COLLECTED POEMS* 1968; Faber & Faber Ltd/Harcourt Brace & Co for extracts from the poems 'Burnt Norton' & 'The Dry Salvages' by T.S. Eliot in *FOUR QUARTETS*. Copyright 1943 by T.S. Eliot, renewed 1971 by Esme Valerie Eliot.

Whilst every effort has been made to trace the owners of copyright material, in a few cases this has proved impossible and we take this opportunity to offer our apologies to any copyright holders whose rights we may have unwittingly infringed.

The Readme *File: Some definitions, units and abbreviations*

This book frequently uses some words which either have a number of common uses, or which are defined differently by various groups. It seems useful, therefore, to set out the meanings that will be adopted here. Also included are some explanations of frequently used acronyms and units of measurement.

Environment is taken to include all those features and processes of planet Earth which are outside the human species. This includes both other living things and non-living matter. So spiders and energy flows would be included. But these may have been modified by human activity and still be included in this usage: thus domesticated animals and the products of nuclear energy generation would be included in 'environment'.

Nature is used to mean all those features and processes outside the human species which are also outside human action. Nature is what precedes human activity; 'a natural environment' is one that has not been altered by human action. It can thus refer also to non-living processes, so that the 'laws of nature' include gravity.

The visual expression in front of us of both nature and environment is *landscape*.

Culture refers to all the ideas, beliefs, values, knowledge and technical capabilities that are the shared basis of action by an individual or community. It is also transmitted from generation to generation and is usually subject to continual change.

All these words can carry depths of meaning that refer to particular times, places and cultures. We have to use them, however, and to put them in inverted commas to show that we 'know' that they have a special historical or political context at this point is not helpful in a text such as this.

Similarly, spatial scales are not defined with any mathematical precision. By *local*, I mean the everyday experienced environment. This will vary according to culture but is likely to be perhaps within a 12 km radius from a home base. The *regional* scale takes us from a local scale up to the size of the nation for small countries like England or parts of countries like Russia or China. More self-explanatory is the *continental* scale, which encompasses the whole of, for example, Europe or North America but only part of Asia, so we have East Asia or South Asia as continental-scale entities.

The use of the boxes in the text should be explained. At the end of each chapter is a box (surrounded by a thick line) which sums up the previous chapter and links it with the next. Intermediate boxes (with a thin line) simply summarize a preceding section, especially where this has been long or complicated. They do not intend to amplify or provide detailed examples as is the practice in some books.

There are a number of abbreviations and acronyms used in the text. These are usually spelled out on their first appearance but it may help to give some of the commonest here:

LDC less developed country, used synonymously with LIE (low-income economy);
DC developed country, used equally with HIE (high-income economy);
NIC newly industrialized country
EU European Union (still seen elsewhere as EEC or EC);
MNC multinational company;
TNC transnational company;
ppm parts per million (by volume): a measure of concentration of one substance in another, e.g. of carbon dioxide in the atmosphere.

Other measures include g = gram; t = tonne; 1 calorie = 4.2 joules. Kilocalories (kcal) are thousands (10^3) of calories. One million is sometimes indicated by 10^6. Less common exponentials include G (10^9), T (10^{12}) and P (10^{15}), where G = Giga, T = Tera and P = Peta.

For the far past, Mya is used for million years ago, and Kya for thousand (kilo) years ago, with a baseline of AD 1950, the same baseline as used in radiocarbon years BP (Before Present).

Most technical terms are explained at their point of first use. A few however seem to benefit from further explanations. These are given in the Glossary at the end of the book and are indicated in the text in **bold** type.

At the end of each chapter is a short annotated passage containing suggestions for further reading. A few works appear more than once since they cater to the material of more than one chapter.

CHAPTER 1

Human societies, 'environmental problems' and nature's storehouses

Any account of the relationships of humans with their surroundings starts with certain assumptions, which are often unstated. Whether or not we make it explicit, most humans have taken it for granted that our non-human surroundings will provide us with life-support in the form of air to breathe, food, water, and resources of all kinds. We have also regarded our surroundings as places in which wastes of all kinds can be disposed. A third category is non-material in the sense that many localities satisfy demands for recreation or for beautiful scenery or the presence of wildlife.

Given the level of world population at present, and the comfortable circumstances in which some people live, this relationship can be regarded as successful. Our environment has been beneficent. But there is another category of relationship which is now demanding a great deal of attention, namely those surroundings or 'environments' as a set of problems for human societies. Rather than accessible sources of materials, convenient sinks for unwanted residuals or places of delight, the environments present problems. Indeed, when the notion of environment is brought to our attention outside the disinterested atmosphere of academia, it is very often as a 'problem'.

By way of introduction to these matters, we have 16 'vignettes' of human–environment interactions at four different spatial scales, each following a general introduction to the notion of the environment (a) as a beneficent store of resources for subsistence and for pleasure, and (b) as a source of problems for human societies. In each of them we can see an interaction between the world of nature and that of human culture.

The beneficent environment

Our surroundings are important to us. The environment which people perceive is there has all kinds of value for them. These values may be material, since their immediate surroundings may yield food, or shade on a hot day. They may be those of pleasure as in a neighbourhood park for children, or a beach. They may even be those of memory, especially for the old. At greater distances from where we live, these values change and the connections are often less obvious to us, but they are still there. We all need the atmosphere to give us oxygen and then to receive carbon dioxide when we breathe out. Our outward connections are variable: in complex industrial societies, though, we interact with our environment at local, regional and indeed global scales (Fig. 1.1). At simpler levels of culture, those interactions have often been clustered at the local and regional scales. One of the great changes in the world has been the opening of those societies and their culture to wider influences.

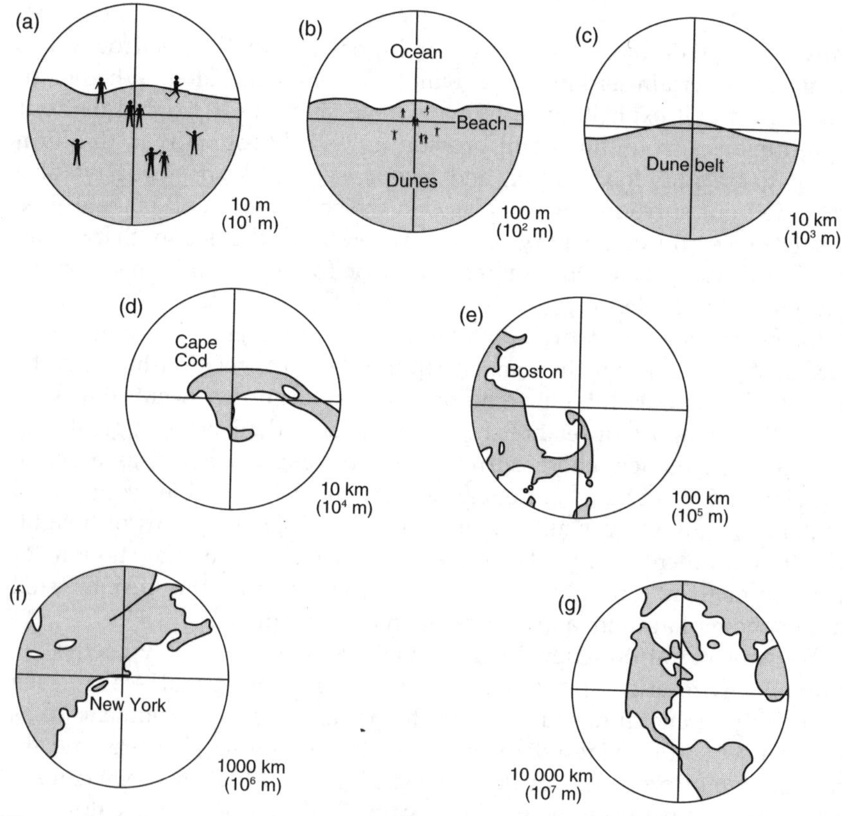

Figure 1.1 Scales of interest in human–environment relations. The smallest scale (a) is that of some people on a beach on Cape Cod (MA, USA). Successively larger scales are shown until the beach is the centre of a 10 000 km radius circle (g). The next level would be that of the whole globe (from Haggett, 1979)

The local scale

If we live in a relatively unindustrialized society, then the local environment will probably yield us food. We can see the crops growing daily and we know what has to happen if there is to be a good yield. Some of these inputs, like rainfall, are beyond our control. Others, such as weeding and bird-scaring, we have to organize. Harvesting and storage are our responsibility, too. So there is a daily if not hourly contact with the weather, the soil and the crops. But if nothing goes wrong, then the environment has been a source of life for us. Less critical than food, but hard to come by in some places, is the environment as a place of rest and recreation. Many European cities have patches of leisure gardens where apartment-dwellers can cultivate some fruit, vegetables and flowers and can sit among some small trees, grass and perhaps an ornamental pool. It is just as much a human-created environment as a field of maize but the purpose is different: largely to bring pleasure rather than necessary sustenance. But it is local and within the daily perception of its users. In this case, though, the food may well come a long distance.

Fresh food

Ask people all over the world about their chief pleasures from their immediate surroundings and a very likely answer will revolve around the virtues of the fish just taken from the water, the fruit eaten straight from the bush or the vegetable with damp soil still adhering to the roots. Although preserved food may be good (and the technical sophistication of preservation in HIEs is very considerable indeed), an immediacy of sensation connects individuals with what they perceive as a beneficent nature in harmony with human effort. So strong is this feeling in the HIEs that getting away from the technical to the elemental is a strong motivator of effort: gardeners will expend time, money and effort in getting a small but tasty crop; others will coo over the produce in a country market and its apparently local derivation; sophisticates will choose the lobster from the tank which they will be served in 20 minutes' time. In general, many better-off people in HIEs will spend time and money getting back to a low-income economy (LIE) stage, though without the uncertainties and discomforts of the LIE food system.

Pleasure gardens

The land use of most communities reveals the presence of small patches near the dwellings which are, even in the most collective societies, under the control of individuals or nuclear families. They are strongly manipulated so they cannot be said to be 'natural', yet their whole purpose is to demonstrate the goodness of nature. This is shown in one way by the presence of food plants. Vegetables, fruit, spices and herbs are the likely plantings rather than carbohydrate crops of basic energy needs. Usefulness may extend to medicinal plants: many introduced plants found their way across medieval Europe from the herb garden of one monastery to another. Pure pleasure has long been another function of the garden: even in poor countries, a backyard plot will have a few trees that yield a pleasing scent or a dense shade. Where money permits, then exotic plants of all kinds find a niche, from imported flowering plants and shrubs to native trees pruned to fit the plot, seen in its extreme form in Japanese bonsai. Where housing styles allow no gardens, then a detached plot may be available, or some of the pleasures may be sought in public parks and gardens, where the scale of planting can be larger and the range of specimens greater. Where larger tracts are owned then the scale of the operation may become larger (Fig. 1.2) but the inspiration is much the same.

The regional scale

If we live in a modernized society, many of our environmental connections are regional. The environment supplies us with many materials but few of them are from local sources. The fruit we eat, for example, may come from the same zone of the globe (like the tropics or the temperate zone) as the one where we live, but from several hundreds of kilometres away. Likewise, the water supply that is piped to our household is probably gathered in an impoundment up to 200 km away. For pleasure and recreation, too, we may travel some distance: perhaps 100 km for a day or weekend visit to a walking and climbing area, perhaps 1000 km for an annual vacation in a different climatic and cultural zone. In a poor country such connections are less likely to impinge on people's lives until, for example, a government agency puts in a road or makes it possible to buy industrially produced fertilizer. Most communities like to get rid of wastes with a solid component, such as domestic garbage, at this scale. Far enough to be out of mind but near enough to be cheap to get rid of is the view of more than one culture.

Fresh water

The environment yields us nothing more basic than fresh water. Every human needs about two litres per day simply to survive. Deprivation of it induces death far more quickly than starvation. Some water may in fact come from the local environment in the sense of a stream or well outside the back door or may even be collected from the roof. But even in LIEs,

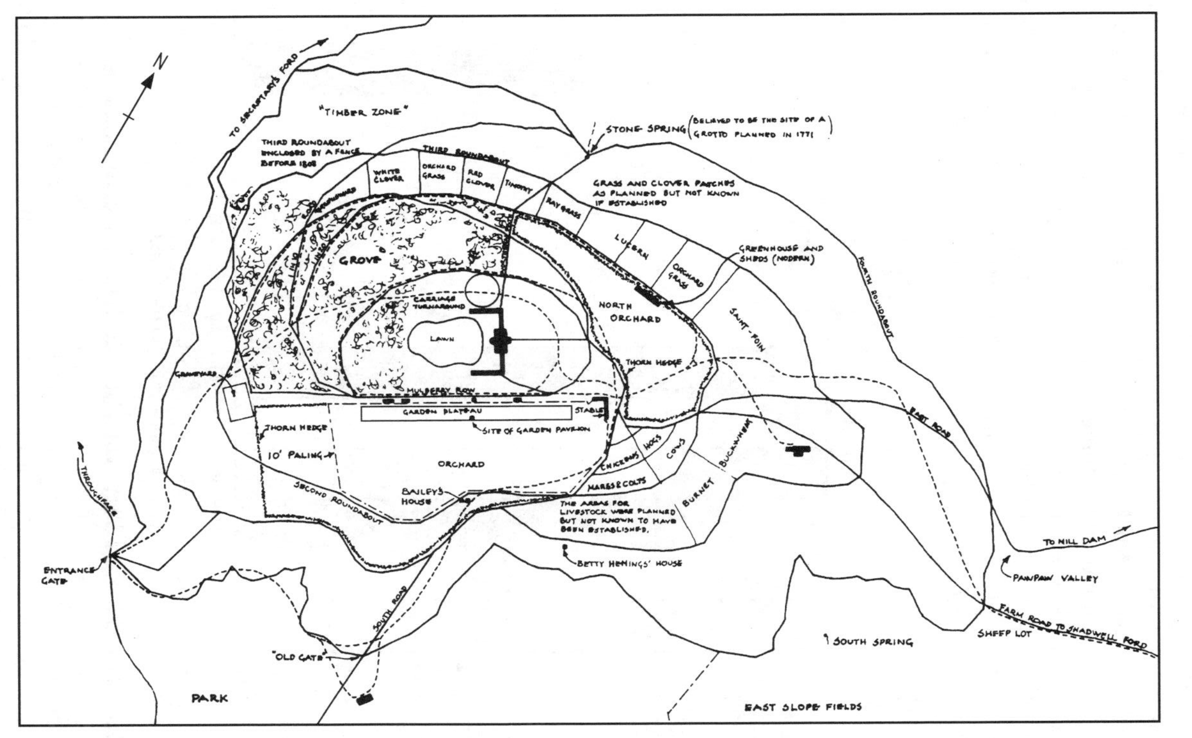

Figure 1.2 Thomas Jefferson's concept for a farm in a garden at Monticello, Virginia, established during the 18th century. (He lived from 1743 to 1826.) Around the house can be seen lawns, a grove of trees and an enclosed garden; beyond that is the periphery of farm and timber production (from Beiswanger, 1984)

water may often have to be fetched a few kilometres, and in the richer nations the tap is usually at the end of a pipe that started some distance away. So necessary are large quantities of water in an industrial society that much will be undertaken to ensure supplies, especially by way of engineering, often of a heroic kind: the Roman aqueducts were just the first examples of such structures. The distance from which water has to be collected is a function of many things but in HIEs the density of an urban-industrial population is often determining, for their demands will soon outrun local supplies from rivers or wells. In LIEs, a dense population will also overuse such local supplies but then somebody has to walk to another source of supply; this is often the job of the women (Fig. 1.3), as is

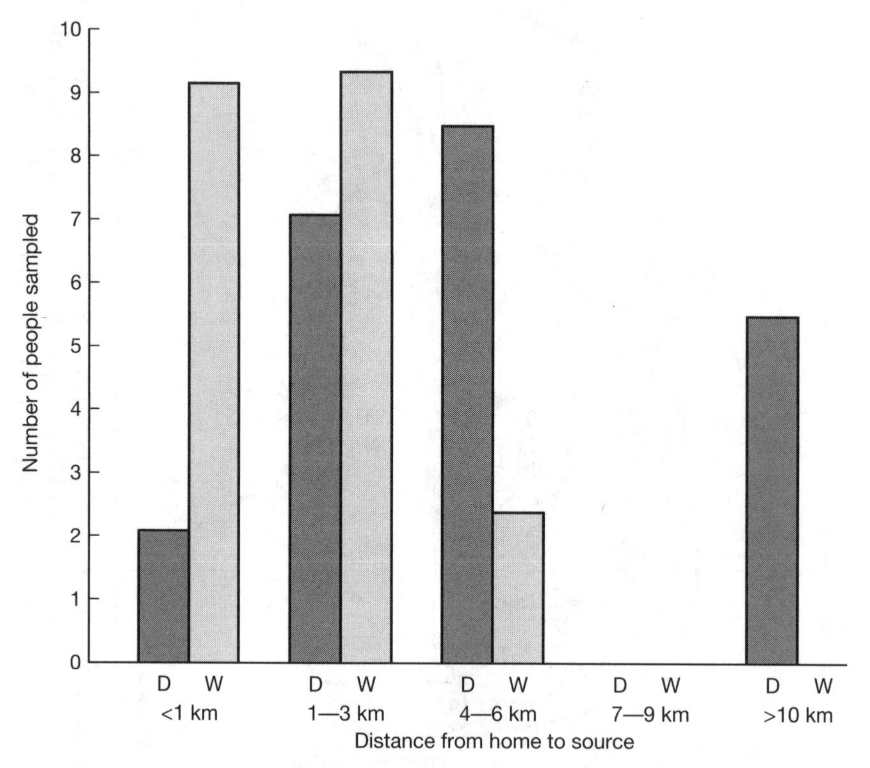

Figure 1.3 The distance that water has to be carried to households in Kitui District in Kenya. The women are responsible for this and some have donkeys to help. A total of 22 people were sampled and this graph records the number of individuals travelling the various distances for water. In the wet season (W), fewer long journeys have to be made than in the dry season (D). The author says that 'women themselves identify water carrying as one of their greatest burdens' (drawn from numerical data in Curtis, 1986)

the gathering of fuelwood if that too is in short supply. Nature here seems to turn niggardly rather than beneficent. Water use is subject to a wide variety of legal systems in different cultures: in Bali the temple priests regulate its use; in the UK it is in the hands of private water companies subject to some government regulation.

Travel for pleasure

In many societies leisure time comes in small and medium-sized chunks and in the latter it may be possible to undertake an all-day trip or an overnight visit. The object of the visit may be to see friends or family or a particular place or even perhaps simply to make a journey for its own sake. The environmental relations are sometimes of a determining nature: groups of Inuit in the Canadian Arctic cannot go visiting if the weather is too bad; family picnics in the UK are sometimes surprised by good weather. Often, the proximity of an area of valued scenery or a cultural monument may be the object of such travel and in each case the environment is the primary source of the enjoyment and recreation, along with the company. Most people view a spectrum of environmental choice without making an explicit distinction between the human-created and the apparently natural landscape; the place appropriate to the occasion is selected. In LIEs, a visit to a religious centre in a rural area might combine all parts of such a spectrum; in HIEs some might opt to spend the day in a forest and others in a cathedral city. In all instances, though, there is a felt need to undertake some travel away from everyday surroundings. The experience of a different, even if not necessarily unfamiliar, set of surroundings is an example of environment showing a positive value.

The continental scale

On a continental scale, the natural and cultural diversity may well be such that we feel no relation to the environment as such: 'the Asian environment' is a pretty meaningless phrase in either ecological or political terms. But in an age of multinational blocs, there may well be environmental concerns and policies that extend over a very wide area. The European Union (EU) has environmental policies designed eventually to encourage 'sustainable' economies throughout its jurisdiction, and the lack of tariff barriers in the EU allows materials to be moved more easily. Continental travel for pleasure is perhaps a very widespread use of resources in developed areas of the globe, since different climates, different scenery or even different cultures are likely to be encountered on such a scale. So of the three scales discussed, the continental scale is the least likely to mean much to an individual, but that is not to say that no linkages exist.

The longer vacation

Within most continents, there are some radically different environments which exert a call to those with the money and the time to visit them; for example, the coasts and the mountains within Europe, North America and Japan. For many years after the 19th century this appeal was focused upon the summer holidays of industrial cultures. Both managerial and working classes had at least two weeks of paid leisure time; now, the winter has been brought into the frame in the mountains, as has the round-the-world tour (Fig. 1.4). In all these cases, though, the environment provides a kind of resource base for the vacation experience, even though a heavily human-dominated set of structures may be superimposed upon it. Thus not only the deserted beach backed by a palm grove is sought but also the wall-to-wall carpet of frying people backed by bars and discos. Some may want to sleep in a wooden, thatched hut aware of the surf grinding the pebbles a few metres away; others want an air-conditioned high-rise. Cultural tourism, too, has its environmental relations. In rural areas especially, a shrine, a castle or a famous garden, together with its landscape, forms the basis of the encounter that is desired.

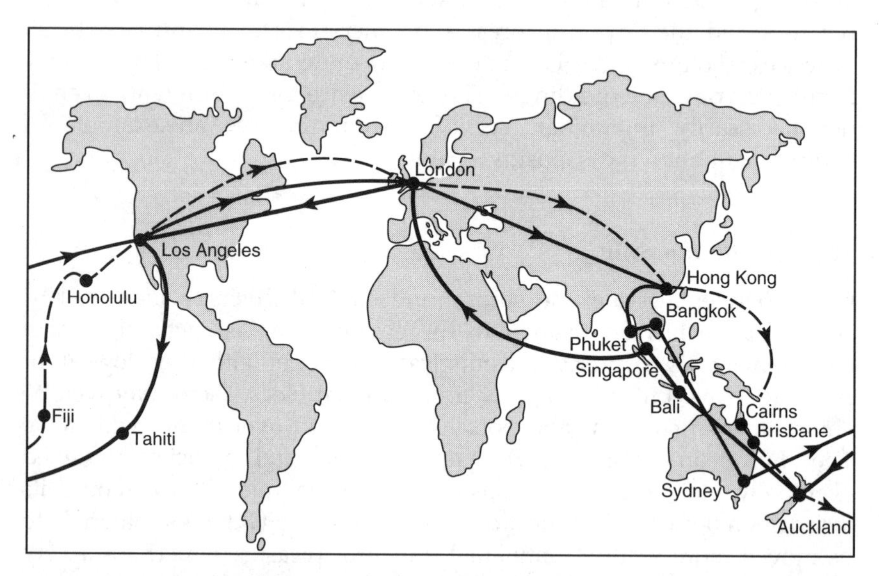

Figure 1.4 'Round-the-world' holidays of 18–20 days length offered by a major airline for 1996. Three different itineraries follow a roughly similar set of paths (from British Airways, 1996)

Food from a different climatic zone

If we cannot go to the exotic areas of our continent, then we like them to come to us. Hence the appeal down the millennia of trade in edible items from faraway places. The inhabitant of a rural area of an LIE may share very little in what is largely a luxury trade but the richer segments of even such societies demand these goods: Parisian bread is flown every day to some of the poorer countries of West Africa. There is in fact a world trade in exotic foods, but it becomes more intense when the distances are that bit shorter. Thus in North America the humble orange is grown in Florida and California; early asparagus came from a few Californian locations. Both are much prized in Boston and New York. In northern Europe now, the strawberry season has been much extended by the outdoor crop from Italy and Spain. These southern nations contribute avocado and kiwi fruit, too, competing with producers like Israel. In the old regime of the USSR and Eastern Europe, there was a short season for the many items of fruit which came from the southern parts of COMECON but there was probably only one delivery to most places: the day the melons arrived, everybody ate melons until they were ill. The extension of trade, especially by the use of air freight, has had the effect of distancing consumers from the environment of producers. When a crop was strongly seasonal, this was a sign of difference in the world: climatic and perhaps cultural. When any item can be had at any time, such perceptions are much harder to sustain.

The global scale

It is difficult to relate in any immediate way to the whole global environment. The industrial countries can call in materials from the whole globe: Japan seeks timber resources in all the world's temperate and tropical areas, for example, and many of us eat food from many parts of the world, including the deep oceans. Energy resources, too, seem to move on a set of global patterns, with oil and coal being transported vast distances in response to financial patterns generated by the instant electronic transfer of money. All of us, however, depend in many ways on the global integrity of climate and in particular on its predictability. Although climate is rarely absolutely the same from year to year, there must be a sufficiently narrow range of variation to know, for example, which crops will grow or how much heating oil to stockpile against the winter. To some extent, there can be globalized pleasure, too, for while most of us may never visit Antarctica (for instance), it gives many people happiness to think of those largely untouched areas of wildlife in such a bleak place; enough concern, for example, for nations to vote to keep it that way in perpetuity.

Wood products

A more literate world is an increased paper-using world, in spite of the advent of electronic media. This adds to demands for wood for construction timber, and other wood products such as particle board and cardboard. The materials produced are usually light and cheap to transport, and technical advances mean that most trees can be converted: the old image of spruces becoming newspapers and mahogany becoming sideboards has vanished. Add the fact that recycling is much practised in the more environmentally conscious HIEs, with impacts upon the economic structure of the industry, and there is a portrait of an important environmental resource at global scale. (Fuelwood, by contrast, is very much a local or regional resource.) Like many such resources, this dissociates the materials from the environments in which they are grown: there is no way of telling where the trees grew from which this book is made and if the paper is recycled then a mélange of sites would have been involved. What we are aware of, probably, is that trees are a renewable resource if managed properly and that certain forest areas are said to have globally regulatory effects on climate. So there is a sense in which trees are a critically important part of our environment at a number of levels. 'A culture is no better than its woods', said the poet W. H. Auden, and many of us have an intuitive feeling that this is exactly so.

Oil

A map of the transport of oil round the world (Fig. 1.5) shows that it is a global resource. Neither production sites nor consumption locations are evenly distributed but it is a fact that it is almost ubiquitously available even though very expensive compared with incomes in some LIEs. It is rare, though, for us to consider that it comes from the lithospheric component of our environment and that it is a non-renewable resource. Even though we realize there must be a finite amount of the stuff in the Earth's crust, we keep hearing of new finds, new efficiencies of recovery, and ways of getting more joules per litre delivered to the wheels of industry. So none of us individually acts as if the material has a finite lifetime which might expire during that of our children: nature's bounty is perceived as virtually infinite. Cheap motor and aviation fuel gives us access to all sorts of places and so, if we stop to think, oil is an environmental gift that underlies both necessities (consider the use of it in food production) and luxuries. LIEs without oil are amongst the poorest parts of the globe and the most vulnerable to environmental stresses of all kinds. It is economics and politics which determine who uses the oil, however—not considerations of need or justice.

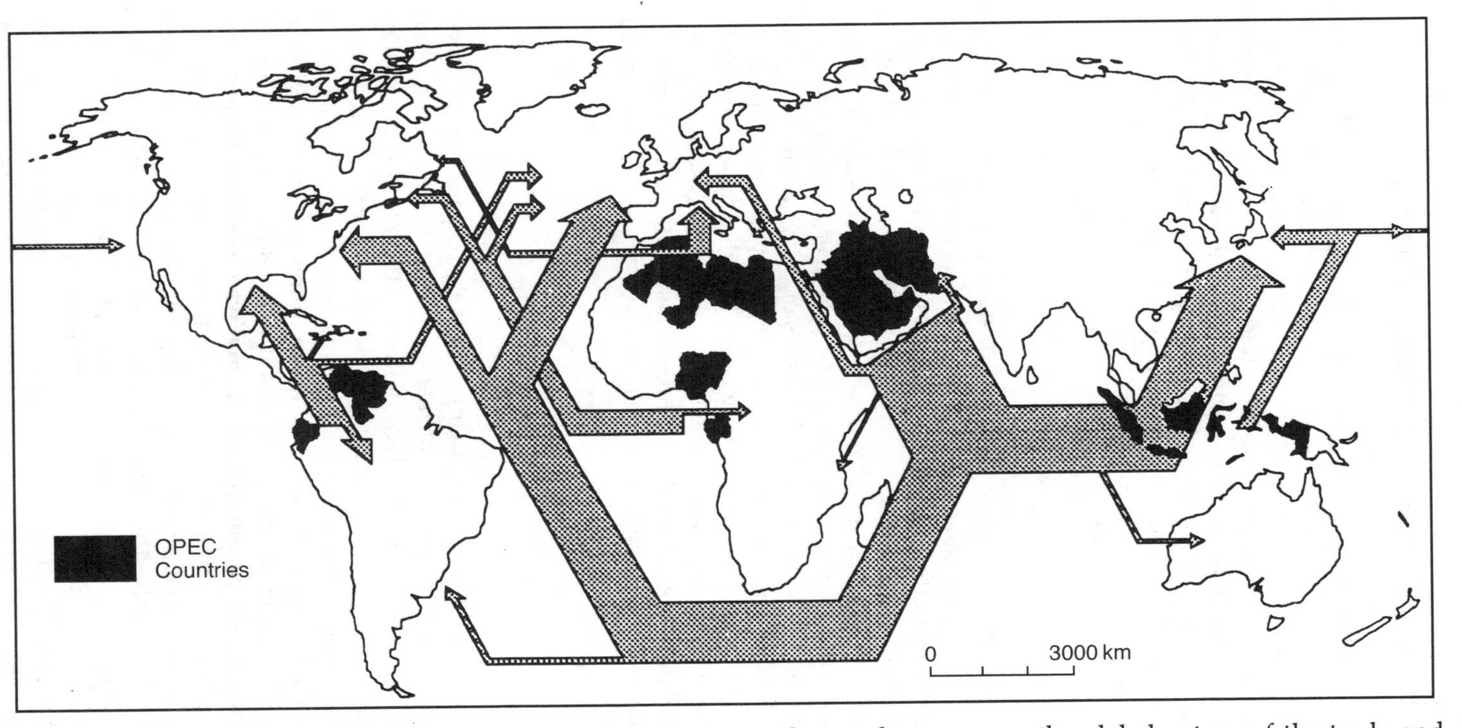

Figure 1.5 Major crude oil export flows during the late 1980s. The emphases are on the global nature of the trade and the dominance of the Middle East as a supplier. The thickness of the arrow corresponds to the share of the trade. The shading-in of OPEC (Organisation of Petroleum-Exporting Countries) reminds us that the price of oil is a political matter as well as reflecting technology and market economics (from Smil, 1993)

The diversity of environmental problems

Even the accounts of beneficence above often hint at a 'downside'. Water sources can be overused, soils too intensively cultivated and animal populations over-exploited. So abundance turns to problem. At this point in our narrative, we shall not attempt to rank these difficulties in order of magnitude or priority. Instead, we shall try to get some idea of the variations in their nature and, especially, their spatial scale, along with some notes on their occurrence in time and place. There will be no attempt to provide comprehensive lists.

The local scale

We are all familiar with some facet of our local surroundings that we cannot directly control (and that we wish were otherwise) but which does not appear to affect people outside a restricted locality. Examples might be a small stream which every two or three years floods into a small number of houses; an abandoned garbage tip which now and again bursts spontaneously into fire; a group of deer or monkeys which forsake the woods for the manicured delights of an agricultural crop; the rise in the level of some diseases after an irrigation scheme is complete. The list could be multiplied over many pages and the causes are equally diverse. Most (but never all) of these events and processes are of recent origin; local knowledge extends to similar traits in the past but they are not seen as deep-seated and inevitable features of the lifeways of the locality.

Waste tips

A very common feature of many people's experience of their environment is the local presence of many kinds of wastes. In industrial countries, most wastes are carried away from settlements for disposal or treatment (Fig. 1.6). This convenience is seldom complete and in poorer countries it is rarely the case. The situation is most obvious to the eye in the case of solid wastes. In LIEs any waste dump is seen as a mine of potential goods but in the rich countries it is simply a large volume of unwanted matter to be got rid of in any way possible. One of the cheapest methods of dumping waste is to fill an existing hole in the ground and earth it over. Even well-managed tips, however, may create environmental hazards: the materials may generate methane as they decay in anaerobic conditions and so fire is possible. If the hole is not lined with an impermeable material, then leachate will enter the local water-table to the detriment of the local water quality. Very serious possibilities exist that illegal (or even legal though undesirable) toxic

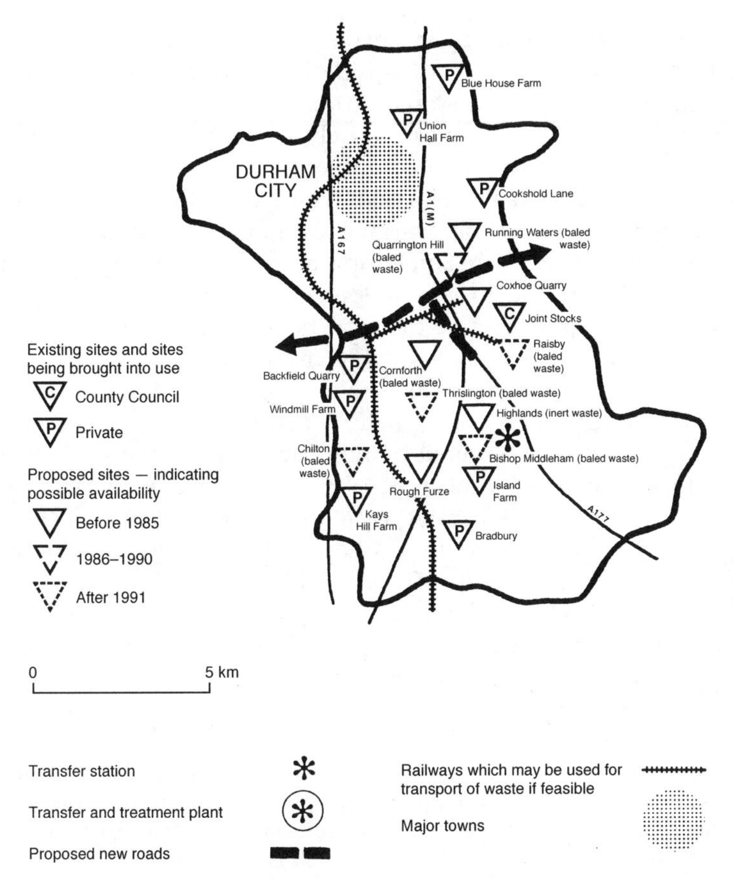

Figure 1.6 Part of a strategy for the disposal of solid wastes in northeast England in the 1990s. Even for an area of de-industrialization, the plan is complex and involves both local government and the private sector (from Durham County Council, 1984)

materials form part of the dump and will contribute to leachates or even be exposed, as has happened with rotting metal drums of cyanide. There are also are the disamenities caused by dust and waste materials, such as paper blowing off a tip, or the presence of rats, which can carry diseases. The pressure to get rid of such wastes has led to many industrial nations having flouted UN Conventions so as to export their solid wastes to LIEs with little legislation and even less hard currency.

Volcanoes

In world terms, 'live' volcanoes are not a very significant feature, with perhaps 50 active at any one time. Equally, we can have little doubt that those who live in their shadow regard them as a negative environmental feature. The only upside is likely to be the tourist influx during times of relative quiescence: Mount Aso on Kyūshū (Japan) has a small town at its base almost entirely devoted to servicing visitors. The obverse is the destruction of property and life that can take place during eruptions. Two features stand out: first, these eruptions have seldom been without some warning even in pre-scientific times, so people can be evacuated. Some refuse to go, of course. Second, a major eruption is usually on a scale that defies the resources of modern technology. The best that can be done even now is to try to divert the strongest lava flows away from settlements or productive land, as was done to some extent on the slopes of Mount Etna (Sicily) in 1991 by the US Army, using helicopters to emplace obstacles in the path of a river of lava. Limited though such interventions are, they are more effective than those taken in 1779 when Vesuvius blasted several villages and threatened Naples with the fall-out from a 3000 metre fiery pillar: in that instance, it was hoped that the blood of the patron saint of the city would liquefy as a sign that he would intervene. Images of the mountain became very popular, it seems, in 1789 as nature's equivalent of what was happening in France. Nearer to home, as Susan Sontag puts it in her novel, *The Volcano Lover* (1992), about Naples at that time,

> Statistically speaking, most disasters happen elsewhere, and our capacity for imagining the plight of those disaster strikes ... is limited. For the time being we are safe and ... life (usually meaning the life of the privileged) goes on.
>
> [Susan Sontag: The Volcano Lover]

There is in this a hint that the role of hazards in our perceptions of the environment may be both wider and narrower than simply the local effects of the (in this case) tectonic processes themselves.

The regional scale

In the unaided perception afforded by our senses, we are aware that there are classes of environmentally based problems that spread beyond our immediate vicinity. Such problems may be lateral extensions of local features, e.g. when a river floods along a 50 km length of its channel. We may become aware that the people in the next settlement also now have more diseases since they hooked into the irrigation scheme, that all the towns near a now-defunct nuclear reactor seem to have an abnormal proportion of children with cancers; a journey to visit a relative some 75 km distant may reveal that many of the lowland agricultural areas have severe gullying from soil erosion and that the upland forests have suffered badly from wildfire. Some of these problems are happening for the first time ever but others stir the memories of the elderly as having happened before, maybe worse, maybe not quite so badly. But we survived and carried on.

Eutrophication of water

The provision of fresh water is a basic need. However, water has other uses: the carrying and dilution of wastes is one of these. Sometimes the wastes carried are not put there deliberately but enter via runoff from the land. This is often the case where lakes, rivers and offshore waters carry high concentrations of nitrogen and phosphorus in solution. Such high concentrations lead to an increase in the numbers of other organisms, which are otherwise limited by the much lower natural levels of those elements in the water. In fresh water, this applies especially to algae, so that high levels of N and P are often detectable from algal 'blooms' that cover the water; any municipal park lake in summer will probably show this phenomenon. This may not be highly deleterious to the ecology: a few bottom-living plants may suffer from temporarily lowered light levels. The greatest change is in the decay of the algal material. This is brought about by bacteria, which in their rapid growth and reproduction use up large amounts of dissolved oxygen in the water. (In warm seasons this quantity is at its lowest anyway.) So animals which depend upon oxygen for respiration are killed: this usually means a high mortality of fish. Together with the products of plant decay it also brings a very bad smell. It means also that the water cannot be used by humans unless treated, so this part of the environment becomes a financial problem as well as a nuisance.

Floods

In the summer of 1993, the Mississippi–Missouri basin of the USA was flooded to a greater extent than any earlier records indicated. In July of that year, some 41 000 km^2 of crops were inundated. This flood was perhaps an extreme example (at St Louis the flood peak was 14.6 m above normal) of an event which happens in most river basins when the channel can no longer carry the quantity of water pouring off the land and so spills out sideways. What turns 'over-bank flow' into a flood is human presence. Floods cause damage to crops, structures and often soils, and bring about the death of plants, humans and other animals. Hence, floods are often labelled as 'natural hazards' or at the very least, 'environmental hazards'. Some regions are exceptionally prone to repeated and devastating inundation (Fig. 1.7). The human response is frequently to try to regulate the river by building levees, flood control

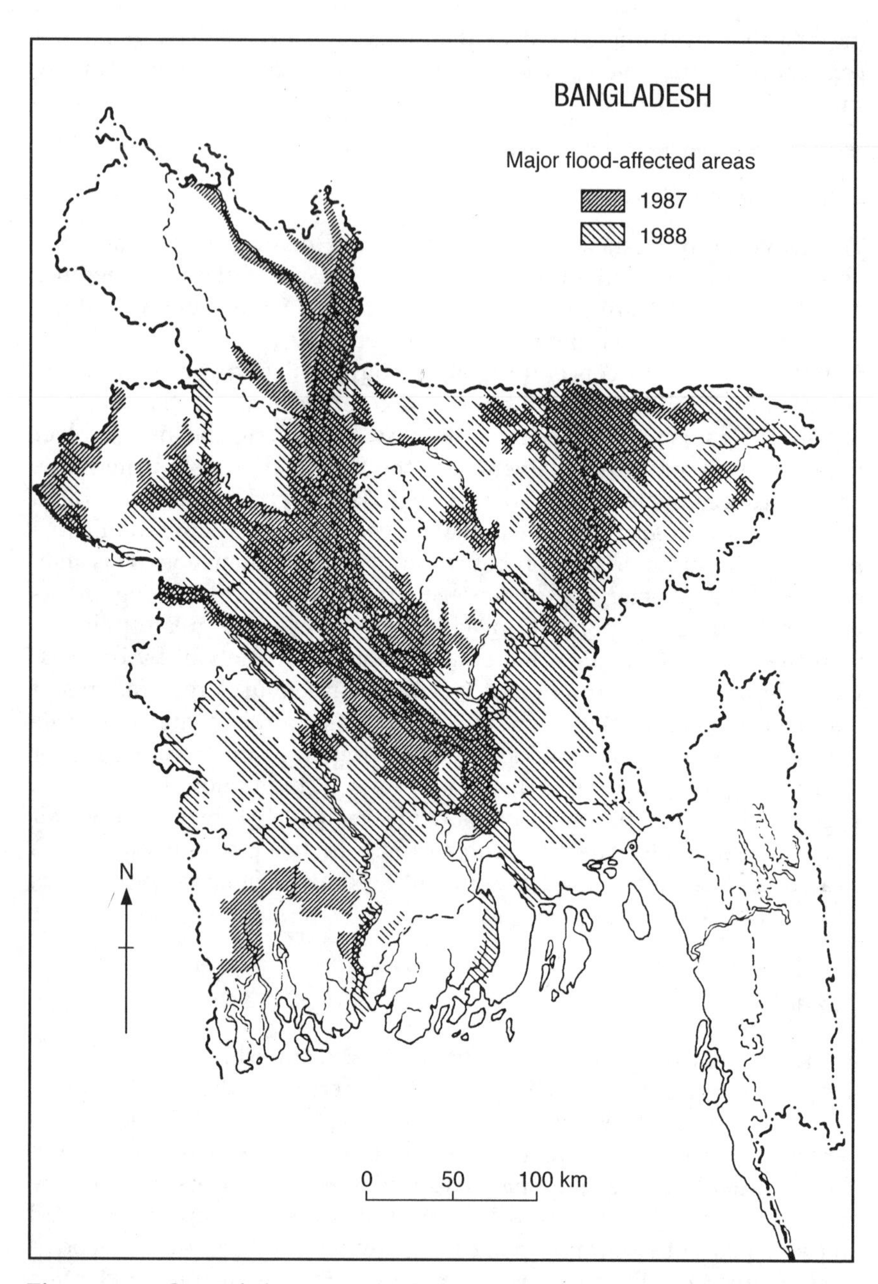

Figure 1.7 One of the regions of the world most affected by floods is Bangladesh. In 1987 and 1988 these were especially severe: this map shows the extent of flooding in those two years, showing clearly the proportion of the entire nation which suffered inundation (from Mandood Elahi and Rogge, 1990)

dams, and straightening channels; in short, engineering ameliorations. But looking at humans and their regional environment as one system, it often becomes apparent that human actions are a partial cause of the floods. Deforestation and urbanization, for example, lead to faster runoff from watersheds and thus produce earlier and higher flood peaks; if a regularized channel is built from point X to point Y then it delivers more water faster below point Y; if the levees between X and Y contain the river then it may burst through any structures below Y. So although the volume of water is provided by nature, its fate thereafter is much affected by humans. To some extent the flood is the evidence of our own failure to understand the connections between the environment and our activities. There is then a strong cultural element in the flood and its environmental linkages.

The continental scale

If we travel extensively, or pay attention to the media, then we cannot fail to notice that very large regions of the globe, such as North America or Africa, are characterized as suffering from environmental problems in what appears to be a homogeneous fashion. (Close inspection may show that this is an over-generalization but that it is not too far from reality.) In the case of Africa, some groups of years produce very light rainfall at critical times and so in rural areas there ensues famine, with large numbers of deaths, diseased people, economic refugees and perhaps conflict. In much of North America, insect plagues break out: spruce budworm in intensively managed forests, grasshoppers in intensively used semi-arid grasslands; other insects spread plagues like chestnut blight, which remove a whole species from deciduous forests. These phenomena are seen as today's examples of long-standing quirks of nature, but careful investigation of both past and present usually shows that more people are now affected for the simple reason that population densities are higher in these areas than they were at the time of earlier drought years or locust plagues.

Acid precipitation

In the 19th century, the city of Durham was the centre of a densely populated region in which coal was mined and burned in large quantities. One result was that the sandstone fabric of the cathedral was so rotted that a layer 4 inches (10 cm) deep needed to be scraped from the whole building. Later investigations have shown that combustion of hydrocarbon fuels puts large quantities of sulphur into the atmosphere and that these eventually rain out downwind. Examples of the effects include shifts in vegetation on the Pennine Hills of north central

England, where it appears that the 19th century saw a shift in the herbaceous vegetation of unenclosed land to species such as cotton-sedge, which are tolerant of soils with a very low pH. An example from our own times is the acidification of fresh waters in Scandinavia under the influence of fall-out of sulphur compounds from power stations in the UK and central Europe. This has proceeded to the extent that many lakes and rivers are virtually devoid of salmon and trout. The continental extent of the sensitivity is demonstrated by maps which show that about two-thirds of North America and at least one-half of Europe are susceptible to acid precipitation, the effects of which have been tracked up to 8000 km (Fig. 1.8). The ecology of its effects once precipitated are much less well known than the physics of its transport in the atmosphere.

War

A number of instances show that modern mechanized warfare can produce environmental consequences over very large areas. If these are not quite continental in scale, they are certainly larger than regional. An outstanding case is that of the Western Front in World War I. Over a large area of Flanders, the impact of trenching combined with artillery barrages was to produce a high-energy environment in which there were few living things apart from men and horses (fed from outside the region), rats and lice, with possibly (as John McCrae's poem puts it),

> The larks, still bravely singing, fly
> Scarce heard amid the guns below.
>
> [J. McCrae: In Flanders Fields]

Time has more or less healed this environment, but there has been less time for recovery in Vietnam, where chemical warfare was added to mechanical and explosives during 1965–1972. Explosives, in the form of huge quantities of bombs, cratered many regions and created a great deal of malaria mosquito habitat. Herbicides have more or less permanently prevented the growth of some coastal mangroves, opening the coast to storm erosion. Inland, their effects are still seen in malformed children who were victims of dioxin poisoning when still foetuses. A further example of relics of war contaminating the environment is that of minefields on land where inadequate charting or slow progress in clearing mean that danger is still present: Poland and the Falklands/Malvinas are two examples. None of this is significant beside what a nuclear war might produce environmentally.

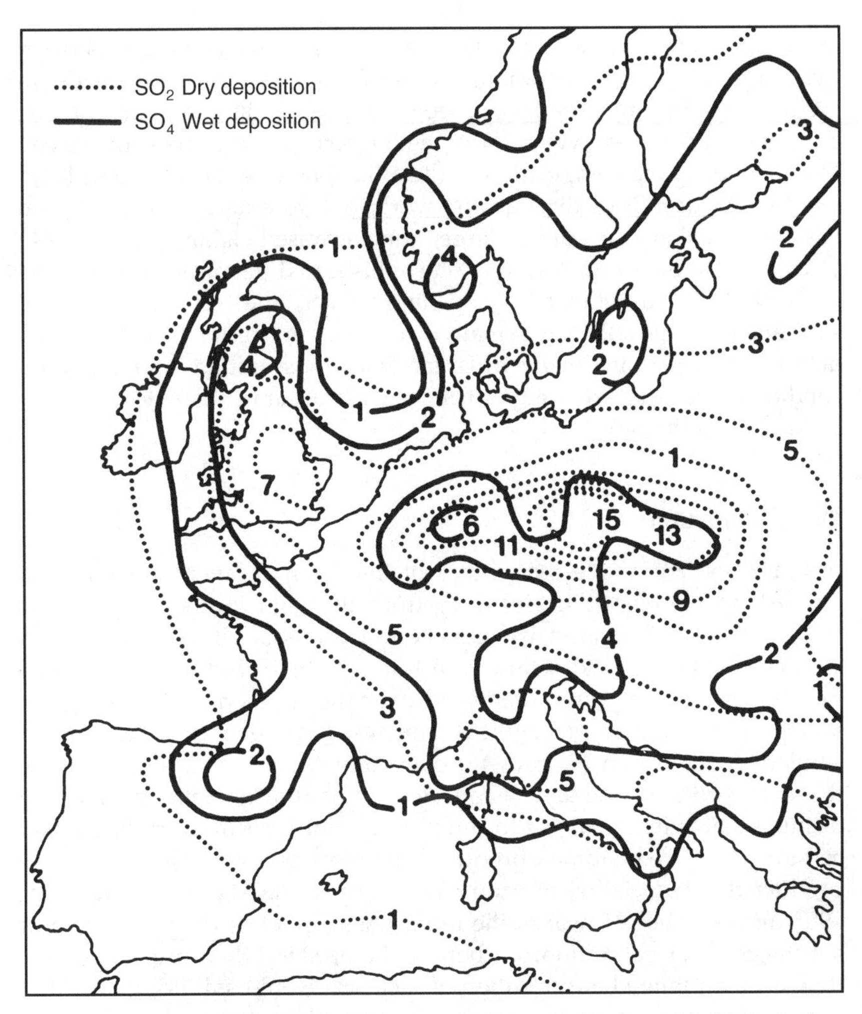

Figure 1.8 The fall-out of sulphur over Europe in the mid-1970s before any serious attempt to reduce emissions. The units are g/m^2. The most seriously affected area is that of central Europe, with the sources being in the former Socialist economies of Eastern Europe such as Poland and the Czech Republic. Although their fresh waters were badly acidified, the Scandinavian countries received relatively low quantities of such acid deposition (from Smil, 1993)

The global scale

The advent of satellite-based remote sensing and of global electronic communication means that we can now comprehend some aspects of the whole globe. So some global changes can now be verified that would otherwise remain identified only at a smaller scale. There are perhaps two sorts of global alterations: (a) those which are world-wide in occurrence but are essentially separate events, like regions of soil erosion; and (b) those which form one interactive and coalescent system, like changes to the composition of the atmosphere or the spread of long-lived toxic substances through the world oceans. Since these latter two environmental systems are so large, human-caused changes are likely to be difficult to detect and we rely on sophisticated instrumentation to tell us that they are indeed happening. Since the concentrations of substances are usually very small, it may be very difficult to tell if any detected changes are really different from those found under natural conditions. This is especially so if we lack accurate knowledge about conditions in the past.

Ozone depletion

About 20–50 km above the surface of the Earth, a layer of ozone (O_3) gas helps to protect the biosphere from the high levels of ultraviolet (UV) radiation that can cause genetic mutations in all forms of life, as well as skin cancer and cataracts in humans. In 1985 it was discovered that this layer was being attenuated over the southern pole during the spring and summer; later measurements have demonstrated that a similar phenomenon is present over the Arctic as well. Since the thickness of the ozone layer is affected by natural fluctuations, the role of human activity is subject to uncertainty, but it seems very likely that certain long-lived human-produced substances have the power to produce the dissociation of ozone. In this light, the Montreal Protocol of 1987 moves nations towards the rapid phasing-out of the production of substances like chlorofluorocarbons (CFCs); since they are long-lived, then even an immediate cessation of their use would still bring lingering effects. The loss of ozone is also implicated in the global warming discussed in the next paragraph. The main use for CFCs is now in refrigeration and one calculation has it that more people would die from food poisoning due to inadequate refrigeration than from skin cancer: most risks have to be evaluated against others.

Global warming

There is no reason to suppose that the glacial/interglacial cycles of the last two million years are over: we are probably living in the second half of an interglacial period. Within that context, however, human activity appears to be capable of affecting the composition of the atmosphere to the extent of making it retain more radiation than hitherto; hence the phrase 'global warming'. A number of gases are emitted that probably produce this effect but about 55 per cent of the outcome is due to the emission of carbon dioxide (CO_2). The concentration of this gas in the atmosphere in the early 19th century was *c.* 275 parts per million (ppm), in the 1890s it was 285–290 ppm and in 1994 it was 357 ppm, which is a level never before reached while humans have been living on the planet. Other gases involved in radiative forcing are nitrous oxide and methane (Fig. 1.9), together with the ozone 'holes' mentioned above. Modelling of these data suggests that if the current rate of emission of 5 Gt of carbon per year is continued then there could be a global warming of 1.5–4.5 °C by AD 2030. One concomitant of this might be a sea-level rise of perhaps 10–30 cm by AD 2030. But these rises need not be linear: rapid changes of climate are known to have occurred in the past and so the atmosphere could 'flip' from one state to another. Unpredictable fluctuations would very likely be the worst consequence for human societies the world over. Because these scenarios are based on models, their predictions contain elements of uncertainty, though in late 1995 the Inter-Governmental Panel on Climatic Change (IPCC) affirmed that there was now good scientific evidence of warming. But any admission of uncertainty (as honest scientists are bound to make when discussing

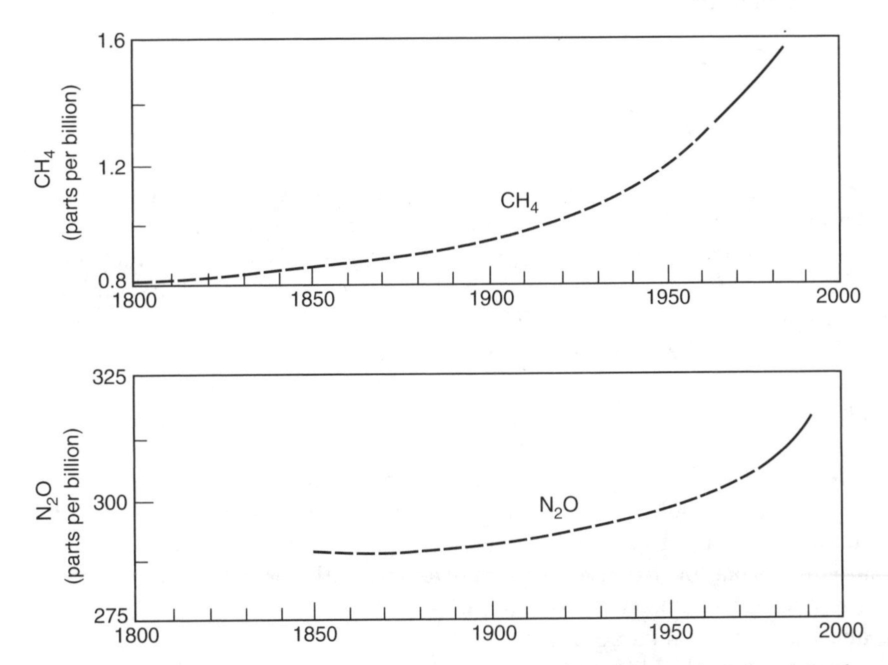

Figure 1.9 The atmospheric concentrations of two gases involved in the 'greenhouse effect'. Dotted lines are reconstructions, solid lines are actual measurements. Complex modelling provides better predictions than simple extrapolation of these trends (from Smil, 1993)

such large-scale phenomena) enables other groups of people to discount the forecast changes as unlikely to come about and hence disregard the need to change human behaviour. Clearly the mere presence of science within a global culture does not carry unquestioned authority.

This first box is an example of a number in the book: its purpose is to tie the material in front of it with what comes next. In that way it acts as a kind of hinge, articulating two sections which are separate but related.

Looking back over this material so far, it is clear that a number of types of approach are featured: there are the measurements of the natural sciences, some statements of the uncertainty of certain future possibilities, the formulations of international agreements, the measurement of the response of local groups to events in the world, and the lived experience of individuals. So the interaction of humanity and its surroundings is not simply a matter of measuring 'nature' and then charting or enforcing the human response to those data. Some understanding of the ways in which we think about humans and environment is clearly the next step, together with how we then communicate that thinking.

Ways of knowing

In the book heading, the phrase 'humanity and environment' is used. This implies that there is some form of separation between the two. (It is possible to believe the 'environment' part exists only in the human mind but we will accept for the first part of this book that what is beyond our own skin actually exists.) There are, however, some fundamental differences in the way that external entity is understood. For example, there are those investigations which claim to be **objective**, i.e. the observer is a detached consciousness (like the proverbial clean slate) and so all the phenomena and processes described are external to his/her mind before they are brought in, so to speak, for description and explanation. Public verifiability is therefore possible: measurements made of an object or a process will be the same whether the observers are in Ōsaka or Austria. This is classically the realm of the natural scientist but social scientists may attempt to produce studies of humans and their societies using this methodology.

An apparently different type of knowledge is labelled **subjective**, because it is internal to the observer and may thus vary according to that observer's individual characteristics. This category of information is central to the humanities and the fine arts: it is not reckoned that everybody should create, or respond to, a novel or a sculpture in exactly the same way. It too extends to the social sciences: for some commentators, lived human experience in all its variety is the only valid starting point.

Furthermore, any study can be **reductionist** in its approach, as distinct from **holistic**. The first term advises us that a complex thing or process can be understood if we learn more completely about the behaviour of its constituent parts. We can fathom the nature of biological cells by investigating molecules; if there is still more explanation required then the atoms which constitute the molecules must be studied. An enquiry into subatomic particles is the next (and currently final) step. In this mode of thinking, for example, human personality will eventually be explicable in terms of biochemical processes. **Holism**, as the word implies, accepts a total whole as the subject of description and explanation. Indeed it asserts that wholes have emergent qualities which are more than the sum of their parts and which cannot be predicted from a knowledge of those constituents. Holism suggests that human personality will never be predictable from biochemistry, just as the freezing properties of water are not predictable from a knowledge of the properties of the oxygen and hydrogen atoms.

Having set out this basic classification, let us now turn to some of the major ways in which humanity–environment relations have been described and interpreted in recent years.

The natural sciences

The word *science* has come to apply to a systematic set of practices applied to a particular set of phenomena: this may include both humans themselves and their environments. For non-human phenomena and humans seen as biological organisms, the term *natural sciences* is used. Investigations which use as much scientific methodology as possible but which look at social entities are described as *social science*. The practical applications of the natural sciences via machines are labelled *technology*, and so embrace professions like engineering.

All the natural sciences in some way deal with the phenomena of the environment as perceived by the human senses but all have become subdivided into specialisms. The scientific discipline that has so far contributed most to the study of environment is ecology, though its position is being challenged in the 1990s by those physicists and chemists who are applying themselves to making global models of the atmosphere and oceans. The usual definition of ecology is the science of the relations of living organisms with each other and with their non-living environment. Ecology today is especially concerned with the dynamics of such interactions in terms of, for example, population ecology, ecological energetics and the cycling of mineral nutrients at a variety of scales. These and other measurable parameters are held together in the concept of the ecosystem, which is a spatial area (at any scale) in which interchange of energy and matter takes place between living and non-living things. One feature of interest is the presence or absence of equilibrium (again at a variety of spatial and temporal scales) within and of an ecosystem. The study of the Earth as a single ecosystem finds one expression in the idea that the biophysical systems of the planet maximize conditions for life: the Gaia hypothesis of J. E. Lovelock (1979).

In their purest forms, the natural sciences are solely an intellectual exercise. For example, it seems unlikely that the evolutionary history of extinct hominids or the five-toed sloth, or the history of the population explosion in kittiwakes after 1945, are of much interest except in terms of intellectual curiosity. In reality, though, the natural sciences are constantly seeking practical and manipulative approaches to the world. The first of these is technology, where this is defined as any humanly produced means of altering the world, be it mechanical, chemical or biological. Technology now has a special place in any discussion of the meaning of environment. We have used it to change so much of our surroundings that it is rarely correct to talk of the 'natural environment' any longer. Beyond that lie interesting and difficult questions as to whether technology changes our whole outlook at fundamental levels beyond the simple 'can we build a road here?' type of question. If one such line of argument is accepted, we live in a technological environment rather than a natural environment, and technology might determine human history.

A second major idea proposes that science shows us how we ought to behave, i.e. that ethical principles can be derived directly from the findings of natural science. This involves the provision of 'hard' information by the natural sciences, to which the rest of society has then to adapt. Science becomes a kind of fundamentalist knowledge, from which we can 'read off' the proper way to behave.

The 'objective' social sciences

The social sciences are founded upon the conviction that the methods of the natural sciences have a primary place in studies of human societies. This view in turn rests on certain assumptions, such as the notion that every event has an identifiable cause and that a particular stimulus will, under given conditions, produce the same response from people. It must also be agreed that there exists an external world of actual behaviour which can be observed and recorded by agreed methods, and that there are indeed detached observers who can scrutinize events and processes without changing those events and processes by the act of recording them. A further assumption is that there is a structure to human society which is capable of being described in terms of regularities; this is perhaps equivalent to the assumption of order in the cosmos which is inherent in the natural sciences. More practically, there is the presumption that the theories developed in the social sciences can be used to change societies in predictable ways; so social engineering parallels civil engineering.

Hence, the aim is to detect regularities in human behaviour which can be used as the basis for prediction. The individual human must therefore be subsumed into studies of patterns of behaviour: the concept termed *behaviourism*. Classical examples of social sciences of this type have been the disciplines of economics and politics. In modern times their study has tended to move apart, whereas in the 18th century for instance they were contained in the same field, as political economy. Economics, deriving etymologically like ecology from the Greek *oikos*, meaning 'a household', is the study of how humans put together distribution methods for any commodity that is in limited supply. In the case of the environment this means either the environment itself (where its totality is demanded, as in the case of wilderness areas or outstanding scenery) or the materials that can be extracted from that environment, i.e. natural resources (Turner *et al.*, 1994). In everyday life, a market-oriented version of economics is now the most important mediator between humans and their environments: the question 'is it economic?' is the significant one in most projects.

Politics *sensu stricto* is the study of the organization and conduct of government and in its objective sense has been concerned with describing how things are and how they ought to be, the particular concern of political philosophy. It is not difficult to see that the possession of power to channel resources is a central feature of any distribution system and therefore as far as environmental constructions are concerned the older term 'political economy' is the most useful. It is also about the search for public support and power by groups who claim primacy for the environment in their hierarchy of values, like the Green parties and their equivalents (Eckersley, 1992).

Other social sciences with an interest in environmental constructions are sociology and geography (Pickering and Owen, 1994). Objective sociological studies have mostly been of the type which delineate in statistical terms the attitudes of groups in society towards particular issues, such as the desirability of the development of civil nuclear power, or which measure the demand for alternative life-styles that might alter the nature of environmental impact. Sociologists are also interested in how societies might change their environmental relations in a given direction, often that of becoming more 'green' (Wall, 1990). Geographers have mostly been concerned with studies of perception and cognition, though investigations of the regional and national distribution of environmental changes which are intimately linked with social values (e.g. the designation of national parks or commentaries on planning issues) are often found.

Many of the social sciences lead directly to ethical prescriptions and some thence to legislation. Litter, for example, has little ecological significance but is recognized as symbolic of an unacceptable social attitude towards our surroundings and so it is often forbidden to create it. The results vary by culture: contrast London and Singapore. So reading off 'good' behaviour from economics, political science and their ilk is not straightforward, since the question is raised of how those disciplines constructed their ideas of 'environment' in the first place.

The centrality of human experience

All that has been said so far makes it clear that the human mind is a node in the interplay of individuals, societies and their environments and this has attracted a great deal of investigation. The natural sciences, it is argued, overlook the cognitive links that join phenomena with their backgrounds and their history and so are always limited in their understanding. Less restricted is a view in which the world has meaning for humans (individuals as well as groups can be considered here) and so each person constructs a world with a set of objects and relationships, rather in the manner of a bubble beyond their skin. Given the technology of communications today, this envelope may of course be quite large, and it is constantly changing in size. The usual term for this sum total of a person's involvement with the cosmos in which she or he resides is the **lifeworld.** It is normal for lifeworlds to overlap and so the method is not confined to individual humans and what they make of their world. Basically, however, this viewpoint discerns environments as parts of personal life histories in which we see people creating themselves.

In one way or another, all these constructions are analytical or explanatory. By contrast the creative arts, in a holistic sense, just 'are'. 'In a work of art the intellect asks questions; it does not answer them.' This is not say that they have no public purpose; some avowedly have no function beyond that of conferring momentary interest or pleasure upon the observer, but many others engage in dialectic with history or politics. Some of the arts contribute more than others to environmental constructions. They may also direct us away from the rather rigid conceptions of self and social identity which derive from the natural sciences and the more objective social sciences. So they may help us to explore ourselves and our surroundings, as do maps, signposts and advertising brochures.

Normative behaviour

Those aspects of human behaviour and of the nature of the world with which we have so far dealt all have one feature in common: they deal with things as they are, or were, or might likely be. **Ethics** is concerned with how things ought to be and how people should behave. Whereas most other constructions are concerned with description, explanation and prediction, this category is about rules, recommendations and proposals: the term **normative** is applied. Such constructions depend upon the disclosure and examination of **value systems** which may not always be arrayed around human needs as distinct from demands. From these origins come environmental custom in non-literate societies and law in those with writing. Most advanced societies now have large bodies of law dealing with environmental matters. There are special problems in dealing with any entity that cannot plead for itself in court, like a tree, or in finding the right arena in which to deal with transnational phenomena like the River Rhine or the Mediterranean.

An overview

All these ways of knowing about the apparent relations of humans and their surroundings have one feature in common: while humans can communicate directly with each other through speech and language, they cannot do so directly with their non-human environments. We can scarcely converse directly with our nearest hominoid relatives, let alone with trees and rocks. We apply words to what we perceive by way of objects and processes, either in the calculated way of the sciences or perhaps in the aftermath of an emotional response. So we investigate the environment via the paths discussed above but above all we hear ourselves talking, since all the frameworks in which we can claim we know anything at all are of our own devising, even pure mathematics. The results of this interrogation then resonate through the various channels of society which are set up to hear and store such information. Writing is the main medium of storage, though nowadays it is supplemented increasingly by images.

In a complex, industrial society these avenues are well delineated and separate from each other. They will include, for example, the legal, educational and religious systems, the economic and political structures, and the media for communications. In each of these channels, information about people and environment will be discussed and transmitted: the word 'resonate' seems highly appropriate (Luhmann, 1989). There are some difficulties resulting from this arrangement:

- Knowledge may make sense in a particular channel but not be recognizable outside that channel since the communication may be conducted in a specialized language, such as that of the physicist accustomed to using mathematics rather than ordinary language, or the lawyer dependent upon very precise definitions of a particular term.

- While the information in a particular channel may thus make a comprehensible sound, it may resonate disharmoniously with that in another channel so that the total is rather like noise, i.e. an unstructured sound. An analogy might be pressing the keys of an organ keyboard randomly: each pipe makes a perfectly tuned sound, but the whole is unlikely to seem rational. So while many individual environmental 'problems' may seem soluble it may be the totality that is intractable.
- For reasons of cultural history, Western societies tend to segregate the resonance into binary states, that is, to formulate two positions like zero and one. Thus we have 'economic' and 'uneconomic', 'legal' and 'illegal', and in the extreme of normative thinking, 'good' and 'evil'. Our Western culture then urges us to eliminate one of these by overcoming it: the verb 'conquer' is much used. This has translated, of course, into the phrase 'environmental problem', which implies that the environment is somehow misbehaving. Since it cannot of itself do anything so laden with human judgements, a more honest view would be to consider those predicaments as cultural problems with an environmental delivery route.

The discussion above virtually eliminates the idea mentioned on p. 24 that there is a simple model in which 'hard' scientific information is accepted by humans who then respond to it in a totally rational manner. Any knowledge is subject to the distortions of the individual channels and to interference from others. There is a further element: we never start from zero. Inevitably the residues of history are present, both as the actual presence of patterns (e.g. of resource use or population numbers) from the past, or the pervasiveness of long-standing sets of ideas. We now need to examine some of these.

A broad picture

In the last section of this chapter, we shall try to get behind the 16 vignettes in a more systematic manner by concentrating on three important themes:

- The human population: clearly the numbers and distribution of humans is relevant to all manner of environmental considerations, such as the use of resources and the impacts thus generated.
- The historical perspective on these material links in terms of what people have in the past thought about them, rather than necessarily what they did.

- The nature of the material links between humans and environments: what is the difference between basic needs and culturally driven demands; what influences does the environment still exert on cultures?

This material then leads on to a consideration in Chapter 2 of the view of ourselves and the world which the natural sciences give us.

The human population

In 1996, the estimated world population level of the Earth was 5.83×10^9, i.e. 5.83 thousand million. The overall growth rate was 1.38 per cent per annum, at which rate the number would double in 51 years. Yet about 10 000 years ago, that level was probably of the order of 4 million. During the intervening period, there has been a great deal of economic and technological change, which has enabled all those people to survive and reproduce, in a world which has itself undergone some natural changes. Some of the most notable contextual factors have been as follows:

- The global warming of the world during the period 10 000–5000 BC, when many of the world's major **biomes** (such as the tropical forests, savannas and grasslands) assumed approximately their present distributions and became stable backgrounds for human utilization. Thereafter, climatic fluctuations were on a smaller scale, though these were not necessarily insignificant regionally, e.g. the 'Little Ice Age' (AD 1550–1850) in Europe.
- The domestication of plants and animals, especially in the major centres of Southwest Asia, Meso-America and Southeast Asia. Control of the genetics of certain species facilitated the intensification of production so that controls on population levels were no longer so tight: more people per unit area could be fed.
- The development of the use of fossil fuels in the 19th and 20th centuries. The advent of coal and especially oil allowed even more intensification of food production and also its extension since land could be transformed with much greater ease than ever before. Demand for goods also meant that many millions could find livelihoods in manufacturing and associated occupations.

None of the above should be interpreted as a simple cause of population growth, since the determination of family size is a far more complex process. Nevertheless, without such developments, the growth of the human population to present levels would probably not have been possible. They are examples of a class of changes which could be labelled as permissive. That is, they are necessary for the response to happen but they do not themselves bring it about in an inevitable fashion. We shall encounter this class of factors several times in this book.

Population growth up to the late 20th century

The genus *Homo* probably evolved in southern Africa about four million years ago, and the species of which we are members, *Homo sapiens*, appears to be a product of the last 40 000 years. It too began in Africa, but quickly fanned out from that continent, especially in the wake of the retreating ice sheets at the end of the Pleistocene, about 10 000 years ago. From that base it has grown to its present (1995) level of 5759 million.

The course of human population growth can be divided into four major phases, the first of which comprises the late Pleistocene and early Holocene, when hunting and gathering was the sole form of human economy. During these periods (culturally named the Upper Palaeolithic and Mesolithic), the supportable densities of people averaged about 2–3 per 100 km^2. One estimate of global population once the major ice-covering of the northern hemisphere had melted is for million in 10 000 BC.

Soon thereafter we can distinguish the beginning of the second major phase, which was made possible first by the movement and expansion of people into areas hitherto unoccupied in Oceania and the Americas, and second by the widespread adoption of farming. This Neolithic revolution allowed growth rates of 100 per cent per century in the last three millennia BC, so a population of 100 million is postulated for 500 BC, with AD 1 seeing a total of 170 million. Thereafter, growth rates seem to have slowed up to give an absolute level of 265 million by AD 1000. The next phase was led by Europe and China and although one of considerable growth overall, it did suffer some checks: the Mongol invasion of China in the early 13th century resulted in 35 million deaths and destroyed much of the agricultural structure. In Europe the plague brought down the population from 80 million to 60 million (Livi-Bacci, 1992). On a world scale, the total in AD 1400 was about 350 million, having been 10 million higher a century earlier, suggesting that some form of check was imposing an upper limit.

Any such curb was lifted in the modern period, when technology, colonization, industrial development, new food-producing techniques, scientific knowledge, and modern medicine all combined to raise the barriers present in medieval times. Growth rates seemed to take off in the 16th and 17th centuries, and especially after AD 1700, again fuelled especially by China and Europe. In the 19th century, for example, Europe's population gained by 135 per cent, of which 20 per cent emigrated to other lands. This phase is different from the others in the sense that it is largely attributable to a decline in mortality and especially to a reduction of deaths from infectious diseases. A world population of 350 million in AD 1400 became 700 million in the last quarter of the 18th century, 900 million in 1800, 1625 million in 1900 and 5759 million in 1995. Of these people, less than 1 per cent live in Oceania, 20 per cent in the Americas, 12 per cent in Africa, 16 per cent in Europe and the rest in Asia.

Population distribution and trends today

The current position is one of two major world types of population: those which are growing slowly or are nearly stationary, and those which are growing quickly (Findlay, 1991). In the first category, the declines in mortality which fed the rapid rates of growth in the 19th century have been followed by reductions in the birth rate. In the second group there are still high birth rates even though mortality is down and so annual growth rates average 2 per cent per year. The slow growth areas comprise Europe, the former USSR, Japan and Oceania (19 per cent of world population) together with North America (5.5 per cent). Faster growing regions comprise Africa (11.5 per cent), Latin America (8.4 per cent) and Asia except Japan (55.8 per cent). In total, the industrialized nations have about 24.4 per cent of world population and the developing countries 75.6 per cent. The world rate of population growth in 1995 was 1.68 per cent per annum, which is a doubling time (δ) of 42 years (Fig. 1.10).

The low-growth areas are eastern and western Europe, the former USSR, Australia and New Zealand, North America and East Asia, including China as a developing country with a growth rate of only 1 per cent. Its average rate of increase was 0.8 per cent per annum: a few nations like the former West Germany were actually losing population but all were at or below the China level of 1 per cent per year. All show a combination of rising levels of material benefits and falling levels of fertility, though the social milieu of the latter in Western Europe, for example, is different from that in China or Romania. The reasons for having fewer children voluntarily involve such factors as the costs in time and money of raising them, the buffering provided against old age by the state or by private pensions, thus reducing reliance upon children, and the revaluation of the role of women in society. The results include ageing populations, with the 65+ cohort being the fastest growing. In developed nations, some 13 per cent of the population is 64+ years old and 22 per cent is <15 years old; in the less developed nations, the figures are 4 per cent and 39 per cent respectively.

The faster growing parts of the world are mostly the less developed and poorer regions. These are found in Southeast Asia, Latin America,

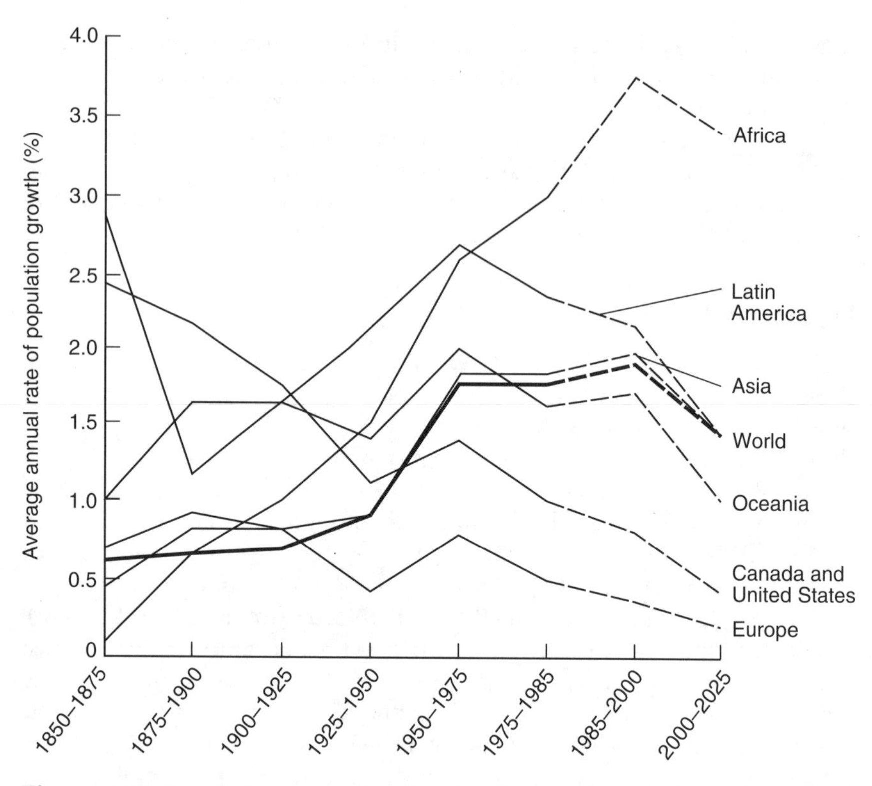

Figure 1.10 Population growth rates for the world and its regions from 1850. Note the upturns during the 1920–1975 period, with some levelling after that. Extrapolations beyond the present use the UN Medium Variant, which models continued falls in fertility. Recall that a 1 per cent per annum growth rate means a doubling in *c*. 70 years, 2 per cent in 35 years, and 3 per cent in 23 years (from UN data)

the Indian subcontinent, the Middle East and Africa, and within these broad categories there are countries with very large populations and very high rates of growth, such as India (194 million, 2.3 per cent per annum), Indonesia (918 million, 2.1 per cent per annum) and Brazil (159 million, 2.3 per cent per annum). Africa has the highest set of population growth rates in the world and within that continent, the sub-Saharan regions are the highest, e.g. Kenya at 4.2 per cent per annum, with a doubling time of 17 years). These fast-growing areas have a world average increase rate of 2.5 per cent per annum; their doubling time is 28 years.

These then are lands of high birth rates and declining incomes. Per capita income between 1980 and 1986 fell by 28 per cent in Nigeria, 16 per cent in the Philippines, 21 per cent in Argentina and 11 per cent in Peru, for example. Further, most of these regions have very heavy burdens of external debt so that much land has to be devoted to export crops to service those loans. Thus grain production per capita fell by 17–25 per cent in several sub-Saharan countries, by 24 per cent in Peru and by *c.* 50 per cent in Mozambique and Haiti over the period 1970–1985, emphasizing the role of political troubles. These populations are still subject to high levels of infectious and parasitic diseases, and cancer is increasing where smoking is becoming more prevalent. Such populations produce heavy impacts on local resources such as fuelwood, soils and water.

In a world which adds to its human numbers by about 192 000 per day (about half the population of Iceland), there is much interest in projections of future numbers. Most of these are educated extrapolations of present trends and hence subject to inaccuracy: most past predictions have turned out to be wrong. Nevertheless, increasingly sophisticated modelling is providing better guides to the range of future numbers, and in this field the work of the United Nations Organisation is outstanding. The factors which will affect future levels of population the most are mainly those affecting people in developing countries since that is probably where 95 per cent of the population growth in the next 100 years will take place. Thus the critical variables will be the cultural changes in those countries which affect the age at marriage, the importance of children in the labour force, the costs of raising children, the social, educational and occupational status of women, and the degree of urbanization. The flow of knowledge and materials to limit family size is also important, but secondary since the decision to be informed and then to use appropriate ways of spacing children is the most important. Modernization of LDCs may initially raise fertility, for better nutrition and hygiene by themselves may well lead to the survival of more infants. The success of family planning in bringing down birth rates when conditions for acceptance are favourable is beyond doubt: Costa Rice, for example, reduced its birth rate from 47/1000 in 1961 to 28.5/1000 in 1985, and the world rate fell from 32.2 in 1970 to 26.0 in 1985.

The UN Medium Variant Projection is one of the most commonly used guides to future numbers. It suggests a doubling of the world's population by 2100. Africa will have the highest growth rates, which will increase the number of people on that continent from 555 million to 2.6×10^9 by 2100. The biggest absolute increase will come in Asia, from 2.7 to 4.9×10^9 by 2100. This variant sees the replacement fertility level (RFL, i.e. zero growth in terms of natural increase) being achieved by 2035; this would bring about a stabilization of the world population by 2100 at a level of 10.2×10^9. (The World Bank goes for the same year but at a level 1×10^9 higher.) If the replacement fertility level of 2.1 children per family is delayed then the absolute number of population is increased and the year of achieving stability is put back, and vice versa.

Thus if RFL came 20 years earlier then there would be 2.2×10^9 fewer people; if it were 20 years later, then 2.8×10^9 more. The difference is today's world total.

A cultural history of humanity and environment

Important though numbers may be, human culture is equally significant. It seems to have been a permanent feature of all human cultures that they have engaged in cogitation about what their environment means to them. The object of such thought is likely to have been called by a term more analogous to our word 'nature' than to the more recent scientific coinage 'environment', which in this sense is mostly a 19th century development. By contrast, 'nature' was commonly used in this context as early as the 15th century. Some of this thinking was transmitted orally from generation to generation and so much has been lost along with the rest of such cultures. Other parts have been written down and so we have a good idea of their lineage, subject always to the selection that has taken place in any historical record (Pepper, 1984; Brown, 1990).

Environmental thought through time

An example of non-literate thought about the relationships of humans to nature is that of the aboriginal inhabitants of North America. This is not merely a historical curiosity, for in the context of environmental problems in North America, there is a renewed interest in the attitudes of the pre-European inhabitants. The current focus stems very largely from the well-publicized statements of a small number of Native Americans which were made about the time that they were being forced to yield their lands to the incomers. That of Chief Seattle (*c.* 1854) is one of the best known:

> Every part of this earth is sacred to my people ... We know that the white man does not understand our ways. One portion of the land is the same to him as the next ... He treats his mother, the earth, and his brother, the sky as things to be bought, plundered, sold like sheep or bright beads.

But it has been pointed out that this speech has been 'improved' a number of times so as to contain phrases acceptable to environmental activists and the Christian religion. Those unconvinced by the environmental piety of the Native Americans point to the wholesale slaughter of bison by driving herds over cliffs or their refusals to co-operate with the Hudson's Bay Company in maintaining a sustainable supply of furs. Nevertheless some detailed studies point to environmental tenderness even in recent times: the Koyukon Indians of the Yukon are one such group. So it is difficult to come away with a single verdict for a continent and this would probably be true of many other groups round the world if we had a comparable depth of studies.

Though the present dominance of Western thought in this field is undisputed, the major literate cultures of the rest of the world have had quite distinctive views of the world and their place in it. The outstanding examples are those of China and Japan; the equivalent resources of India and of Islam are without doubt available, though less explored. In China, for example, the example of Taoism (*tao* = 'the way') sets a pattern in which the stream of all things in humanity, nature and earth can form an essential harmony, provided they are allowed to follow their own pathways:

> Knowing when to stop averts trouble.
> Tao in the world is like a river flowing home to the sea.

Just how this can be squared with the realities of 6th century BC China and its agricultural expansion at that time is never quite explained. Nevertheless, the non-interventionist tenor of Taoism has had some revival in recent times, especially among those seeking alternatives to the dominance of the Western view of the world. This characteristic has flowed over into those parts of Japanese culture most influenced by Buddhism. So the Japanese character for 'nature, environment' is read as 'self-thusness' in terms of anything which is following its own ways rather than being directed from outside. This suggests that no distinction between humans and environment can be drawn. Most of us find it difficult to see this being an operative concept when standing at 6 p.m. near Shinjuku Station in Tokyo. In Tokugawa times and in an apparently eternal agricultural economy, it may have been different, as Bashō's haiku suggests:

> One whole paddy field
> Was planted ere I moved on
> From that willow tree

But again, there is a resource of thought which may repay more thorough exploration both now and in the future.

There is little controversy over the opinion that today's western view of the world is suffused through and through by the discoveries of modern science and technology. Before their existence, however, many societies had thought about their place in nature. As we would expect, the thinkers of Classical Greece devoted time to the topic, as did the Hebraic societies, medieval Islam and the early Christian West. As early as the third millennium BC in Sumeria, the text known as *The Epic of Gilgamesh* describes, as one commentator puts it,

> ... Enkidu, the 'natural man', reared with wild animals and as swift as the gazelle ... seduced by a harlot from the city ... and led on by stages, learning to wear clothes, eat human food, herd sheep, and make war on the wolf and lion, until at length he reached the great civilized city of Uruk.

This appears to be an allegory of the transition from hunting and gathering to agriculture, though not attended in this culture by a Fall, as it is set down in the Hebrew Bible or Old Testament. The Hebrew Bible itself has been arraigned as permitting the exploitation of nature (as set out in Genesis 1:2) but closer analysis provides for a notion of human domination over nature being given only to the righteous and faithful, whereas the transgressors were punished with natural disasters.

Of all the Greek traditions of thought, that of the Stoics has been most influential. Simplified, their argument was that the Earth was the best place for survival. It was absolutely complete in every way and also beautiful. Human cultivation combats disease and struggles with the wild, exercising a control over excesses which might have otherwise occurred. The Earth was clearly designed for humans. The reverse idea, that large parts of the planet are inhospitable or useless to humanity, is associated with Epicureanism. The Epicureans contended that too much area was taken up by mountains, forests, rocks and the sea, so that there is no evidence for purposeful design. Early Christian thinking tended to adopt the Stoic position, for obvious reasons. But both Christian theology and Greek cosmology regarded nature as something to utilize: no natural object was sacred and transforming it carried no danger of divine retribution. Further, not even the higher animals were worthy of moral consideration: being cruel to animals was not wrong in itself.

A later critical development was the birth of modern science. Two names stand out: those of René Descartes (1596–1650) and Francis Bacon (1561–1626), though this ignores the contributions of men like Copernicus, Kepler and Galileo. The major legacy of Descartes was the view that nature could be conceptualized as a machine that could be understood by reducing it to the seven basic physical qualities (mass, length, time, etc.) plus the numbers 0–9. For Descartes, human thinking was not reducible in those terms and so a separation of mind and matter, subject and object, was introduced. Bacon was equally worried about the ways in which 'facts' might be separated from the humans who were the channel for them. His approach was governed by an initial demand that many observations should be made, after which laws governing relationships could be formulated. The agenda for collecting this reliable knowledge was that such information equalled power over nature and hence human progress. The ultimate goal for Bacon was no less than the reconstruction of the Garden of Eden before the Fall. From these dissimilar conceptual roots has grown much of today's science and technology, based upon 'real' phenomena, i.e. those which can be observed by the senses or their extension via instrumentation. Laws can then be constructed which describe regularities in the behaviour of phenomena and allow prediction of what will happen. Prediction can then be used to control. The more deterministic the system, the better the predictive power and the more 'successful' the science is judged to be, and so it occupies a higher place in the pecking-order of esteem among the scientists. The physics of materials thus ranks higher than ecology, both of which outrank the social sciences, for we are describing a downward trend in predictability. The practical outcome of much of the theory and of the discovery of laws has been a technology in which the good machine is one whose behaviour is entirely predictable.

Environmental thought today exhibits many continuities with the themes of the last few paragraphs but has at once become (a) more diverse as the number of disciplines of study has built up since the 19th century and (b) more holistic as people have sought to look at humanity–environment as a whole. But the general themes, those of beneficence and of niggardliness, and of plenty and of problems, remain. We shall need to return to these topics in due course.

Links between humanity and environment

The sociologist Abraham Maslow (1968) has pointed out that all humans have basic needs: some of these are purely biological, such as food, water and shelter; others are mostly emotional, like affection and sex, and yet more are social, such as the need to have an occupation and the desire for esteem from others. What is interesting in the present context is the resource and environmental implications of these needs. Even if only the most basic of these were to be satisfied, the demands upon environmental flows for the food, water and shelter of 5000 million people would be quite strong. When esteem by others requires such acquisitions as a Porsche or a handbag of whale-flipper skin, then the potential for manipulation becomes very great indeed.

Basic needs

It is possible to find comparative measurements of human interactions with the environment. One of these is that of energy consumption

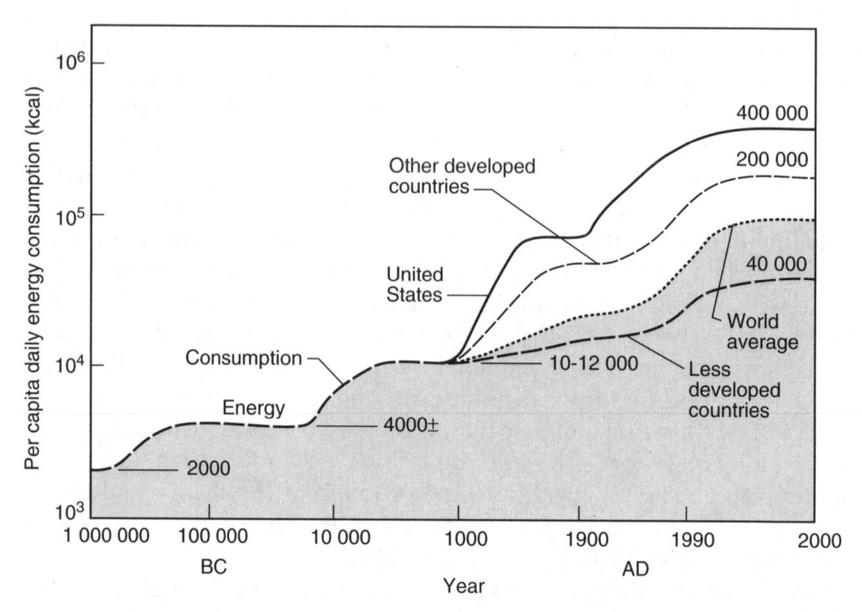

Figure 1.11 The growth in energy use over a very long period. The divergence between the HIEs and the LIEs in energy consumption is very marked. Each surge in energy availability confers a greater ability to manipulate the natural and human-dominated environments (from Simmons, 1989)

from commercial sources, since possession of commercial energy sources generally means access to materials from elsewhere or the ability to manipulate the local environment (Fig. 1.11). The basic survival requirement for an adult is in the region of 2000–3000 kcal/day (though less is possible for short periods) but in order to bring a certain dignity to life, then clean water, fuelwood, food, basic medicine and basic education are needed. Provision of these takes about 27--37 $\times$ 10^3 kcal/cap day. The average consumptions in some LIEs are set out in Table 1.1, where it must be recalled that biomass energy is the chief source but is not included in calculations of commercial energy. More complex calculations of the environmental linkages of nations exist, based on population, life-style and technology, but in the end they tell us little more about the relative material standards of different parts of the world than the energy consumption data. They may, however, attempt to disentangle the contributions of population growth and life-style to environmental change. For LIE farmland area, for example, in the years 1961–1985, population growth was responsible for 72 per cent of environmental change, and increase in consumption 28 per cent. In developed countries, the population

Table 1.1 Energy consumption in some LIEs, 1991

Place	Per capita commercial consumption (GJ)	Per capita traditional fuels (MJ)	Percentage of traditional fuels	
			1991	1971
World	60	3702	6	5
Africa	12	7275	38	42
Mali	1	5627	89	90
Zambia	20	6812	26	52
China	23	1724	7	7
Nepal	1	10 247	93	97
El Salvador	10	10 950	43	61
Brazil	23	13 328	36	51
Albania	26	4628	15	21
Fiji	14	16 175	53	60
Papua New Guinea	8	15 083	65	78

Source: WRI (1995) *World Resources 1994–95*. Oxford University Press, 1995, Table 21.2

growth explained 46 per cent and consumption patterns 54 per cent. For global CO_2 then, population growth comes out in the 40–44 per cent bracket.

Cultural demands

The differences in material standards between the richer and poorer parts of the world have clear consequences for environmental relations. Many of these, as indicated in the vignettes, are on very extensive spatial scales: the outreach of rich nations like Japan or Canada for resources is truly tentacular (Simmons 1991). In global terms, the high-income economy (HIE) group of nations now uses about 58 per cent of the world fertilizer production, 75 per cent of the oil and 86 per cent of the natural gas. From this it generated 91 per cent of the world's industrial waste and 93 per cent of industrial effluents, as well as 74 per cent of the carbon dioxide emissions from fossil fuels. This is all from a base of 22 per cent of the global population. So each of us in a developed country uses typically 12 times more oil than our counterpart in a less developed nation and produces 40 times as much industrial waste, including 75 times more hazardous waste. In the 1970s this could be quantified for the lifetime usages of a Western individual (in West Germany as it then was): 460 t of sand and gravel, 99 t of limestone, 39 t of steel, 1.4 t of aluminium and 1 t of copper. Their provenance will have ranged from

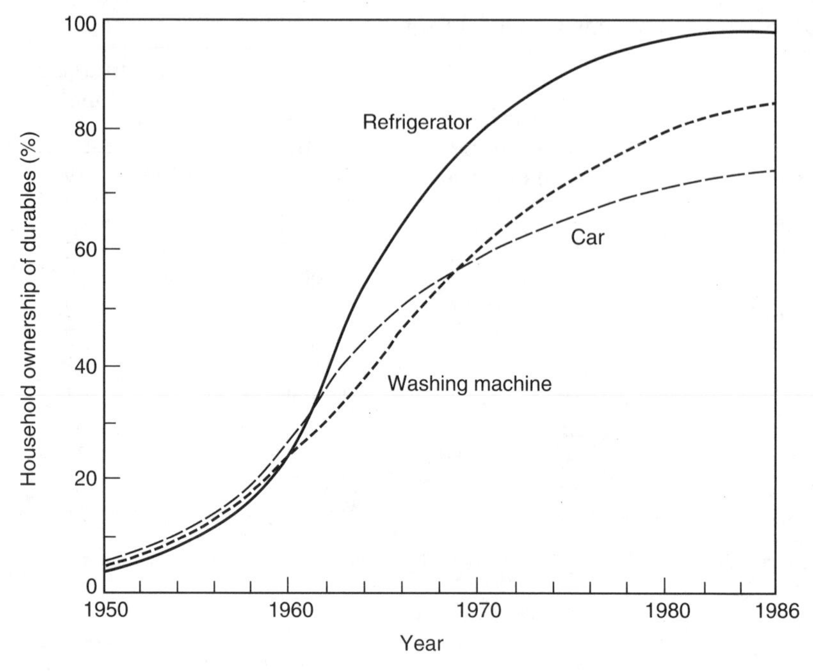

Figure 1.12 The rise in affluence as measured by the possession of consumer durables in France between 1950 and 1986. Consider the increased energy and water demands which are implicit in these curves (from Smil, 1993)

the very local (e.g. gravel), to possibly another continent (e.g. copper), although increasing amounts of this will be recycled, as has happened with steel. The environmental impacts of the industrialized life-style can be seen in the instance of the motor vehicle, where 1993 calculations suggested that over 10 years, each car produces 44.3 t of CO_2; 4.8 kg of SO_2; 46.8 kg of NO_2; 325 kg of CO and 36 kg of hydrocarbons along with 1016 m^3 of polluted air. This means that in the lifetime of each car, every one of them produces three dead trees and 30 'sick' ones. Each car is responsible for 200 m^2 of tarmac and concrete; in Germany the parking and driving requirements command 3700 km^2 of land, which is 60 per cent more than is allocated to housing. Finally, disposal of the car produces 26.6 tonnes of rubbish. Altogether, the external costs per vehicle were about £2400 (DM 6000) per annum in the 1980s (i.e. pollution, accidents and noise, after income from taxes on fuel and vehicles), which is equivalent to a free pass for all forms of public transport or 15 000 km of first class rail travel. Figure 1.12 shows graphically how the affluence of one Western country has risen in the last 40 years.

The interlinkages today

This chapter has had one underlying agendum: to suggest the complexity of all the matters concerning the interactions of humans and their surroundings. The types and quantities of resources used are enormous, populations are very diverse spatially and culturally, and all is set in a historical context. There are other features, too, of the whole:

- The human species still possesses the capability to wage nuclear war and thus to destabilize climate, ecosystems and all properties of human life and culture.
- The scale and speed of human actions today are without precedent and hence any 'traditional' approaches to problems are likely to have only local significance.
- Unpredictability and uncertainty are increasing, so that old ways of thinking which related to a foreseeable world that was somehow more 'external' to a way of thought are no longer practicable.
- Interdependence is growing, including connectedness of political, biophysical, economic and cultural phenomena, together with interaction between local and global scales (and those in between), which means that (a) new vulnerabilities emerge but (b) new forms of synergy and co-operation become possible.

This means that any thinking about the world and our place in it has to adapt to this evolving set of systems. Some of the characteristics of our thought might be:

- The capacity to include complex biophysical, socio-economic, historical and political factors in one framework.
- The capacity simultaneously to consider processes at different spatial and temporal scales: the local and the global, the fast and the slow.
- A way of discussing the dynamics and the structural changes of natural and human-directed ecosystems at different scales which evolve in ways which are non-linear and not in equilibrium.
- Ways of interfacing accounts of phenomena which are measurable in the scientific sense with those which are qualitative and which call for judgements based on ethical rather than (say) numerical grounds.

None of these is easy by itself, let alone altogether simultaneously. Yet as we discuss some of these lineaments in this book we must all the time consider how well they help in such tasks.

This second kind of box (enclosed with a thick line) finishes off a major section, such as a chapter. It is mostly concerned with summing up the previous chapter and pointing forward to the next. It is intended as a device for linking across the necessary divisions in a text which in an ideal world would be seen as a whole.

Chapter 2 will consider the main findings of the natural sciences which appear to be germane to our task. We need to be reminded of what kind of information the sciences give us: it is not related to a single human consciousness but is verifiable between observers. But it also comes from a social setting in which prediction and then control seem to be the real agenda rather than simply the disinterested accumulation of 'facts'. We might use the pictures of Earth from space as a metaphor: the closer in we get to the object then the higher the resolution of the quantitative measurements that can be made (e.g. of the area of different land uses) and the better the chances of applying reductionist methods of investigation, such as looking at the expansion and contraction of one part (e.g. the deserts) and seeking explanations from just those measurements. We can also be reminded that it was humans who built the machines that make the images and process them (highly apparent in false-colour pictures from satellites) and so all the facts that are collected are preceded by some kind of idea about what it is we are looking for anyway and also, possibly, why. Since the sciences are so diverse, we shall have to be very selective in reviewing their findings here.

Further reading

The initial vignettes cover such a wide set of processes that no single source could amplify them all and so they are not referenced. Thereafter, readers wanting further detail of the Gaia hypothesis should start with Lovelock (1989). There is a plethora of recent books on environmental economics but the one showing the greatest interface with ecology is Barbier *et al.* (1994); a similar outpouring in politics can be accessed through Eckersley (1992). Wall (1990) is more about commitment to change at social and personal levels. Luhmann (1989) is not easy (though first try the translator's introduction) but is rewarding. The history of population is told in a straightforward way by Livi-Bacci (1992), with some forecasts; Findlay (1991) has more comment and more linkage to our themes. The Open University text that contains Brown's (1990) essay on changing human attitudes is also full of other interesting material. Pepper (1984) is especially good on post-19th century times. Some data on current consumption levels can be found in Simmons (1991) but seek more up-to-date material of this kind in the campaigning journals of bodies like FoE and Greenpeace and in the *New Internationalist*.

Barbier EB, Burgess JC and Folke C 1994 *Paradise Lost? Ecological Economics of Biodiversity*. Earthscan, London

Brown S 1990 Humans and their environments: changing attitudes. In Silvertown J and Sarre P (eds) *Environment and Society*. Hodder and Stoughton, London: 238–71

Eckersley R 1992 *Environment and Political Theory. Towards an Ecocentric Approach*. UCL Press, London

Findlay A 1991 Population and Environment: Reproduction and Production. In Sarre P (ed.) *Environment, Population and Development*. Hodder and Stoughton, London: 3–38

Livi-Bacci M 1992 *A Concise History of World Population*. Blackwell, Oxford

Lovelock J 1989 *The Ages of Gaia*. OUP, Oxford

Luhmann N 1989 *Ecological Communication*. Polity Press, Cambridge

Pepper D 1993 *Eco-socialism. From Deep Ecology to Social Justice*. Routledge, London and New York

Pepper D and Colverson T 1984 *The Roots of Modern Environmentalism*. Croom Helm, London

Wall D 1990 *Getting There. Steps Towards a Green Society*. Green Print, London

CHAPTER 2

Data from the natural sciences

Our early admission of the word 'culture' reminds us that all human activities are subject to mediation by our various cultures. Even such a basic human need as food is heavily influenced in its acceptabilities by what is culturally acceptable. Some people will eat raw sea slugs but not blue cheese for instance. In contrast, science as an activity is one of the least prone to variations between cultures but it has grown in particular historical circumstances. To most it is of itself a cultural practice therefore, and not some unalterable cosmic 'given'. Because of the way in which results in the natural sciences are achieved by agreed and rigorous procedures, however, they have a special standing when applied to the non-human elements of the universe. This applies to many of the physical and behavioural traits of human beings as well. But science rests on certain prior assumptions about the nature of the cosmos and so it consists of theories as well as factual statements. The relation between the two is often closer than at first imagined and this chapter will give some attention to both, though the current statements about the natural world will receive the more detailed treatment.

The sciences' view of the world and ourselves

Science suggests that it is possible to prepare detached accounts of what things are like in nature (Chalmers, 1982). We need to know about the kind of information relayed to us by the natural sciences for our attempts to predict the future, a task often undertaken in the service of improving the degree of control we can apply. The types of information brought to us by the natural sciences in all their diversity and specialization are manifold but they tend to have in common the characteristic of being reductionist. This means that the immense complexity of whatever was under investigation was dealt with by breaking down systems into component subsystems, or complex mechanisms into constituent parts, so that available techniques of investigation could be deployed with greater precision. The complex and interacting movements of elements like nitrogen, sulphur, phosphorus and carbon through the pools which hold them temporarily (such as the oceans, the atmosphere, the land and fresh waters) must often be disaggregated into separate cycles (see the later material in this chapter) to make their measurement possible. What must be asked at the end of this chapter is the degree to which these and analogous reductionist models lose the kind of validity that makes them useful as generators of new ideas or as guides to practical action.

The interrogation of nature

Our information about nature comes mostly from one of two sources. The first is direct experience. When I walk our dog by the river early in the morning I sometimes see a kingfisher. What I 'know' of that bird in that moment is direct and little mediated by other knowledge. Yet

> After the kingfisher's wing
> Has answered light to light, and is silent, . . .

there is a plethora of other information, from its position in a Linnaean classification scheme that places it in the family Alcedinae and the genus *Alcedo*, to maps of its distribution in England and Europe and bookfuls of information on its feeding and nesting habits. This all boils down, first thing in the morning, to knowing that it is quite an uncommon bird and that my day is the richer for the sight: fewer students are likely to be snarled at that day. The source of the other data, of course, is mostly the natural sciences: the patient accumulation of observations made according to procedures which ensure that if done elsewhere or at a later time will yield data that are genuinely comparable. Note though that here as in many environmental cases, controlled experiments are less likely to be performed than in investigating new chemical compounds, for example.

Explanations of system function

Because environmental systems are often reckoned to function on a large scale, they are rarely suitable for experiments unless reduced to smaller-scale subsystems. We cannot manipulate experimentally the climate of a whole city and its surrounding area very easily but we can mimic the formation of photochemical smog (PCS) in a way that provides a good explanation of how it is formed in cities with certain climatological and topographical features. But this will not tell us the full story of the effects of PCS on plants or people, for example, and we ought not to ask for volunteers to expose themselves to different intensities of PCS in the laboratory. The use of other animals is highly controversial.

The complexity and spatial scale of environmental systems has not deterred the natural sciences from undertaking many types of investigations. To each environmental system, a number of the classical natural sciences have been applied. In the case of the atmosphere, for instance, the science of physics deals with the movement of air masses and leads to the application of mathematics in describing those dynamics; chemistry can be applied to the nature of the aerosols which might form the nuclei for cloud formation and rainfall. The combination of applications of parts of these sciences brings about a new crystallization of knowledge which we label meteorology (if it deals with daily weather phenomena) or climatology (if it deals with the pattern over years). The oceans have been subject to measurements by geophysicists, by chemists and by biologists and their combined knowledge becomes oceanography. In the freshwater environment, the outcome of the same combination is hydrology. The ice-masses are the objects of research by physicists, by climatologists and by geomorphologists and their syntheses are called glaciology. The land surface has, obviously enough, been investigated more than other environmental systems and the list of interested sciences is long: climatology, geomorphology, biology, pedology and geology constitute only a partial list. The outcome in a synoptic sense is physical geography but in recent years there has been an emphasis on describing and explaining dynamics rather than only portraying static patterns and so the science of ecology has been called upon to provide models for the whole physical land system: 'landscape ecology' is one of the syntheses that has emerged. At the scale of the planet's systems, ecology seems one of the most fruitful sciences, partly because of its breadth of concerns (Fig. 2.1).

Each of the basic sciences and each of the resulting system-oriented studies makes use of different kinds and intensities of data. The terrain of England, for example, has had its soils mapped in minute detail, so that maps at a scale of 1:50 000 can be published, giving a detailed taxonomy of soil types; an accompanying memoir contains further detail of the chemistry and water relations of each of the soil types recognized on the map. Contrast this with the input to global climatic models (GCMs), where the basic unit of mapping is a rectangle of 1° latitude by 1° longitude and not all of these have the basic climatological data needed for world-scale climate modelling. We might also contrast an England where almost every patch of semi-natural vegetation has had detailed studies of its ecology made at regular intervals in this century with many parts of the Third World, where even a basic inventory of species present has not yet been made.

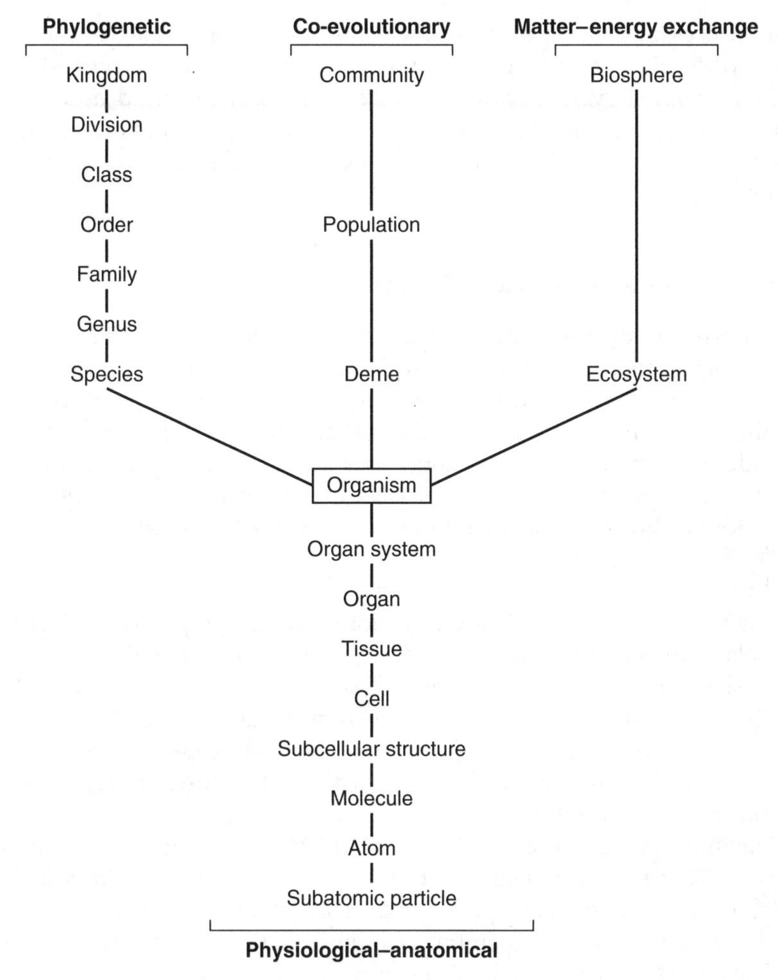

Figure 2.1 The scales of ecology. The investigations of this science start with the very small but divide at the level of the individual organism. Thereafter, the larger scales vary according to the questions being asked – whether about classification (phylogenetic), the formation of communities (co-evolutionary) or ecosystem function (matter–energy exchange). If physics were the central science then another top layer, that of the cosmos, would have to be added (from Pickett *et al.*, 1994)

Such variations in gathering information are paralleled by differences in the amount of processing which the data receive. In some investigations, the data are used in a virtually raw state: an emergency survey of insect populations as part of an Environmental Impact Assessment will go into the section on biotic diversity with probably no more processing than a count of the total number of species and a list of any rarities. By contrast, the output of a satellite image is a refined piece of high technology involving the skills of those who build and launch satellites together with those who program the instruments to send and receive digital signals which can be manipulated into explicable patterns.

The communication of scientific data

When we survey the output of all these sciences and their resultant technologies, there is a bewildering gamut. For the specialist there may be computer files of digital data which can be transmitted between continents within seconds; at the other end are snatches of half-considered stories set to very loud rock music and jolly icons in a 20-seconds-per-item teenage TV magazine. In between, almost everything is to be found, but two considerations are worth while singling out at this stage:

- The versions of scientific investigations that are prepared for decision-makers in bureaucracies and politics. These are inevitably shortened versions of the original (many reports and even books have 'Executive Summaries') and tend to reduce the complexity of any set of systems or problems. In particular, they may cater to a consumer's demand for assurance by diminishing the uncertainty of any conclusions or predictions.
- The importance of the word in any explanations. Equations, diagrams and pictures are frequent aids to interpretation but the bottom line, so to speak, is dependent upon words. The use of language, therefore, is always crucial in any communication of scientific endeavour. Many discoveries are couched in terms of metaphor.

The main and inescapable lesson is that the communication of science is always selective: at each stage between the formulation of a project and the appearance of results in print, selectivity is applied; in the later stages, to quote only the most obvious example, editors of journals are always looking to shorten papers. So all accounts, even when mature investigative techniques are used, become representations of the apparently external world which was the initial object of the inquiry: the map is not the territory.

The cultural context of science

In spite of its many intellectual and practical achievements and the degree (unparalleled in any other branch of human affairs) to which it is an international endeavour, natural science is not always devoid of cultural content: it is not in the ideal position of a detached but all-observing consciousness floating above the objects that it studies (Ziman, 1980, 1994). In the worst instances, a field of study is not pursued, because of a political motive: the Nazis abandoned certain branches of physics because they were 'Jewish science'. In the post-1945 USSR, the results of investigations had to conform to Marxist–Leninist theory if they were to get an airing: Darwinian evolution was for a long time unacceptable, just as Lamarckism was politically correct. (Ironically, Marx had wanted to dedicate the first volume of *Das Kapital* to Darwin.) By contrast, governments will put money into science for political or military ends, thus skewing the output to certain types of knowledge: the billions of US dollars spent on the SDI ('Star Wars') programme, with virtually no useful results at all, is only an obvious example of a daily process.

The ease with which societies are allowed by their populations or sustaining élites to spend money in such ways has led social scientists to suspect that there is a hidden agenda in the case of the natural sciences. Part of this is scarcely hidden at all: science and technology seem to lead to material abundance. But beyond that statement, there is sometimes discerned a deeper cultural project focused upon the domination of nature. Sometimes this is in the name of immediate prosperity for all or profits for a few, but even more it is in the name of wanting to impose human domination over everything which is not obviously and immediately human. This may be driven by what appears to be a strong hominid trait to form hierarchies, in which the cosmos conforms to a line management model of the God–man–woman–children–animals–other living things–non-living things type. Within this, it is often argued that the key element is 'man', with the science that drives this model being an essentially androcentric activity. Science as a cultural activity is seen as a outgrowth of the male psyche; there could be, logically, a feminist science.

> Science, hence, is part of human culture. Even the most strenuous attempts will not make it a totally detached, free-floating activity. Data are gathered in the light of pre-existing theories or even policies.

Asking ourselves questions

Our own species is not immune from the dissecting views of the natural sciences. More than one eminent biologist has suggested that their mission is to explain human individuality in terms of the molecular composition and metabolism of the body. At its extreme, perhaps, is the statement of the sociobiologist E. O. Wilson (1975) that 'the organism is only DNA's way of making more DNA.' In his view, genes are the basic source of power over behaviour and the brain exists merely to carry out the instructions of the genes. Therefore, if a piece of human behaviour such as altruism appears attractive to us, then it must have a role in ensuring the perpetuation of a particular set of genes. This view is strongly reductionist since it holds that the sum of the parts is simply the sum of the parts. Nevertheless the success of reductionist sciences is rarely questioned when it comes to physiological matters, for the understandings which have given us much modern medicine are well agreed and much appreciated. Rigorous experiments and a well-tested body of theory have given us drugs to combat infectious diseases, for example.

Human culture

In all its global variety, human culture can be subjected to the same analyses as other phenomena. One school of thought has regarded human cultures in the same light as animal behaviour, namely as a set of adaptations to the natural environment which ensure the survival of a species. Culture is thus part of an evolutionary process and indeed in the case of *Homo sapiens* is assumed to have replaced physical change as an expression of continued adaptation. If aspects of current human behaviour seem maladapted to the natural environment (examples such as rapid population growth or technologically based warfare are often quoted) then the conclusion is sometimes drawn that our species is destined for relatively early extinction. There is indeed no apparent reason why this one species should last for ever (no other species can be traced throughout the entire fossil record, after all) but our attitude to 'environmental problems' is underlain by the desire to ensure the perpetuation of our species. Otherwise, why not burn the candle at both ends and the middle and go out in a pyrotechnic display sooner rather than later?

One reason for apparent maladaptation is assigned to culture lag. In this interpretation, many human behavioural traits were implanted in the hunting and gathering stage which has occupied at least 90 per cent of our evolutionary history. So male aggressiveness, which is a positive feature in a hunting culture, has no socially approved outlet in an industrial society. Equally, we are good at responding to immediate stimuli analogous to the sudden sighting of a prey species, but less good at the perception of long-term threats. The latter is called the 'boiled frog syndrome' since it is said that if a frog is dropped into a pan of boiling water it will jump out, whereas if placed in a pan of cold water which is gradually heated up then it will never perceive that the moment has come to get out, and so will perish.

Such cultural arguments are also underpinned by a discussion of the range of our perceptual systems, since our neurophysiology is at the base of what we receive from around us and hence of how we construct

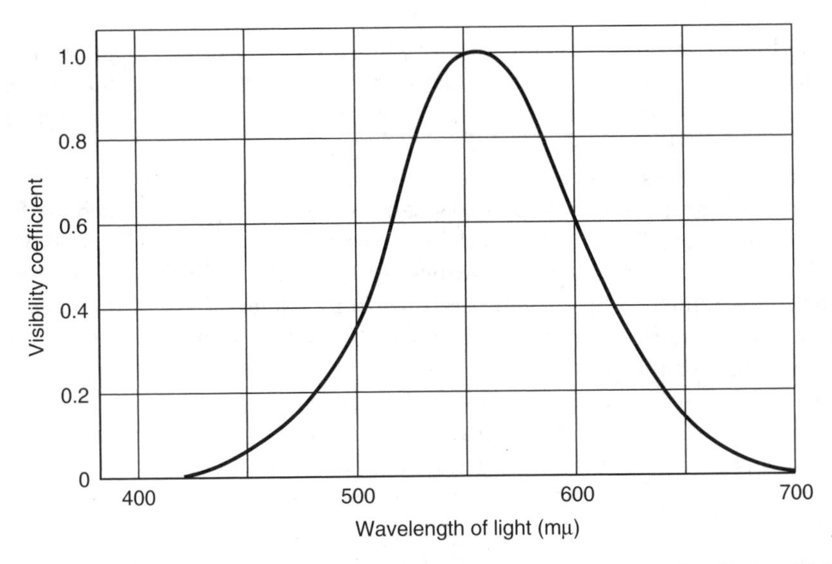

Figure 2.2 The sensitivity of the human eye when in the light. This shows that we receive only a portion of the light wavelengths that are present in the world; the same is inevitably true for other senses (redrawn from Gregory, 1972)

it in our brains. It is a commonplace that our vision, though excellent for 3-D imaging, is limited at low light levels compared with many other animals. Even in the light, it is restricted in the range of its receptivity (Fig. 2.2). Similarly, our hearing cannot consciously detect frequencies above a certain range. There are, presumably, many such stimuli on the planet which we cannot detect but which may affect our behaviour. Low-level ionizing radiation is a well-documented example: there is no way in which we know that we are being exposed to low levels of radioactivity from either natural or human-induced sources unless we extend our perceptions with the aid of instruments. Less well understood (and hence even more controversial) is the effect on human health of magnetic emissions from high-voltage overhead power lines.

Spatial scales

In the last 50 years especially, attention has increasingly been paid to those natural processes which are truly global in nature. Such mechanisms are not simply those which occur at any place in the world but which are actually connected with each other. Soil erosion, for example, may happen anywhere in the world but any one occurrence has no significance for the others. On the other hand, the injection of a long-lived compound in aerosol form, like a CFC, has global significance since it is diffused by atmospheric processes throughout the entire gaseous envelope of the planet. Most of the rest of this chapter is concerned with the global scale.

Studies of these processes are usually of two types: those which have been built up from a multitude of smaller-scale investigations, such as the silt load which the great rivers deliver annually to the oceans, and those which were planned as global studies. An example of the latter would be the measurement of ocean temperatures from satellite data. In the case of world-wide studies, the usual outcome is the construction of a model of the distribution of the studied feature (such as carbon or copper or water) in the form of a diagram which displays reservoirs together with the fluxes (flows) between them. Such reservoir–flux models are often made for smaller scales as well but do not cope well with short-term dynamics: they are of little use in weather forecasting for example.

It is thus possible for the natural sciences to construe human culture in a rather narrow light as a particularly human form of animal behaviour. The actual diversity of culture too can be studied as if it were the ways of ants or baboons, and described in much the same terms. In some of the examples in the rest of this chapter, 'human impact' upon the rest of the world is often seen in terms of the changes wrought by an alien but behaviourally homogeneous species.

The science of the global environment

Science undertakes the formidable task of studying the great functional systems of the world (Fig. 2.3), e.g. its landforms, the circulation of the oceans, the surges of the atmosphere and the distribution of plants and animals, and then attempts eventually to relay that information in generalized formats which an individual mind can comprehend, as if they sought, with William Blake, 'to see a world in a grain of sand,' partly as an intellectual endeavour but also in the full realization that human societies now have the power to deflect the courses of some of those great cycles.

World-wide studies

Until the advent of rapid communications, the accumulation of a global or world-wide view of a phenomenon was a slow business. A scientist would himself travel and formulate some ideas and then engage in a protracted correspondence with other like-minded men (usually *sic*), supplemented by what was published in learned journals and occasional

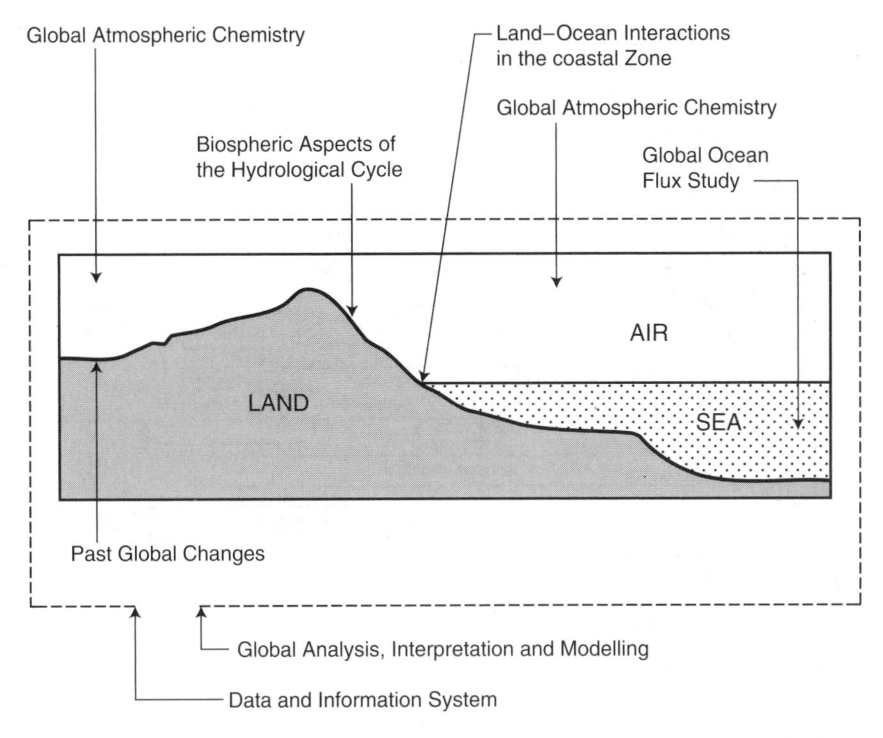

Figure 2.3 A way of interrogating the global systems and their linkages. Each system is studied separately and in conjunction with one or more of the others, and at intervals through time. In fact each of these boxes refers to a major international research programme

books. Charles Darwin (1809–1882) is a good example (Desmond and Moore, 1991). Nowadays, the communication is far more rapid: the number of printed sources is immense and is now supplemented by the resources of the Internet. The digital computer has contributed to these changes by making possible the storage and processing of quantities of data (mostly numbers but not entirely so) on scales not credible until the last 20 years at most. In the natural sciences, as in finance, cheap electronic communication has led to the formation of a global system of almost instantaneous information transfer, so that there is a human-made global 'cycle' to add to those of nature, such as the water and carbon cycles.

Nature of the models

As stated above, most of the models constructed, at whatever scale, are of the reservoir–flux type and sometimes the words 'pool' and 'transfer'

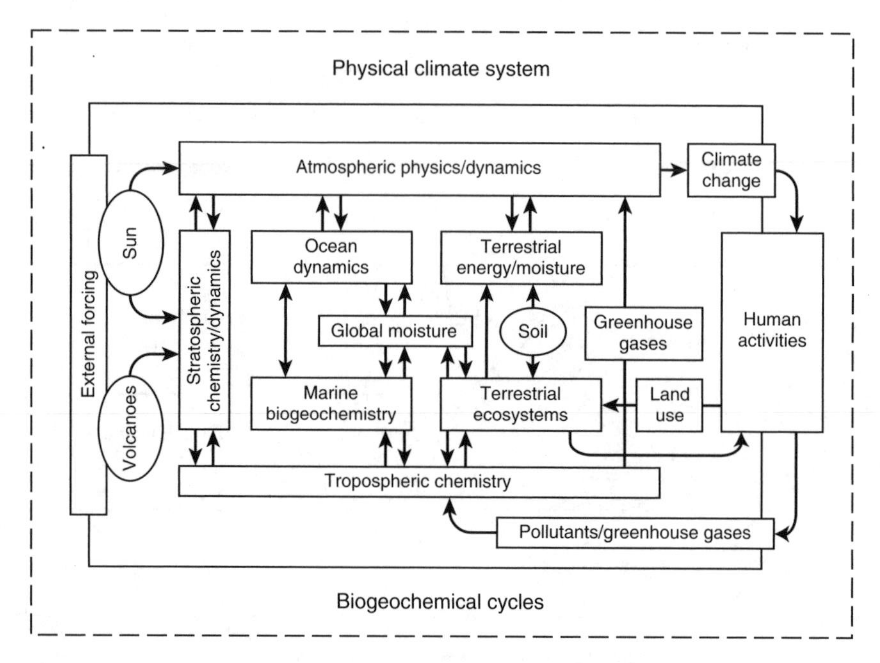

Figure 2.4 A pool and transfer model of the connections between biosphere and the climate system; it is possible to measure flows of elements and compounds between boxes, whether the boxes are natural or contain the results of human activities (from International Geosphere–Biosphere Programme)

are used (Fig. 2.4). These have a number of positive features. In particular, they give a good overview of the whole process at the chosen scale: the quantification can include not only the reservoirs and the fluxes but the residence times of any given element in one of the pools. They also show gaps in knowledge rather well. Given such an overview, it is usually possible to assess the relative roles of nature and humans in the cycles. There are also negative features of such models. Each reservoir tends to be a black box; it is easy to interpolate the missing numbers without actually making any measurements and the quantities tend to be averaged in a way that may ignore important spatial or temporal variation. Their 'overview' capability may also detract from the realization that they are not very good at conveying data about changes over time.

General results

The investigation of the various cycles as separate entities (e.g. the oceans, nitrogen, sulphur and carbon) has fortunately not obscured the

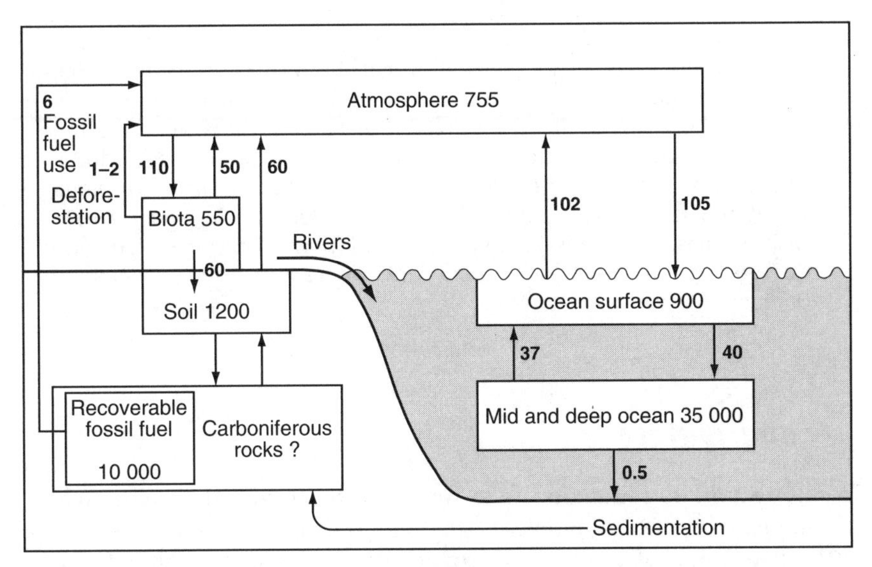

Figure 2.5 A simple model of the global carbon cycle showing the size of the main pools and the annual fluxes between them. The immense quantity that resides in the oceans is perhaps the greatest surprise to most readers of such diagrams. The reservoirs are in Pg of carbon and the flows in Pg of carbon per year (1 Pg = 10^{15} g = 1 billion tonnes) (from Middleton, 1995)

realization that many systems are interconnected. There appear to be, for example, linkages between ocean temperatures and most other global climatic fluctuations; the glacial/interglacial cycle is in harmony with variations in the concentration of atmospheric CO_2; and it has long been realized that carbon cycles between organic and inorganic pools (Fig. 2.5).

Work in the last 50 years has brought about the realization that some of the global cycles can be affected by human societies in more than simply marginal ways. This has applied especially to the soils and life-forms but also to climate at varying scales. Some, naturally, are unchanged at the global scale: the hydrological cycle happens at a magnitude beside which the interventions of humans are so puny as to make very little difference outside the regional scale. Some cycles, however, are clearly changed (the increasing concentration of trace gases in the atmosphere is a good example) but we do not know for certain the effects either upon the predictability of the system's behaviour or more directly upon human individuals and societies. We can only make models of these relationships, and admit all the uncertainties.

Two major points emerge: the first is that the movement of elements round the globe takes place on a variety of spatial scales but the truly global is often one of these. Appended to such considerations is the time factor and most of the models deal with periods of three months and more, not the dynamics of every day. The second point reminds us that the label often applied to these cycles is 'biogeochemical', which implies that life is involved in an element's progress from one reservoir to the next. Inevitably, this increases the chances of significant human involvement.

Air and water

In this section, the intention is to focus upon the global systems which are modelled as the dynamics over time and space of the atmosphere and of water. Thus we shall cover recent thinking (hardly *current* thinking since any account will be out of date as soon as it goes to press) about long-term climatic changes of the glacial/interglacial type as well as shorter-term natural changes. Following on naturally comes the central role of the oceans, both their physical characteristics and their level relative to that of the land masses. Finally, we remember that a large quantity of fresh water is in the form of ice and that the accumulation and melting of the world's major ice masses are closely bound up with phenomena such as climatic change and sea-level fluctuations.

Global changes of the last two million years

More specialized sources must be consulted about the details of geophysical fluctuations during the geological period known as the Pleistocene. This is reckoned to have started between 2.5 and 1.6 million years ago (Mya). It ended 12 000 years ago (12 Kya) with the beginning of the Holocene, in which we now live. For both, however, some reminder of the major alterations in climate and other geophysical features is needed as a context for the human tenure of the Earth. Then, as now, two major sets of processes can be discerned:

- Long-term changes happening in cycles of perhaps 100 Ky, subject to shorter-term accelerations over periods of *c.* 10 Ky or less. If scientists had been present, then such changes would have been largely predictable.
- Unpredictable catastrophes such the impact of a meteorite. In the present context, the most significant to humans are probably major volcanic eruptions; a series of them even more so.

The Pleistocene

The transition from the Pliocene to the Pleistocene was marked by world-wide cooling and the first evidence of this is seen in the traces of glaciation at about 2.3 Mya in the northern hemisphere (Goudie, 1992). Thereafter, there is good evidence for a series of fluctuations between (a) glaciations spreading equatorwards from the poles and (b) warmer interglacial periods. The evidence comes from terrestrial deposits, ice-cores and deep seabed deposit cores and the main findings of each agree, even though a variety of e.g. dating methods are applied to different materials. In higher latitudes, glaciation is still being experienced, with waxing and waning of the ice limits in response to short-term climatic fluctuations. Beyond the ice, there is a zone of periglacial processes where the soils are unstable because of summer melting of a thick layer of ice in the subsoil (permafrost). Figure 2.6 shows the maximum extent of ice and permafrost at the height of the latest glacial maximum, called the Wisconsin glaciation.

Middle latitudes are now free of ice except on high mountains. During the Pleistocene they experienced a series of alternations of glacial periods, when they were covered with ice-sheets, and interglacials that were warmer than the present non-glacial period, with hippos and elephants present at 50 °N, for instance. Strong cold winds blowing across the ice-margins deposited thick layers of loamy material now known as loess. The length of the glacial/interglacial cycles cannot be calculated exactly but in middle latitudes, the interglacial periods seem to have lasted between 10 and 130 Ky with a bias towards the longer intervals, and the glacials seem to have lasted 95–175 Ky, with short periods of remission known as interstadials. The difference in yearly average temperatures between the full glacial and mid-interglacial times was about 8 °C in north temperate latitudes, but interstadials saw ameliorations of only about 2–4 °C. In the equatorial and south temperate zones, the difference between coldest and warmest was about 4 °C.

In low latitudes, the cold stages nearer the poles seem to have been cooler and drier; ice was present on higher mountains in the 'glacial' episodes, and forests retreated to lower altitudes. Drier conditions meant that savannas replaced closed woodland in some regions, and there is evidence that only towards the end of the Pleistocene did the lowland Tropical Moist Forests (TMFs – the tropical rain forests) assume their recent distributions. With climatic changes, the levels of lakes fluctuated considerably.

Globally, one of the most notable changes of the Pleistocene was the response of sea-level. This is affected by the volume of water in the oceans (itself a function of (a) its temperature, since it expands when warmer, and (b) the quantity of water which is bolted up as ice) and by

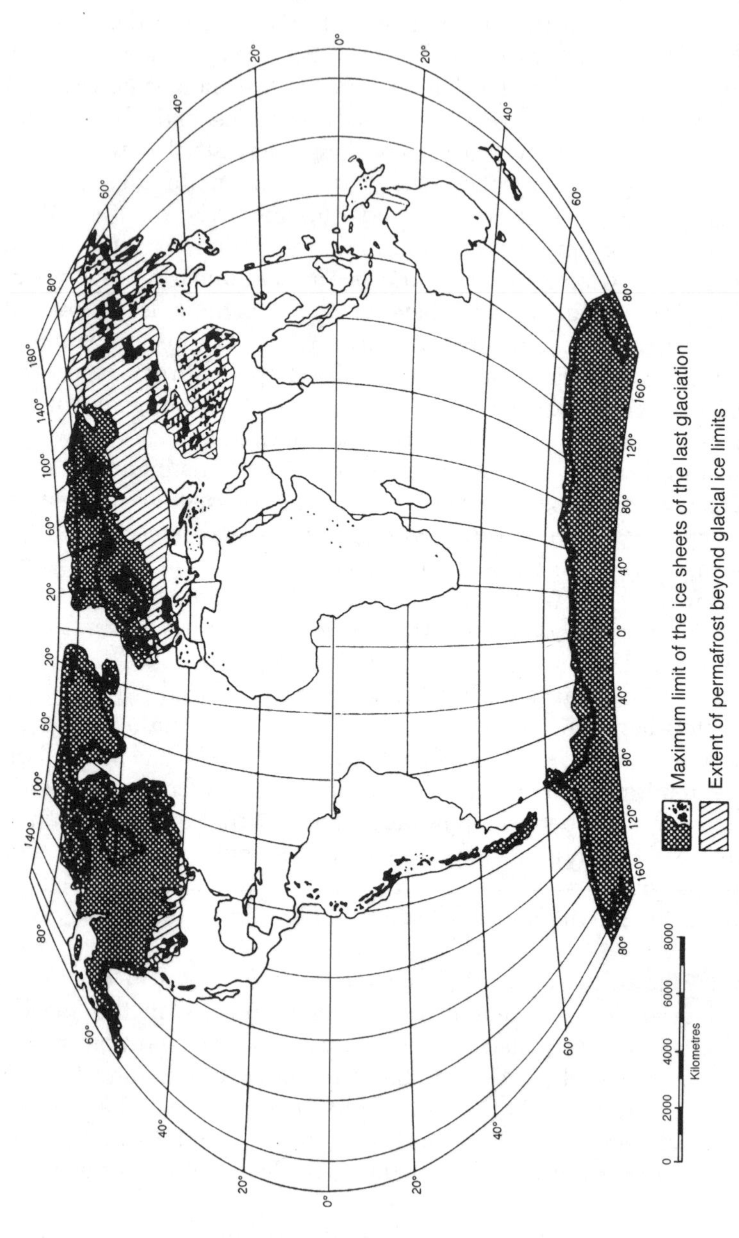

Figure 2.6 A map of the greatest extent of ice and permafrost during the latest glacial period of the Pleistocene. The ice sheets were typically 2–3 km deep and often stripped rock to a depth of 120 m. Note especially how much of Eurasia and North America was affected (from Williams *et al.*, 1993)

the movements of the land. Depressed under the weight of millions of tonnes of ice, the land springs back when the ice retreats.

The Pleistocene was, therefore, a world of change with cycles often of amplitudes of 100 000 years. These cycles brought in their train other changes, such as variations in the salinity of the oceans, but equally importantly the periods of low sea-level allowed the migration of plants and animals and the subsequent evolution of new genetic variety. These two million years also saw the evolution of our own species.

Another unique species

The biological evolution of the genus *Homo* is subject to an intensity of investigation and debate applied to no other group: we very much want to know where we came from and when. Scientific reputations are made (and lost) on the basis of a quantity of evidence which for any other species would ensure its relegation to the back rooms of obscure museums. Debate with 'creation scientists', who do not believe in evolution at all, adds heat if not light to the vigour of the inquiry.

It seems as if a first *Homo* can be distinguished at about 2 Mya, with its ancestors classified as various species of *Australopithecus*, a genus which appears to have died out after 1 Mya. The first stone tools are seen about 2.5 Mya, and the first hominid to migrate widely out of the African regions where the genus first evolved was *H. erectus*. A linear descent brought about *H. sapiens* from about 200 Kya. This was of *sapiens* the *neanderthalis* subspecies from 200 Kya until its demise *c.* 30 Kya, and in the *H. sapiens* or modern form from perhaps 90 to 40 Kya. With the end of the Pleistocene, certainly, *Homo sapiens sapiens* colonized most of the newly available land, with the exception of Antarctica (mostly still ice anyway) and some of the remoter islands of the Pacific. Australia, however, was colonized about 40 Kya and the Americas between 20 and 12 Kya.

Australopithecines were probably scavengers, with hunting accompanying the 'arrival' of the genus *Homo*. The first controlled use of fire in the landscape as well as at the hearth is credited to *H. erectus* and there is a steady improvement in the variety and efficiency of the tool kits whose remains have survived. There appears to have been an explosion of inventiveness after 40 Kya and the famous cave art of France and Spain appeared about then.

It is impossible to believe that the environments of early hominids played no part in their evolution, but certainty is evasive. Glacials being cool and dry might have reduced the area of forest and thus favoured species able to adapt to savanna, as seems to have happened with *Australopithecus*. Glaciation may also have isolated populations able to withstand cold conditions – circumstances put forward to account for the physiognomy of the neanderthals. But evolutionary outcomes in humans, even pre-modern ones, may well be the result of cultural choices rather than environmental selectivity; we are unlikely ever to have much reliable knowledge for the years before 10 Kya (Bilsborough, 1992).

The Holocene

The latest of the glacial stages of the Pleistocene terminated between 15 and 10 Kya. Then the global rate of warming is estimated to have been 2.5 °C per millennium, with overall global temperature rises of 5 °C and, in the ice-covered areas, 8 °C. The global CO_2 concentration also rose (from 190 to 270 ppm), suggesting a linkage between global temperatures and carbon dioxide accumulation, though not proving any causal links, one way or the other (Roberts, 1989).

The improvement of temperatures was accompanied by rapid rises in sea-level but the rate of both slowed down until, in the centuries around 9000 before present (BP), global temperatures reached levels from which they have since fallen. In temperate latitudes, temperatures were then typically about 2 °C above those of the present. The time of the highest temperatures seems to have varied regionally: in the Antarctic between 11 and 8 Kya but in the Arctic between 5 and 4 Kya, for instance. The period of highest Holocene temperatures (called the Hypsithermal) lasted until *c.* 5 Kya, after which there was a global decline in temperatures until very recently. Early intimations of cooling included the lowering of treelines in southern Sweden by 45 m around 5.3 Kya. Thereafter, the accumulation of evidence allows better resolution of periods of climatic change (Bell and Walker, 1992), so that we can discern the following:

- A 'little optimum' between AD 700 and 1300. Southern Greenland, for example, was colonized by Norse communities by AD 985 but abandoned in the mid-14th century.
- A 'little ice age' between AD 1550 and 1850. In the Alps, this brought glacier advance, avalanches and flooding; in Norway it was a period of many requests for tax relief because of natural disasters. Elsewhere, crop failure and lower productivity were common. A similar phenomenon can be detected in Peru in the years AD 1490–1880.
- Recovery from 1860 to 1920, then a plateau (or even some cooling) until 1980, after which we had the warmest decade of the entire instrumented period: the 1980s experienced the five global-average warmest years out of the last 100, for instance. This has also been a phase of increased variability of climate: 1910–1970 saw a high frequency of dry years.

Any of these identifiable intervals can be interrupted by volcanic eruptions. If the emissions are on a large enough scale, like those of Bali in 1981, then a global fall in temperature of 1 °C will ensue; locally this may be as high as 1.5 °C and both may last up to two years. Several other examples can be traced in different kinds of records.

The secular detail is not perhaps the point here. What is important is that climate is not necessarily stable on either a decadal or a regional basis. Either may demonstrate the kind of variability that has significance for human affairs. True, such significance is lower in an age of technology, but consider that the prevention of some of the examples quoted above, like avalanches, landslides and lower global temperatures, is beyond the capability of the most advanced technology, even in those regions of the world with access to such comforts.

Where does this place the present concern about global warming due to human activities? Given the historic link between global temperatures and CO_2 levels, the constant input of dust and aerosols from economic activities, and the glacial/interglacial sequence, it looks as if we might be past the middle of an interglacial whose course we are altering by means of the radiative forcing which is known as the 'enhanced greenhouse effect'. While it might look an attractive way to combat the onset of the ice, the outcome of radiative forcing in terms of its effects on sea-level and coastal storms, together with the high probability of increased variability of climate (and hence much more difficulty in forecasting both weather and climate), makes it much more prudent to allow nature to take her undeflected course. This is why the Rio Convention of 1992 on climatic change was signed and why it was intensified by the subsequent Berlin meeting in 1995, which set emission limitations and objectives for reduction within specified time limits for developed countries.

> The story of the Pleistocene and Holocene is one of change and variability. Some changes appear to have had a roughly regular set of cycles, other are less easily foreseen. The evolution of human culture has certainly been unpredictable. The complex systems resulting from the interaction of human cultures and the natural environment are therefore unlikely to be deterministic like those of classical physics and so will not be very amenable to precise forecasting of the 'what happens if we do this?' kind.

The oceans

Earth might better be called the water planet: viewed from space, it is the blue of the oceans which dominates the picture. The oceans cover approximately 71 per cent of the globe (3.61×10^{14} m^2 out of a total surface of 5.1×10^{14} m^2); the volume of the oceans occupies about 1370 million km^3, a mere 0.02 per cent of the planet's mass by weight. The oceans are different from the other components of the hydrological cycle in having a high dissolved mineral content and hence are termed 'saline'. This salinity has been constant globally (though there are regional variations) for perhaps 200 My at *c.* 35 parts of salts per thousand (‰). Occupying such a vast area, the oceans not only exert a strong influence on such immediate features as sea-level but also other great systems such as climate and weather.

Links with the atmosphere

The atmosphere contributes CO_2 to the oceans, where it mixes in the top layers of sea water; some mineral ions also become nutrients for marine life in the same way. The oceans yield sulphur to the atmosphere to the extent of 39 million t/year, mostly as dimethyl sulphide (DMS). This is produced largely by one group of marine phytoplankton and enters the atmosphere in aerosol form. It then acts as a source of condensation nuclei for water vapour and increases cloudiness and rainfall downwind. The production of cloudier conditions makes for a cooler sea and hence the production of less DMS: a negative feedback effect. It has been suggested that such a process might damp down any global warming from the enhanced greenhouse effect but the price would be lower marine productivity, i.e. fewer fish to be caught.

The long-range controls exerted by oceans are seen clearly in the El Niño phenomenon. This is an oscillation in the southern Pacific that happens every 2–7 years. Around Christmas time (hence the name of 'the child' in Spanish) the cold upwelling off the coast of Peru is capped by warmer water, with a temperature up to 5 °C above the normal. This has several regional effects, including rises in sea-level all round the Pacific, but the dominant result is lower plankton productivity and hence the failure (now exacerbated by over-fishing) of the Peruvian anchoveta fishery. At a wider scale, it has been shown that El Niño years are drier than normal in southeast Africa, the western Pacific (including Australia) and northern South America. Most striking of all is the realization that the Indian monsoon yields less rain, so that there is a direct correlation between El Niño and agricultural productivity in India (Fig. 2.7). This coupling has led to the term ENSO (El Niño–Southern Oscillation) for the whole event. The realization that it can cause both flooding and droughts on different continents as well as hurricanes has led to a calculation of the costs of the impact of an El Niño event (Table 2.1). A mirror image effect which causes a temporary chilling of the world and the opposite regional effects to El Niño (and is hence called La Niña) sometimes but not invariably follows an El Niño year.

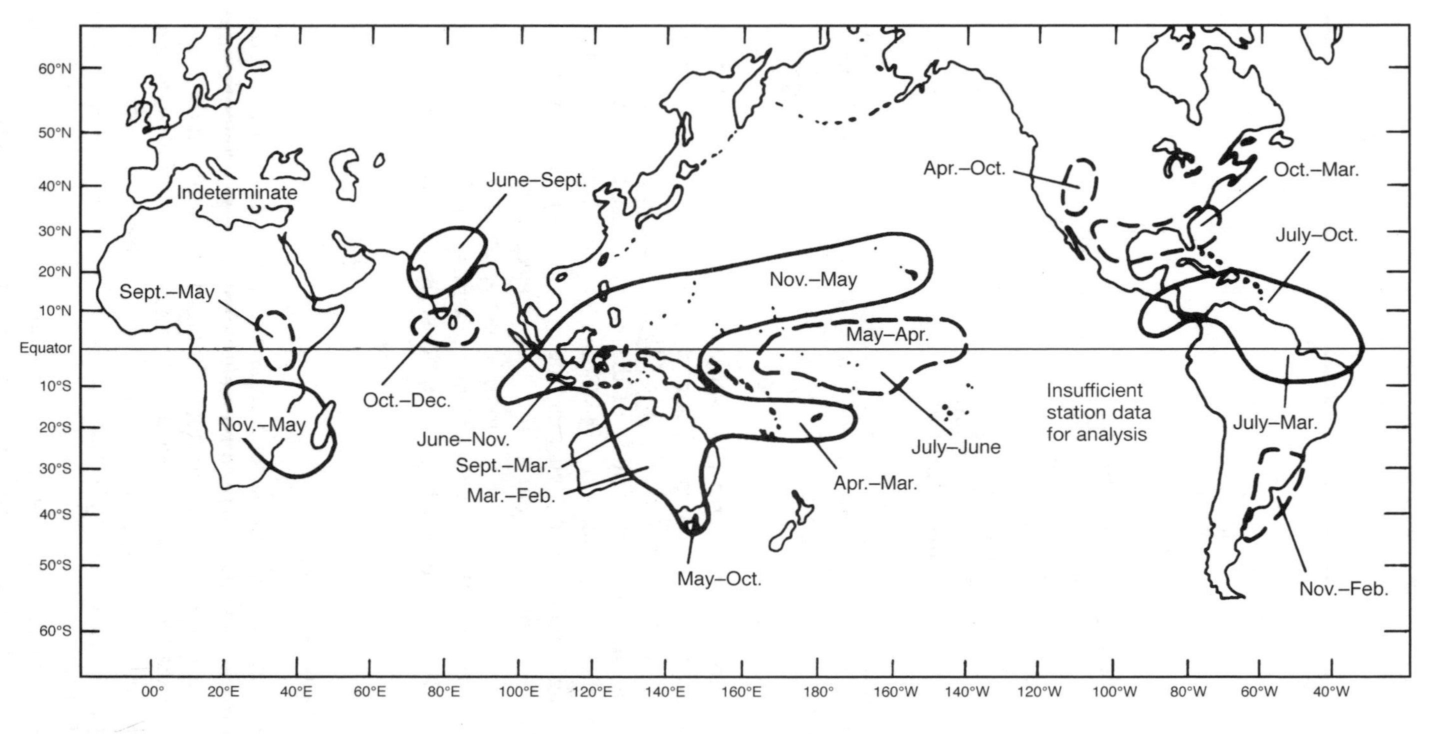

Figure 2.7 A map showing the wide-ranging effect of El Niño episodes and also something of the complexity of their effects. Solid lines enclose the areas in which rainfall is increased; dashed lines enclose areas of decreased rainfall. The months affected are also shown (from Philander, 1990)

Table 2.1 Impacts of the 1982–1983 El Niño

Hazard	Location	Damage in million US$
Flooding	Bolivia	300
	Ecuador & N Peru	650
	Cuba	170
	US Gulf	1 270
Hurricanes	Tahiti	50
	Hawaii	230
Drought and fires	Southern Africa	1 000
	S India & Sri Lanka	150
	Philippines	450
	Indonesia	500
	Australia	2 500
	S Peru, W Bolivia	240
	Mexico, Central America	600
TOTAL		8 110

Source: NOAA [USA] (1994) *Reports to the Nation: El Niño and Climate Prediction.* Washington DC: University Corporation for Atmospheric Research p. 22.

There are many possible explanations for the phenomenon and it has been much modelled, so that prediction of its occurrence has reached a high level of success. One fascinating interaction with other natural phenomena is the apparent cooling of the Earth associated with major volcanic eruptions like Krakatau in 1883 and El Chichón in 1982. There is some disagreement as to whether the massive amounts of debris ejected into the atmosphere (50 million tonnes ejected up to 50 km into the atmosphere in the case of Krakatau) in fact causes global cooling or whether for example the aerosols cause a warming of the stratosphere (Thompson, 1995). But ten large eruptions in 111 years provide a distinctly limited set of data.

Sea-level changes

One of the fears in the present concern about 'environmental problems' is that rises in sea-level will adversely affect many economic and residential zones of the world. So not only are the current trends in sea-level and consequent levels of tides and storm surges carefully watched, but data from the past are analysed to see if there are correlations between past climatic changes and sea-level fluctuations (Emery and Aubrey, 1991).

One problem is that the causes of sea-level rise and fall are multiple. Temperature is without doubt important: sea-level will rise about 60 cm for each 1 °C increase in global temperature, because water expands when heated. Mountain-building leads to local uplift of the land surface and may also have world-wide effects. There is also geoidal eustasy, in which a variation of ±180 m on the surface of the geoid reflects this irregularity. Isostasy reflects the bouncing back of the land and ocean floor surface after it has been under the weight of ice during the Pleistocene. Eustasy is the volume of water added by the melting of ice: if the Greenland and Antarctic ice-caps were to melt completely, sea-level would rise by 66 m. During the Pleistocene there were considerable variations in relative sea-level: during glacial periods, sea-level was often down to *c.* 120 m lower than at present and may even have been 175 m lower. During the Holocene, sea-level has risen in most parts of the world as ice-melt has added to the contents of the ocean basins, but the interaction with other processes has produced a mosaic of rises and falls of different magnitudes. On the east coast of southern England during the last 10 Ky, the very fast eustatic rise after the end of the Pleistocene has been replaced by a much slower process in the last 2000 years, but only in the initial phase is the rise apparently regular. At other times, the curve has a number of irregularities and not all rises in sea-level caused the advance of the coastline, presumably because they were followed by periods of temporary retreat. A map of current trends (Fig. 2.8) suggests that more places are experiencing rises than falls but it is not a simple picture: in Great Britain, for example, there is a 'hinge line', with relative land emergence to the north and land subsidence to the south (Tooley and Jelgersma, 1992). The picture can be further complicated by catastrophic events such as the earthquake off the coast of Chile in March 1985, which yielded a coastal uplift of 33 cm and disrupted the life of intertidal biota.

In summary, sea-level change at present is running at about one-tenth of the fastest rate in the late Pleistocene (about 10 times the average of the last 50 My) but more of the change is due to tectonics and human impacts than to climatic change. However, any effects of the latter are in addition to the others, not replacements for them (Tooley, 1994).

The lesson is one of both stability and change: the oceans are always a potent governor of the weather and climate systems of the planet but are subject to fluctuations at the margins. At the geological time-scale these can be massive but at human time-scales they are small. However, they are large enough and near enough to population concentrations to evoke considerable interest and more than a little concern.

Ice and fresh water

The water planet's stores of water and the movement of water between them are called the hydrological cycle and are best

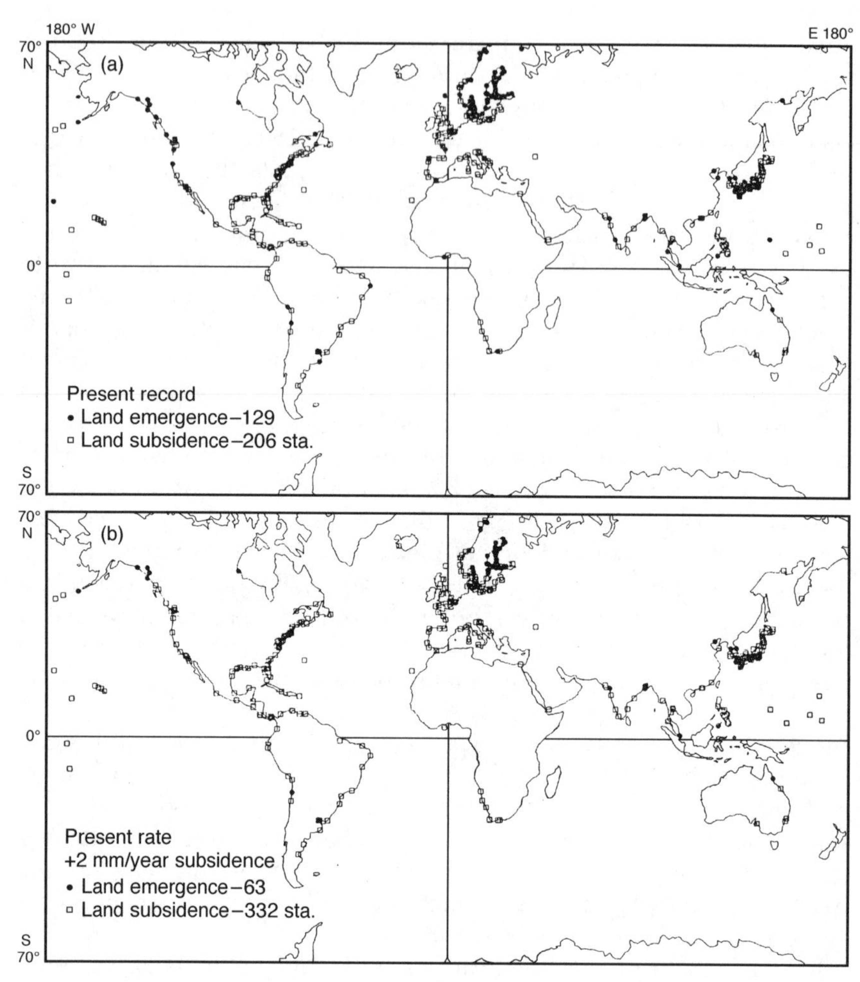

Figure 2.8 The global changes in sea-level as recorded in tide-gauges. (a) present record, (b) present rates +2 mm/year, which increases the number of stations where submergence would be found. At present trends, this could be achieved by AD 2025 (from Emery and Aubrey, 1991)

summarized as a diagram (Fig. 2.9). The volumes of water in the great stores, such as the oceans and the ice-caps, are connected by a number of flows, such as evaporation, precipitation and runoff. As discussed above, the role of the oceans is crucial to this system; of considerable importance, however, are the great ice-caps of the world, since they are closely tied to climate. The runoff from the land to the oceans is of great significance to human societies because it is their main source of fresh water.

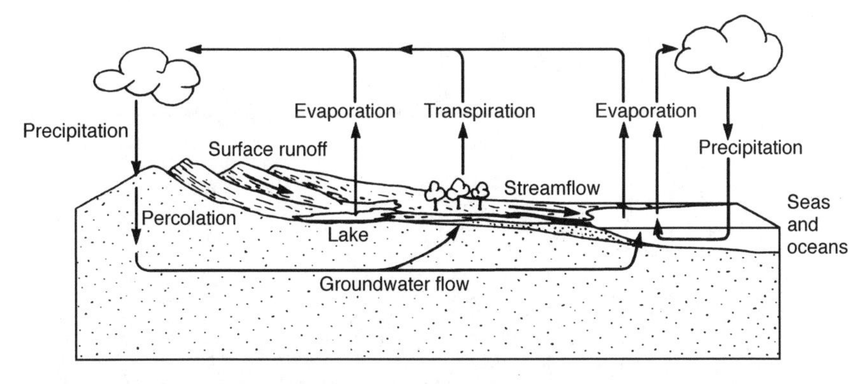

Figure 2.9 A simple qualitative diagram of the hydrological cycle, showing the movement of water through the various atmospheric, biospheric and lithospheric systems

Ice-caps and glaciers

At present, about 1.9 per cent of the world's water is immobilized as ice, mostly in the two polar ice-caps and thereafter as mountain glaciers (Table 2.2 and Fig. 2.10). The polar regions account for about 84.5 per cent of the surface area of the ice, and Greenland for another 12 per cent. For most of the last three million years there has been much more ice in the world than at present (Fig. 2.6 shows the maximum extent during

Table 2.2 Global ice distribution: maximum Pleistocene areas compared with the present

Region	Maximum ($\times 10^6$ km^2)	per cent of total	Current ($\times 10^6$ km^2)	per cent of total
Antarctica	13.2	28.0	12.65	84.5
Laurentia	13.8	29.3	0.23	
North American cordillera	2.5	5.3		
Siberia	3.7	7.9		
Scandinavia	6.7	14.1	0.005	0.03
Greenland	2.2	4.6	1.8	12.0
Other northern hemisphere	4.0	8.6		
Southern hemisphere excluding Antarctica	1.0	2.2	0.026	0.17
TOTAL	47.1		14.97	

See also Figs 2.6 and 2.9.
Source: Sugden, D. and Hulton, N. (1994) Ice volumes and climatic change. In N. Roberts (ed.) *The Changing Global Environment*. Oxford: Blackwell, 150–72

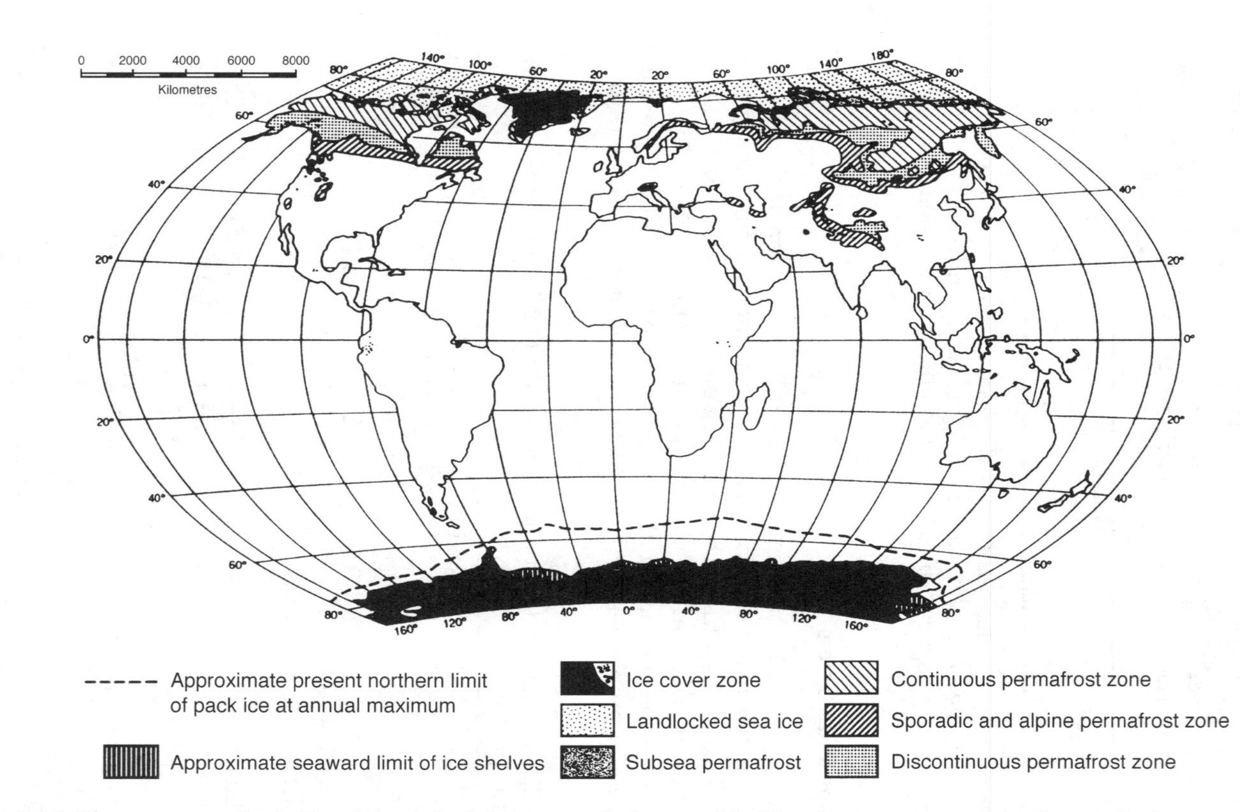

Figure 2.10 The present glacial and periglacial zones of the world. The ice-cover zone is the major reservoir of fresh water on the planet (from Williams *et al.*, 1993)

the later part of the Pleistocene). The evidence suggests that the capability of ice and snow to reflect solar radiation (the albedo, which is 80–90 per cent for ice and snow compared with 18–25 per cent for grass and crops) induces positive feedback. That is to say, once an extensive ice-cap is formed, at least 90 per cent of the incident radiant energy is reflected and so more ice formation is encouraged. The opposite is also true: as warming proceeds, warmer surfaces absorb more energy and the process is speeded up. Ice-sheets, in spite of their massive volumes, can apparently form and break up quite quickly.

This feature has been closely examined in the light of predictions about global warming. Calculations about the rise in sea-level if the ice-caps melted, or surges due to break-up of the west Antarctic ice-sheet, have fuelled speculation. While it is true that a global rise of 6 °C over 200 years would probably melt the Greenland ice-cap and add another 22 cm to global sea-levels, the picture is more complicated. Modelling of the behaviour of ice-sheets in very cold but dry areas like Antarctica suggests that they will increase in mass under conditions of global warming. At the same time, calving could increase depending upon the topography beneath the ice; it seems that sub-ice topography is critical in determining calving patterns. There is then uncertainty about the response of ice-sheets to other global changes: there are regional differences in reactions (Sugden and Hulton, 1994).

Such a variety of responses is expected from the periglacial zone, characterized by the year-round presence of frozen ground called permafrost. Some 20–25 per cent of the northern hemisphere (and indeed 50 per cent of the former USSR) is underlain by permafrost (Fig. 2.9). The northern hemisphere currently contributes *c.* 25–40×10^3 t/year of methane (CH_4) to the atmosphere; methane is a greenhouse gas with 20 times the effectiveness of carbon dioxide in retaining heat. Since many GCMs predict that higher latitudes will experience above-average rises in temperature under global warming, then the notion that the emissions would increase to 45–65 t/year under a temperature increase of 4 °C causes some concern. A complicating factor is the role of the vegetation: if CO_2 levels increase photosynthesis, then will more luxuriant vegetation impede melting of the permafrost and the release of methane, currently rising at the rate of 1 per cent per annum? Such uncertainties make prediction difficult enough to be replaced by the word speculation.

The hydrological cycle

The water planet contains about 13.8×10^{18} km^3 of H_2O. This is arrayed (see Fig. 2.10) in a number of reservoirs, between which the water moves at a variety of rates; the whole set of reservoirs and fluxes is known as the hydrological cycle. By far the biggest reservoir is the world ocean, for some 97.9 per cent of the total water is found in these seas. Not surprisingly, the biggest fluxes are evaporation and precipitation to and from the atmosphere and oceans, well out of sight of most humans, though not nowadays of their instruments. Only 2 per cent of the total water is in fact fresh water, lacking or very low in dissolved salts, and most (75 per cent) of this is ice, with a great deal of the remainder being in the soils and rocks of the crust. So the proportion of the fresh water which is in rivers and lakes and so easily available to humans as a resource is very small: *c.* 0.33 per cent of the total fresh water. This too exhibits certain concentrations: the River Amazon alone is responsible for 15 per cent of the total discharge to the world oceans from the continents. The atmosphere contains perhaps 100 000 times less water than the oceans but the cycling is fast. The residence time of the water in the atmosphere averages at 11 days, whereas in the oceans only the surface layers are so quickly moved: other water bodies in the oceans may be 250 years old (Chorley, 1969).

The movement of water through the hydrological cycle is driven by the energy of the Sun. This is so in the long term since variations in the incidence of solar radiation are the probable cause of ice ages and hence the quantity of water that is solidified as ice. It is also the case in the short term since this energy sets in motion the atmosphere and is the fundamental cause of weather patterns. In particular, precipitation allows life to exist on the land surface, where the land biota contain 1100 km^3 of water with a residence time of about 17 days. It is the water on the land surface (including the soil water, which amounts to twice that in the atmosphere) that forms the main resource for human societies and which is the scene of most of their interventions to try to control and improve the supply of the resource.

Ice and water are, obviously, closely linked. The importance of water, however, is immediate and this is reflected by the short time-scales at which it cycles through some reservoirs. Ice, by contrast, changes much more slowly. The volume of water it represents and its albedo effect nevertheless prevent us ignoring its significance at the global scale.

The land

The land surfaces of the world are wonderfully various. Though the deepest ocean-floor trench (at 11 516 m in the Mindanao Trench, east of the Philippines) exceeds in depth the height of the Himalayas, the diversity of extra-glacial landforms, soils and biotic communities between sea-level and 8848 m is very great. So too are the dynamics of these terrestrial ecosystems: change is constant even under natural conditions, from the short life-span of a mayfly to the century of the elephant, from the boulder-moving flash-flood to the imperceptible creep of soil down a gentle hill slope.

Much affected by the vegetation cover (whether that be natural or humanized) is the rate of denudation of the land surface. This is the rate at which mineral materials, in the form of soil or rock particles, are removed from the land into the runoff and eventually into the sea. The weathering and erosion of bedrock is fundamental to the operation of biogeochemical cycles that affect the atmosphere, the oceans and the accumulation of sediments. It is affected by tectonic processes, the height of the land above the sea, climate, the susceptibility of rock types to erosion and, as mentioned above, the land cover. In spatial terms, the humid tropics are very important since 25 per cent of the land surface yields 65 per cent of the total dissolved silicon, 38 per cent of the ionic load and 50 per cent of the solid load that enters the oceans.

The processes most involved are those of chemical and physical weathering. The chemical weathering weakens the rock and then physical processes complete its disintegration. But there are regional variations in the importance of the two: chemical weathering is more important in warm, moist zones and in areas with a continuous vegetation cover. Physical weathering, by contrast, is dominant in cold and dry lands and also where terrain is steep and the vegetation cover is more likely to be patchy. It appears that vegetation cover reduces short-term erosion by sheltering and anchoring soils but it also produces bioacids that contribute to the break-up of rock.

Slopes at all altitudes are crucial elements in the transport of materials downwards: slope form may be determined, for instance, by the relationships of weathering and material transport by gravity or water. Slope processes can be designated as 'transport-limited' when the supply of material from weathering exceeds the ability of the transport systems to carry it onwards; the other case is 'weathering-limited', when the supply of weathered material is less than the capacity of transport processes to bear it away. The removal is not necessarily directly to the sea, of course: every landscape contains a mixture of eroding areas, stores and sinks (Summerfield, 1991).

All this results in a great variability of denudation rates and input to the seas. The highest rates come in hot, humid lands with high uplift rates: Taiwan is probably the highest in the world, with 13 000 tonnes/km^2/year of solids and 650 t/km^2/year of dissolved material being removed from the land surface to the surrounding sea. While denudation is a process which exhibits a great deal of natural variation, it can be greatly speeded up by human activities, where it is identified as soil erosion. The activities most involved are deforestation, cultivation and construction.

Throughout the last 10 000 years, however, what has presented itself most readily to human senses has been the combination of all these elements of the planet in the form of systems characterized by their soils, and plant and animal life. Science has enabled us to aggregate all the local observations into a world-wide scheme of classifications which shows these integrating units, which are called biomes.

The major world biomes

The most extensive ecosystem unit which it is convenient to designate is called the biome (Archbold, 1995). This consists of a dominant life-form (e.g. deciduous trees, grasses) which extends over an area that corresponds to a particular distribution of soil types and climate (Fig. 2.11). The map is not one of contemporary reality, for human activities have transformed many of the areas plotted on it. It is rather a map of what the world would be like if human activities suddenly ceased and all the resulting plant and animal colonizations were telescoped in time. Since climate is such an important controlling factor in these major world zones, the impact of any global warming is of special interest to ecologists and biogeographers concerned with, for example, the immigration of new species and the extinction of existing ones.

The climate is the fundamental control on the primary biochemical process of the biomes. This is the fixing of solar energy as plant tissue, known as **net primary productivity** (NPP). Light levels are rarely limiting but water and temperature often are, so that the productivity varies from that of the deserts and open oceans to that of the tropical moist forests.

Deserts

Deserts are found in areas of the world where the average precipitation is below 250 mm/year and is often irregular; dew fall during cold nights is the only other source of water. Much of the precipitation is evaporated rapidly by the high levels of solar radiation. The biotic response is firstly in scantiness of vegetation, so that there is usually more bare ground than plant cover, and secondly in marked adaptations of both plants and animals to enable them to survive long periods of drought or the lack of access to free water.

We need not be surprised that NPP is so low (usually of the order of 90 g/m^2/year, but as low as 3 g/m^2/year in some rocky and sandy places; compared with temperate deciduous forests at 800 g/m^2/year)

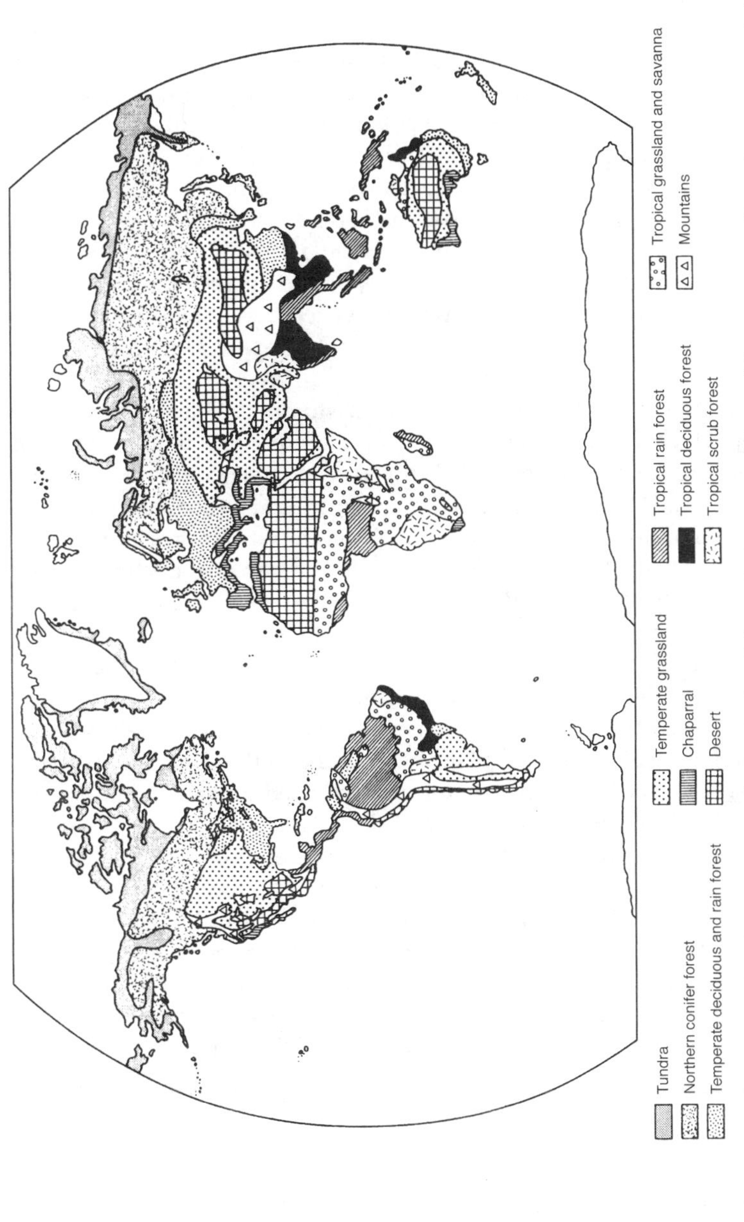

Figure 2.11 The major world biomes as characterized by their vegetation. The zones often have associated animal communities and soils as well, and are clearly related to climate as well as major topographies. Maps like this do not however deal with the alteration of the land cover by human activity (from Simmons, 1979)

and that much of it is underground away from the shrivelling rays of the Sun. Of the above-ground biomass perhaps only 1 per cent will be green since such tissues are also potential areas of water loss.

The outcome in terms of plant communities is a vegetation closely related to water; as more water becomes available, the density of plants increases. Relatively moist areas, such as the Mojave Desert in California, have a vegetation of low (0.5–1.0 m) shrubs (such as creosote bush) set 2–3 m apart as the 'dominants', with other shrubs interspersed. At intervals, clumps of low (0.2–1.0 m) succulents such as *Opuntia* are found. When the rains come (regularly in spring in this case), the ephemeral plants grow and the desert blooms. At the other extreme, the dune lands of the Sahara have sparse populations of only a few plants, such as the drinn grass (*Aristida pungens*), and in the rocky and stony deserts there may be large areas with no vegetation at all except where a little finer soil material can accumulate and support a grass or a small shrub.

Animal life is by no means absent: a number of adaptations have evolved which facilitate the survival of the creatures in the prevailing conditions of heat and drought. Most species are nocturnal, since they minimize water loss by being active at night and probably avoid some potential predators as well. Any animal that has a moist or porous skin, such as the desert forms of worms and slugs, is always nocturnal. Some animals are active in the day but even they mostly leave the ground surface by burrowing, climbing plant stems or flying when the temperature reaches 50 °C; only a few grasshoppers, beetles and spiders are active during the very hottest conditions.

Even though the deserts are sparsely inhabited, any human activities that affect the above-ground plant biomass have the potential to change the ecosystem radically. Grazing of domesticated animals is the obvious method. Pastoralism of animals such as the camel can be a stable system in deserts provided that movement is frequent. In deserts used for recreation (as in North America) severe damage to the ecosystem can result from vehicles which damage the plants, especially succulents; for instance, the use of dead cactus 'skeletons' for fires robs the soils of organic matter as well as destroying the scarce cover for arthropods such as insects and spiders. So although the deserts scarcely look like fragile biomes, they are certainly capable of modification by humans.

The tundra

Tundra is found mostly in the northern hemisphere, where it is the most northerly formation, since beyond it lies either the Arctic Ocean or permanent snow and ice. An analogous set of ecosystems is found on sub-Antarctic islands and the fringes of the Antarctic continent.

In the northern hemisphere, the boreal forest (see p. 79) gives way northwards to a zone of scattered clumps of trees in a low open vegetation and then finally to a broad belt of treeless vegetation in which shrubby willows, birches and alders are the highest vegetation. The transition appears to coincide with a mean daily temperature of at least 10 °C in the warmest month, but where there is a growing season of more than three months. The ground, however, remains frozen all year (**permafrost**), except for the top few centimetres, which thaw during the summer, when water that is frozen for the rest of the year becomes available to the plants.

The growing season is not so short that higher plants such as herbs, grasses and low shrubs cannot grow, and is extended by the long hours of daylight during the summer. An NPP average of 140 $g/m^2/year$ is achieved, which is low but higher than that for deserts and the open ocean. With higher plants, much of this productivity is below ground; for example, the ratio of root : shoot NPP at Point Barrow, Alaska, is 1 : 1.2 (a temperate deciduous forest ratio would be in the order of 1 : 3.0). The litter layer acts as a considerable store of nutrients because of the very slow rates of decomposition.

The animal life is conspicuously adapted to the climate in one way or another. Warm-blooded animals which live above the snow are generally covered in warmth-retaining fur (e.g. the musk-ox, *Ovibos moschatus*; Arctic hare, *Lepus arcticus*; wolf, *Canis lupus*; and caribou or reindeer, *Rangifer tarandus*). Migration is also practised, with the yearly round of the caribou herds (which are responding to the availability of food) a much-studied feature. The caribou drop their fawns on the tundra during May, having wintered at the forest edge. They then range up to the shores of the Arctic Ocean, frequenting high ground to avoid the flies. They are constantly attended by wolves, which keep the herds moving, which helps to avoid overuse of the forage. Many Arctic animals have oscillating populations; that of the lemming is the most famous. More stable are those animal populations that are linked to the food webs of the sea. Seals, and to a lesser extent walruses, are obvious examples, together with various species of sea birds, many of which (such as skuas and gulls) can feed off either sea or land. At the top of this chain is the polar bear (*Thalarctos maritimus*), now a threatened species and the subject of a circumpolar international convention.

The impact of humans in such environments is bound to be strong, for with such a short growing season plants have low recuperative and colonizing powers. The hunter–fisher cultures like the Inuit were always low in density and lived almost exclusively off animals such as seals, caribou and fish, supplemented seasonally with berries, birds' eggs and chicks and small whales. But the Arctic is now exploited for oil, gas, minerals, tourism and military purposes and the impacts on the tundra can be profound, not only by industrially based users but by modernized natives as well. Road construction, fire and oil spills all increase the depth of thaw and the amount of surface subsidence. Many of the human influences have come together in their effect on the barren-ground caribou (*Rangifer tarandus arcticus*) of Canada, which declined from several million before European contact to 672 000 in 1949 and 200 000 in 1958. Excessive hunting was one cause, especially in the years 1949–1960, when very low numbers of calves were born. It is difficult to enforce catch limits because of the terrain and because many of the hunters are Treaty Indians who cannot legally be obliged to observe hunting regulations. Also, the exploitation of the Canadian north has caused many forest fires in the lichen-rich forests where the caribou winter and so their food supply has diminished.

Temperate grasslands

If we imagine a kind of temperate-zone tundra, with low vegetation dominated by grasses, and with trees only along water-courses, then we have some idea of what natural temperate grassland looks like. Usually the rainfall and snowfall (250–750 mm/year) are too low to yield enough water to support forest but are above the level of desert, and so the grasslands are seen as an intermediate form between the forest and the desert. Although lightning-set fires may play a part in producing vegetation mosaics, it is more often the case that the fire results from human activity. Early European explorers in the northeast woodlands of North America recorded large grassy openings in the forest due to fires which the native Americans had started, largely to aid hunting.

The average biomass of these grasslands is 1600 g/m^2 and the average NPP is 600 $g/m^2/year$. However, the shoots are only the obvious part of the vegetation, for the roots of most species will penetrate deeply (up to 2 m) into the soil and so the majority of the plant biomass is in fact below ground: 2000 g/m^2 in some places.

The animal ecology of the grassland has some distinctive characteristics. A few species of large mammals tend to be dominant, as with the buffalo and pronghorn antelope in North America, the wild horse and saiga antelope in Eurasia, antelopes in southern Africa, and guanaco in South America. These large herbivores tend to be herd animals, which affords them some protection from predators (e.g. wolves and coyotes) in the open terrain. All these components are linked to produce the interactive grassland ecosystem. The selective grazing of large mammals largely controls the composition of the plant community, making it more likely that an individual plant will survive if it grows steadily rather than rapidly to the point where it stands out as an obvious target for a grazer. The grassland sod holds the high quantities of nutrients and the organic matter, which helps to retain moisture during long periods of drought and prevents the erosion which occurs when the turf mat is broken.

The effect of human use has been, in general, to break open these tight linkages, either by pastoralism to the point of grazing out the palatable species and leaving a lot of open soil whose mineral nutrients are then vulnerable to leaching, or by ploughing and exploiting the rich stores of mineral nutrients and humus, sometimes with irrigation as an aid. Stable agriculture can be achieved although both the Great Plains and Khazakstan steppe, for example, have seen crop failure and soil loss due to unsuitable agricultural methods. So virgin grasslands are rare since most of them have been altered by pastoralism of domesticated animals, replaced by agricultural ecosystems, or converted to a different species composition through the use of biocides (chemical weed and/or pest killers) or mechanical processes such as brush removal, seeding with leguminous species, or simply through the invasion of new (including exotic) species following utilization by human societies.

Tropical savannas

The tropical savanna, the first tropical biome to be considered, is in many ways intermediate between a forest and a grassland and indeed the term has been applied to a variety of vegetation formations from a nearly closed canopy woodland to a grassland with thinly scattered bushes. Common to them all is a continuous ground layer dominated by grasses. The productivity varies according to the density of the trees: the average of 900 $g/m^2/year$ conceals a range of 1500 $g/m^2/year$ in closed savanna (i.e. nearly woodland) to 200 $g/m^2/year$ where the savanna is more like a desert scrub. The savanna biome is formed in a wide belt on either side of the Equator in areas with a tropical temperature regime. Total rainfall will vary from 250 mm/year on the desert fringes of the savanna, to 1300 mm/year where it abuts true tropical forests, but characteristically there is at least one dry season. The obvious characteristics of typical savanna vegetation are trees and grasses. The former exhibit a great taxonomic variety and are usually 6–12 m in height, strongly rooted and with flattened crowns. They exhibit drought-resisting features, including partial or total seasonal loss of leaves, water-storage modifications and reduced leaves and, in addition, are usually fire-resistant (pyrophytic) in having a thick bark and thick bud-scales. The grasses are often long, reaching up to 3.5 m in height and thus providing ample fuel for dry-season fires. Only the underground parts of the grasses survive the dry season.

Where humans have not over-hunted them, and where fire maintains a variety of habitats, the savannas can support a very diverse fauna. The savannas of East Africa, for example, support the greatest variety of grazing vertebrate life in the world, with over 40 species of large herbivorous mammals (such as African buffalo, wildebeest, zebra and many antelopes) and up to 16 species grazing together, apparently in the same habitat. A wide variety of scavengers and predators is supported by this fauna. Whether fire is the major determinant of the savanna biome's ecological characteristics is controversial. There are areas of South America where savanna-type vegetation is found without fire, and where work on ecological history indicates spells of open grassland before the coming of humans. In Africa, however, opinion seems to favour the idea of the savanna as a delicate balance of the outcome of climate, soils, vegetation, animals and fire, with fire as the key agent whereby humans have created the biome.

Sclerophyll ecosystems

This set of systems is named after the major adaptation of its dominant low trees and shrubs, i.e. the possession of thick leathery leaves with waxy cuticles. This is a way of adapting to a climate in which there is a long dry season. This vegetation type, known as chaparral in California, maquis or garrigue in southern Europe and mallee scrub in Australia, is typically the vegetation of areas of 'mediterranean' climate.

The dominant trees and shrubs are usually 3–4 m high and form a close-set and sometimes impenetrable scrub. In the Mediterranean, the wild olive (*Olea europea*), carob (*Ceratonia siliqua*), evergreen oaks (e.g. *Quercus ilex*), pines such as *P. pinaster*, arbutus or strawberry tree (*Arbutus unedo*), heathy shrubs of the genera *Erica*, *Ulex* and *Genista*, and herbs of the Labiatae family and the genus *Thymus*, are the commonest. In California, many species of bush-like oak are found, as well as species of *Ceanothus*, the chamiso bush (*Adenostoma*), mazanita (*Arctostaphylos* spp.) and several pines. In Australia the equivalent mallee consists of *Eucalyptus* scrub 2–3 m high. In an average area of the biome, a biomass of *c.* 6000 g/m^2 results from an NPP of *c.* 700 g/m^2/year: one sample site had an above-ground biomass of 315 t/ha under *Quercus ilex* and 518 t/ha of organic matter in the soil, connected by 3.8 t/ha/year of leaf fall. It appears that many of the plants grow quite fast on soils which are low in phosphorus and so conserving mechanisms exist: a fine mat of roots penetrates the litter and enzymes within the roots accumulate the phosphorus until it is needed for rainy-season growth. The litter does not penetrate far into the soil in the form of humus and so in the soil profile the unobscured iron, coloured red, gives the name of *terra rossa* to the characteristic soils in the Mediterranean.

The abundant food and good cover of these scrub lands allow a plentiful animal life. In southern California, 201 species of vertebrates were counted, 75 per cent of which were birds. Mammals now tend to be dominated by ground squirrels, the wood rat and the mule deer (*Odocoileus hemionus*) although before the heavy impact of humans, predator species such as the wolf and mountain lion, and **diversivores** (animals able to eat a wide variety of foodstuffs) such as the grizzly bear (*Ursus arctos horribilis*) were more common. The role of small mammals in the ecosystem is hinted at by the fact that wood rats (*Neotoma fuscipes*) have been known to consume the entire acorn crop of one species of oak in a particular year.

In such climates, it is not surprising that fire is a normal occurrence in the biome. In the San Dimas forest of California, lightning set eight fires in 75 years and burning by the native Americans is well documented. Most of the species are adapted to fire (e.g. after it, eucalypts give off numerous stems from the stump, like a coppiced tree) and no doubt have been selected over many thousands of years of its occurrence. The fires seem to stimulate the germination of some seeds, to reduce to ashes much vegetation and litter and hence speed up the process of mineralization of organic matter; and also to destroy **phytotoxic** compounds (poisonous to other plants and perhaps also to bacteria and other soil organisms) secreted by plant roots, which interfere with litter decomposition and the processes of nitrogen fixation in the soil. Its role in succession is illustrated by an Australian study which showed that an unburned *Eucalyptus* forest was experiencing no regeneration of the Eucalypts but that they were being replaced by *Casuarina*, *Banksia* and *Acacia* species.

Chaparral and similar forms of vegetation can clearly be regarded as biomes in which natural fire has played a key role, even though its frequency has been increased by human activity. Fire, though, is not the only form of human impact; centuries of grazing of goats and sheep in the Mediterranean, terracing for cultivation, management for high deer populations, conversion to grassland using biocides, and agricultural use including irrigation, coppicing and urbanization have all hit the sclerophyll scrublands at various times. Some are quite recent, such as home-building in the Santa Monica mountains near Los Angeles; others are ancient, such as pastoralism in the Levant since time out of mind.

Boreal coniferous forests

The boreal coniferous forest biome is dominated by trees. In spite of long and cold winters with considerable snowfall and soils frozen to a depth of 2 m, the evergreen conifers thrive. They are helped by a summer which, although of short duration, has a long day length and at least one month with an average temperature of 10 °C.

In such a physical environment, the evergreen conifer has particular adaptations suited for survival: provided water is available it can photosynthesize all year round, and the needle-leaves help to resist drought when water is locked up as ice and when strong winds increase transpiration rates; the shape of the crown enables it to shed snow so that branches are not broken off by the weight. Large trees up to 40 m high are common and these dominate the structure of the biome: in a continuous stand relatively little light penetrates the canopy so that a lower layer of trees is uncommon. There is usually a continuous ground cover, the components of which vary according to local conditions of drainage and light: at the dry end of the spectrum a lower cover of lichens and mosses may be found; where there is more groundwater the vegetation will include low heath shrubs such as crowberry; and where it is very wet bog-mosses such as *Sphagnum* will cover the ground, often forming open bogs in very wet hollows.

The size of the dominant trees and their ability to photosynthesize all year round if conditions permit mean that the NPP of this forest is sometimes not far short of forests further to the south. In northern Japan, for example, the NPP of fir–spruce forests averaged 2000 $g/m^2/year$, whereas deciduous forests to the south averaged 2160 $g/m^2/year$. But the average for the whole biome is much lower at 800 $g/m^2/year$. Up to ten times the annual litter-fall may accumulate on the forest floor, so that a litter biomass of 100–500 kg/ha is found. (In deciduous forests the equivalent measurements are five times the annual fall and 100–150 kg/ha.) The slow mineralization of this material is short-circuited by fire, which under natural conditions is probably quite frequent and runs along the forest floor consuming the relatively low amounts of litter accumulated since the last fire. Human-introduced fire-protection policies allow debris to accumulate and so a big fire may then result, with the fire running up the trunks (aided by resinous drips) and igniting the crowns. Of the vertebrates, rodents are a characteristic group and generally survive the winter under the insulating cover of the snow blanket. The beaver (*Castor fiber*) is typical of this biome and is perhaps analogous to the elephant in its alteration of the local habitat. Of the larger mammals, various sorts of deer are characteristic, including the moose (*Alces alces*), which is largely an animal of the secondary vegetation produced by fire. The diversivores are typified by bears, including the common brown bear *(Ursus arctos arctos)* as well as the remnant populations of the North American grizzly (*Ursus arctos horribilis*). Small carnivores such as the lynx (*Felis lynx*) and wolverine (*Gulo gulo*) prey chiefly on the rodents, as do the owls and hawks; the characteristic large carnivore is the wolf (*Canis lupus*), which is an important predator upon the populations of deer, caribou and moose.

Humans have used this biome ever since it was formed. It was the basis for many hunting and gathering cultures based on deer, moose and fish. Now it is the focus of the world's softwood lumber industries, whose activities alter the ecology considerably, sometimes temporarily and sometimes over a long period, especially when the frequency of fire is increased beyond that to which the forests are undoubtedly adapted. Also, much of the nutrient supply of the ecosystems is tied up in the biomass and so rapid-cycle exploitation and whole-tree harvesting may reduce the productivity of this biome permanently.

Temperate deciduous forests

To the south of the boreal forest is found a more productive forest type, but one in which there is not growth all the year round; the trees lose their leaves (the **deciduous** habit) in the winter, so that there is a dormancy period as far as the dominants are concerned. The deciduous habit of the dominant trees can be seen as a form of dormancy which is a seasonal response to low energy levels from the Sun and the winter freezing of water. In spite of this, NPP averages at 1200 $g/m^2/year$. The biome is dominated by trees of 40–50 m height. Their leaves tend to be broad and thin (compared with the leathery but narrow leaves of tropical genera) and a large number of them produce nuts and winged seeds rather than pulpy fruits. These trees are occasionally the habitats of climbers such as ivy (*Hedera helix*) and wild vines (*Vitis* spp.); **epiphytes** such as mosses, lichens and algae grow on the trunks. A dense canopy is usually formed and the amount of light percolating through determines the character of the lower layers of vegetation, as does the leaf-mosaic density of the individual species of dominant tree.

Below the trees a shrub layer may form, especially where light penetrates in gaps in the forest canopy. Often the seasonality of the forest is reflected in the ground flora, which exhibits two assemblages: an early spring group which puts out leaves and flowers and which sets seed before the dominant trees have come into leaf; and a summer group which can tolerate the lower light levels of the canopy in full leaf. A lower ground layer of mosses and, on very dry sites, lichens may also be found, and the presence of this stratum is less obviously dependent upon the summer light intensity.

The animal communities too are responsive to the climatic regime. Migration to warmer climates for the winter is common among insect-eating birds such as the warblers; hibernation is found among those less capable of long-distance travel, such as the black bear. Others (e.g. deer) remain active all year, digging through the snow for food when the above-snow browse is exhausted.

The litter layer is the site of a diverse flora and fauna which are responsible for the release of inorganic mineral nutrients back to the soil. There are two main stages in this process. First, the primary decomposers (millipedes, woodlice, beetles and earthworms) attack the litter and break it down into smaller particles. In the second phase, these fragments, together with the faeces of these litter animals, form the food for the secondary decomposers (mites and springtails), which further comminute organic material. Wet material is broken down by bacteria, fungi and protozoa and so more or less complete mineralization is achieved: there is little long-term accumulation on the surface, although some centimetres of humic material is nearly always present.

Western industrial civilization grew up in this zone, preceded by a long period of agriculture. There have been two main effects: the first of these is obviously the clearance of forest and its replacement by agriculture. Secondly, the remaining forests have been heavily managed for an immense variety of purposes, from timber production through forage for domesticated herbivores to amenity in terms of visual pleasure or hunting. The upshot is that it is unlikely that any deciduous woodland in Europe is 'natural' (i.e. unmanipulated by humans), although it may be 'primary' in the sense that woodland has grown on that site from time immemorial (i.e. at least since the Dark Ages). So if a forest is to be called 'natural' it needs to have a certificate of its pedigree which shows that for both prehistory and historical times all the techniques of the ecologist and the historian have failed to reveal any trace of human effect.

Tropical evergreen forests

Conditions most favourable to high productivity seem to have developed in lowland equatorial zones. The constant high temperatures and year-round precipitation have allowed the development of large trees with rapid growth rates, high-speed litter breakdown and fast nutrient cycling.

These forests occur in climates which have both high and constant temperature and humidity, with precipitation of over 2000 mm/year and at least 120 mm in the driest months. The vegetation is dominated by trees of a great variety of species: in some parts of Brazil there are 300 different species of tree in 2 km^2. The trees are typically tall and are structured into three layers. The highest or emergent layer consists of the tallest trees (45–50 m high), which are scattered but project through the lower canopy layer (25–35 m high) that forms an almost continuous cover, absorbing some 70–80 per cent of the incident light. When there are gaps in this lower layer, the normally sparse understorey tree layer may become dense. The effect of these tree strata is to absorb all but a few per cent of the incoming light and so shrubs and ground vegetation are not normally found except in gaps in the forest and at its edges and near rivers.

Animal life, though very low in productivity compared with the trees, exhibits the greatest taxonomic variety of any biome. The richness of food resources available and the relative constancy of environmental conditions seem conducive to such a state. Like the trees, the animal communities are stratified: the emergent layer is inhabited mostly by birds and insects which live their whole lives in this arboreal habitat. Below them, the canopy layer houses the highest variety of animals in the form of tree-dwelling monkeys, sloths, ant-eaters and small carnivores. They rarely descend to the ground, but in the understorey layer the animals may range down from the trees to the forest floor. Ground-dwellers are less diverse than arboreal types but include deer, rodents, peccaries and wild pigs.

The productivity of the rain forests seems to depend upon mechanisms which keep the nutrients in the organic components of the cycle so that they are not leached out of the inorganic phase by the abundant rainfall. The rate of litter-fall from the forest canopy is high but there is a humus turnover of 1 per cent per day so litter does not accumulate: if mean temperatures are above 30 °C, litter is broken down faster than it is supplied; at 25–30 °C, supply and breakdown are about equal. The main agents of litter breakdown seem to be fungi in mycorrhizal associations with the tree roots, so that mineral nutrients are passed directly from the decaying litter to the roots of the trees for uptake. Thus loss of minerals to the runoff is minimized, even to the point where soil animals are forced to feed on the fungi rather than the litter. Earthworms, for example, do little mixing of the soil and so the organic upper horizon is sharply marked off from the mineral soil beneath.

The outcome of high solar input, abundant rainfall and rapid nutrient cycling is a very high NPP, with the mean for rain forests estimated at 2200 $g/m^2/year$ of dry matter; multiplied by the area of the biome, we get 37.4×10^6 t/year, which is far higher than for any other terrestrial biome.

Various governments and lumber companies are removing the forests at a current rate of 11 million ha/year so that of an original area of 16 million km^2 of rain forests (about 5000 years ago), only two-fifths are now left: at this rate of destruction, another 30 years will see the demise of the biome. Apart from its scientific interest (for example, there are estimated to be over 25 000 species of flowering plants in the rain forests of Southeast Asia), such a biotic profusion must inevitably be a reservoir of great economic and genetic potential.

Islands

It seems appropriate to bridge the descriptions of terrestrial and aquatic biomes with a discussion of islands. Here we must remember that the sea is likely to be a strong biological influence, even on apparently terrestrial fauna and flora. In the Arctic, for example, many small islands bear numbers of animals that could not possibly be nourished from their sparse tundra, but which depend on the sea for their nutrition, either directly or indirectly. Islands tend to exhibit certain definable biological characteristics: they have a low species diversity compared with the nearest continental masses, a diversity which gets progressively lower away from the continents, especially along island chains.

Island ecosystems will, of course, vary with the usual environmental factors, and the relief and substrate are major determinants since the term island encompasses Hawaii, with the active volcano of Mauna Loa at 4170 m, and low atolls never more than 15 m above sea-level. The island ecosystems are not necessarily unique in terms of their structure. For example, the island of Hawaii (3200 km from the nearest continent and 720 km from the next island group) possesses natural formations of evergreen rain forest, evergreen seasonal forest, savanna, grassland, scrub, alpine tundra and near-desert. These are arrayed in a general altitudinal sequence similar to continental tropical mountains, and the plants often exhibit similar forms and in some cases similar species.

Usually the effect of humans upon island ecosystems has been strong. This is because humans have brought with them animals and plants that are competitively superior to the native species, or because the paucity of the island's species diversity meant that there were unoccupied places in the ecosystems. Also, since islands often have many endemics (i.e. species found only on one island or island group) in their flora and fauna, it is often easy to bring about extinction – the example of the dodo is perhaps the best known.

The seas

The outcome of both the opportunities for and constraints upon life in the seas is a highly variable NPP. In essence, the near-shore environments such as estuaries and coral–algal reefs are highly productive (since in part the movement of water brings in food and removes wastes so that the organisms spend less energy acquiring food and getting rid of wastes), the continental shelf areas (especially where water wells up from the ocean floor bringing a nutrient supply) being next in productivity, and the open oceans rather less productive.

These differences in NPP suggest that one of the main factors limiting growth in the seas is the supply of nutrients: where nutrients are plentiful and in the surface zone of water into which light can penetrate (the **euphotic** zone, < 200 m deep) productivity can be high (there are no terrestrial ecosystems with such a deep photosynthetic zone). Elsewhere, meaning for most of their extent, the seas resemble a very wet tundra.

The primary producers of the oceans are predominantly phytoplankton down to a depth of 60 m, although large algal seaweeds may have a very high NPP, especially in middle latitudes. The first consumer level is that of zooplankton, which feed either directly on phytoplankton or on detritus derived from them. Then there are the carnivorous zooplankton. Thereafter the many trophic levels of the long food chains of the seas, including fish, comprise carnivorous animals; there are few large animals which are strictly herbivorous. The larger consumers of the sea-bottom are sometimes free-moving (as with some flatfish and lobsters), or may be fixed (sea anemones, bivalve molluscs), or may burrow into the substrate if it is sand or mud (burrowing anemones, bivalve molluscs, gastropods, echinoderms and Crustacea). Bacteria are also found in large numbers in the surface sediments of the sea-floor, where they are thought to play a similar role in nutrient release to that in the litter layer of terrestrial ecosystems.

The active swimmers of the shelf zones comprise fish, the larger Crustacea, turtles, marine birds and mammals. Although individuals may range over a wide area, they are still limited in their distribution by the barriers of temperature, salinity and nutrients that affect the other components of their ecosystems. Even if not directly affected, they are tied to their food sources. Because of the small size of plankton, fish which eat them are important links in the food chains over the continental shelves: the herring family (herring, menhaden, sardine, pilchard and anchovy) are very important in this and in the Pacific some sardines are virtually herbivores. As the fish get larger their food sources may change to smaller fish, so that long food chains with tertiary and quaternary carnivores may be found.

The open oceans beyond the continental shelves are populated entirely with open-water and sea-floor organisms. The oceanic phytoplankton are predominantly very small and the zooplankton are the permanent kind without the addition of larvae of other groups. Even though plankton productivity is low, it eventually supports a characteristic fauna of oceanic birds such as petrels, albatrosses, frigate birds and terns which only come to land in order to breed; most of these are bound to a particular type of surface water even though their movements appear to be unconstrained. Also completely independent of the land are sea mammals such as dolphins and whales.

Compared with many terrestrial ecosystems, the dynamics of marine populations are so little known that the results of human interference are often hard to elucidate, apart from spectacular examples such as the decline of whales from over-hunting. The effect on the seas of nutrient enrichment from sewage and fertilizer runoff, for instance, is complex and affects both species composition and productivity in a number of ways; analogous are the fate and effects of long-lived substances such as chlorinated hydrocarbon biocides (e.g. DDT) and polychlorinated biphenyls (PCBs) although the toxic effect of these upon carnivores such as sea-birds is now well documented. Equally important, though not immediately quantifiable, are the effects of reclaiming estuaries, salt marshes and other intertidal communities which house the larval stages of many marine organisms. Best known of all is the ability of modern humans to reduce populations of whales and fish to very low levels. Nobody now thinks that the oceans are so vast and so teeming with life that human effects are confined to a narrow zone round the shore, and that they are an inexhaustible supply of food or a bottomless sink for wastes, but putting such realizations into action is proving a slow process.

A very productive intertidal environment is the estuary, with its set of mud flats and salt marshes. With tidal effects, strong currents, high turbidity and variable salinity, the estuary imposes a high degree of stress upon its organisms: a wide tolerance of variability in salinity is necessary, for instance. Plants are salt-tolerant (halophytic) and include grass- or rush-like genera such as *Spartina*, *Salicornia* and *Scirpus*, as well as green algae such as the sea-lettuce *Enteromorpha*, while the mud is covered at low tide with diatoms and blue-green algae. Few phytoplankton are found because of the high turbidity.

There are compensating features: dissolved oxygen levels are high because of the turbulence; the intermixture of salt and fresh water acts as a nutrient trap, keeping river-borne nutrients in the estuary for a long time in spite of the river current. In the tropics at least, nutrients also come from the sea and especially from deep waters below the euphotic zone, where they have not been depleted by the phytoplankton. The tide does a lot of work in removing wastes and transporting nutrients and organic matter so that the permanent biota can be sessile and do not expend energy on excretion and food gathering. The outcome is an average NPP of 1500 $g/m^2/year$ (compared with continental shelf figures of 360 $g/m^2/year$ and open ocean of 126 $g/m^2/year$), a quantity similar to tropical seasonal forests. The productivity of the algae and higher plants goes mostly into a decomposer chain in which bacterial breakdown is an important stage, as is the activity of molluscs, worms and other detritus feeders. Bivalve molluscs (e.g. clams, cockles and mussels) in particular exhibit a high productivity: in northern Europe figures of 200 g/m^2 of dry meat biomass of mussels have been reported. A properly managed mussel bed should provide 2000 g/ha/year live weight, about 50–100 times the yield from beef cattle on grassland. The visible carnivore fauna, including the birds which find this ecosystem such an important feeding ground, live off these sessile animals. Importantly for resource use by humans, two-thirds of the commercial fish species of continental shelves spend their larval years in estuaries, and others must pass through them on their way from ocean to upriver spawning ground. The estuaries are therefore important nurseries of commercial fish.

The level terrain of these ecosystems makes them easy to reclaim for many kinds of industrial purpose, and estuaries in general are very prone to environmental contamination. Unspectacular though they are, these places deserve a high degree of environmental protection, not merely because of their bird populations but because of their contribution to the whole marine environment short of the open oceans.

It must be re-emphasized that the map of biomes is partly conjectural. As was said at the beginning of this chapter, it is largely a map of what the world would be like if human activities were suddenly removed and all the ensuing successions were telescoped in time. Alternatively, it might be a map of some time in the past before human manipulation had become significant but after most of the major post-Pleistocene climatic changes and subsequent vegetational adjustments had taken place – perhaps about 1000 BC. But even then, large areas of Southwest Asia must have been altered by agriculture and pastoralism and large tracts of Southeast Asia converted to cereal growing. It is most certainly not a map of contemporary reality. Most of the tundra, for example, is still there and little altered but most of the deciduous forests of Eurasia have gone and the lowland tropical forests of the Zaire and Amazon basins are shrinking fast. This leads to one more reservation about the reality of the biomes: in the case of those which have been altered by agriculture and pastoralism, research has been done on relict areas either accidentally or deliberately preserved. We do not know exactly to what extent their ecology differs from that of the biome in its pristine state.

Connectivities and wholes

Energy from the Sun flows through the Earth's systems and is then radiated back to space. It is a one-way flow in the sense that concentrated solar energy able to do work is radiated to Earth, and dispersed heat unable to do any work is radiated back. Whilst inside the planetary envelope, this energy entrains water to form the hydrological cycle. Many chemical elements in dissolved, solid or gaseous forms are also moved. These elements are usually present in both living and non-living material and indeed make life possible. They are different from the flow of energy since these elements are never lost to the planet and are endlessly recycled: their flows are called **biogeochemical cycles.** The scales are often huge and the residence times very long (Table 2.3), but some human activities have come to be sufficiently high in magnitude to divert parts of the cycles and so produce actual or anticipated problems for other life-forms or for some human societies (Butcher *et al.*, 1992).

Table 2.3 The main components of biogeochemical cycles

Global compartment	Forms of chemicals
Atmosphere	Gaseous
	Particulate including aerosol
Biosphere	Gas
	Solid
	Liquid (solute and suspended)
	Living matter
Lithosphere (crust)	Solid
Hydrosphere	Solution
	Particulate
Noosphere	Symbolic in written and electronic forms

The most important of these cycles are those of carbon, nitrogen, sulphur and phosphorus together with some metals such as mercury and copper. The simplest way to consider these matters is in a series of reservoir-and-flux models but this is of course a classically reductionist mode and fails to point out that many cycles are interlinked. One attempt to suggest that the feedbacks of these cycles and the forcing functions within them add up to more than the sum of the parts is the **Gaia hypothesis**, which will be considered as an example of a holistic approach to the planet's systems and their inhabitants.

Biogeochemical cycles

The following condensed accounts will depend upon box-and-arrow models and it is worth repeating that (a) these do not convey change through time, and (b) the data on which they are based are still accumulating, so that some linkages may well be underestimated in their importance.

The carbon cycle

Carbon is the key element of life on this planet. The study of the carbon cycle therefore involves all the life on Earth as well as inorganic reservoirs. As with many biogeochemical cycles, a range of time-scales is involved, from millions of years for crustal processes to seconds for the interchanges between sea and air or in photosynthesis. We should note, therefore, that most models select the parts of the process that operate on similar time-scales.

Elemental carbon occurs as amorphous carbon, as graphite and as diamonds and in over one million compound forms. At present, the most intense focus is upon carbon dioxide (CO_2). In the atmosphere, CO_2 is controlled by the rates of erosion of rocks and by vulcanism, which is balanced by the sedimentation of carbon (especially as calcium carbonate, $CaCO_3$) in the deep oceans. There is 39 000 Pg (1 Pg = 10^{15} g) of carbon in the oceans and 725 Pg in the atmosphere, but the current focus has meant that the atmospheric reservoir of CO_2 is the best-known part of the carbon cycle, with accurate measurements of CO_2 dating from as early as 1957. An annual increase of *c.* 0.5 per cent per year has been plotted since that time. Carbon dioxide is the main form of carbon in the atmosphere, along with smaller amounts of methane (CH_4), carbon monoxide (CO) and other gases. However, after being about 200 ppm at the end of the last glaciation, the level was more or less steady over the last 10 Ky until the 19th century. The pre-industrial concentration of CO_2 was *c.* 270 ppm but this had risen to 316 ppm in 1959 and 357 ppm in 1995.

In the oceans, there are four forms of carbon: dissolved inorganic and organic carbon (DIC and DOC), particulate organic carbon (POC), and carbon present in the biota. A crucial point is that the process of sequestering carbon on the ocean floors is controlled entirely by living organisms, which are responsible for the rain of organic carbon and $CaCO_3$. Some 90 per cent of the organic carbon falls out on the continental shelves, where the majority of the marine life is found. Between the atmosphere and the oceans, 80 Pg/year pass each way and a mere 5.2 Pg becomes sediment. On land, the terrestrial biota are the sites of the fixing by photosynthesis of CO_2 as plant tissue and the terrestrial biomass contains some 560 Pg of carbon, of which perhaps 90 per cent is in forests and some 60 Pg is in litter awaiting decay and the subsequent release of the CO_2. The lithosphere has the largest reservoir but the smallest fluxes: the crust has about 20×10^6 Pg of carbon but the river loads and vulcanism which symbolize its mobilization do not exceed 1.0 Pg/year. The land's contribution to atmospheric carbon loads is mostly as CO_2 (at a rate of 180 Pg/year) but 1 per cent of the flux is methane, of which 25 per cent is accounted for by a post-1940 increase in the world's domestic cattle population (Table 2.4). Fires contribute 7 Pg/year.

The hints given here about the human effect upon this cycle will be amplified and gathered up with others after the rest of the set of cycles have been described.

The nitrogen cycle

The inventory of nitrogen on Earth (Table 2.5) shows the importance of the atmosphere and the planetary crust. The relevance of nitrogen, however, is more often its association with living tissues and with other elements such as carbon, sulphur and phosphorus. In fact, living organisms and human activities are the main pathways through which

Table 2.4 Sources of methane (Tg of CH_4 per year) emitted to the atmosphere. Those marked * are primarily 'natural' though they may have been affected by human actvity

Source	**Total**	**Range**
Coal mining, gas drilling	80	45–100
Landfills	40	20–70
Animal guts[a]	80	65–100
Biomass burning	40	20–80
Natural wetlands*	115	40–200
CH_4 hydrate destabilization*	5	0–100
Termites*	20	2–100
Fresh waters*	5	1–25
Oceans*	10	5–20
TOTAL	505	222–965
Proportion of recent growth attributed to human activity	65	Might be greater

[a] The main source of recent increase has been domestic cattle.
Compiled from various sources

Table 2.5 Major nitrogen reservoirs (values in Tg, where $T = 10^{12}$)

Reservoir	**Value**
Crustal	6×10^8
Soil organic matter	6×10^4
Soil inorganic matter	1×10^4
Biomass (terrestrial)	1×10^4
Oceanic	
inorganic	6×10^5
organic	2×10^5
biomass	8×10^2
Atmosphere	
(N_2)	4×10^9
(N_2O)	1.1×10^3
others	3.0

Source: D. A. Jaffe (1992), The nitrogen cycle in S. S. Butcher *et al.* (eds), 263–284

nitrogen moves. Fluxes between the very large reservoirs are very small: 95 per cent of all N_2 flux is between biosphere and soil, and biosphere and water (Fig. 2.12).

The main movements of nitrogen involve three processes. The first is nitrogen fixation in which atmospheric N_2 is fixed in a compound such as ammonia (NH_3); the second is nitrification, which is the oxidation of

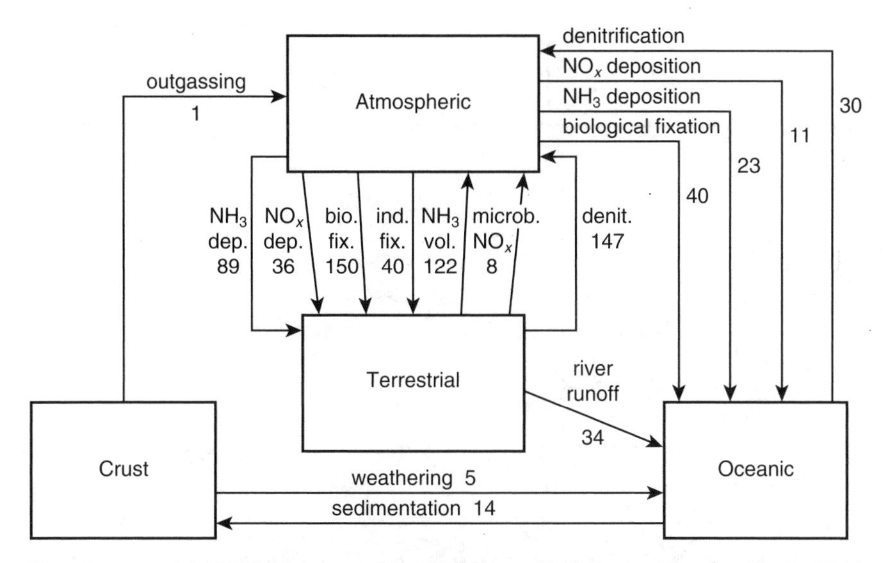

Figure 2.12 A relatively complex diagram of the nitrogen cycle in Tg/year (1 Tg = 10^{12}). There is a net transfer from the atmosphere to the other reservoirs of 81 Tg/year; human-directed fixation of nitrogen is about 100 Tg/year (from Jaffe, 1992)

ammonia to NO_2 or NO_3^- by an organism in order to get energy; the third is denitrification, which is the reduction of NO_3^- to gaseous nitrogen as N_2 or N_2O. All three processes are carried out by micro-organisms. A major pathway is the emission of ammonia to the atmosphere from animal dung, including that of domestic animals. This returns in the form of precipitation or as dry deposition. Best known of all in the cycle is the transfer to the atmosphere of technologically produced NO_x, which produces about 22 Tg/year of nitrogen out of the 54.0 Tg of reactive nitrogen emitted or produced in the atmosphere. Abiotic nitrogen (i.e. in non-living form) is relevant to tropospheric aerosol formation and to both stratospheric and tropospheric ozone levels. Nitrogen oxides (NO_x) in the stratosphere have a residence time of 1–30 days and help to produce ozone (O_3). In the troposphere they tend to break down ozone.

The reservoirs of nitrogen other than the planet's crust are dominated by the atmosphere, which contains 99.9 per cent of the non-crustal nitrogen, in the form of N_2 gas. Compounds such as ammonia are scavenged out in <1 year. On the land surfaces, most of the nitrogen is in dead organic matter, with 4 per cent in living biomass; inorganic nitrogen comprises only 6.5 per cent of this total. In the oceans, 95 per cent of the inventory is dissolved N_2, which in turn is only 0.5 per cent of the total non-crustal nitrogen.

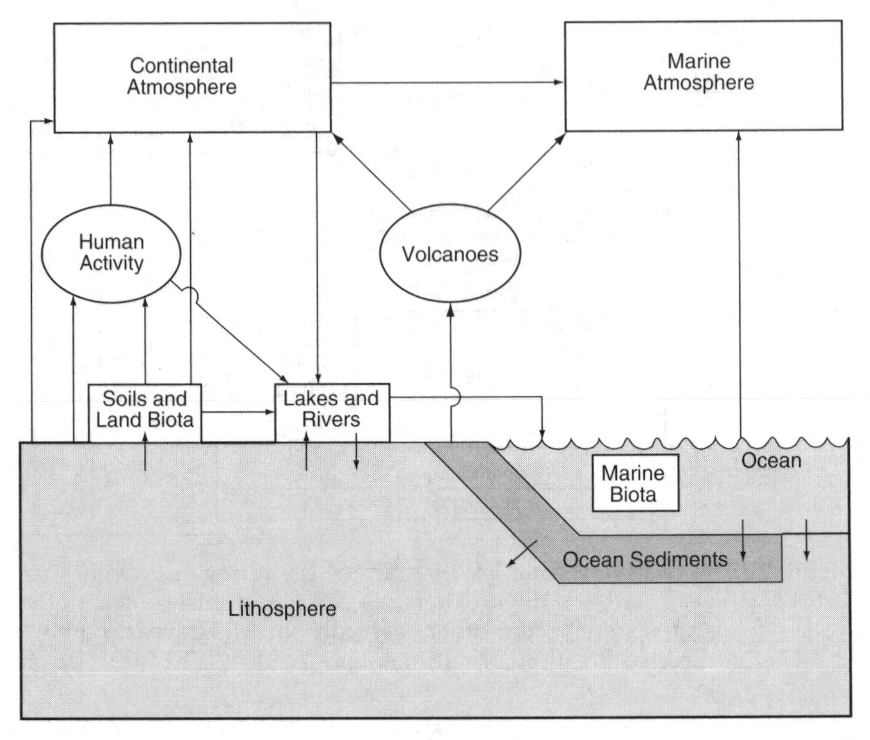

Figure 2.13 Major fluxes of the biogeochemical sulphur cycle, with human activity included as a single box (from Charlson *et al.*, 1992)

The sulphur cycle

Sulphur is a key element in living matter since it gives structural integrity to protein-containing tissues. Elsewhere it is a major contributor to acidity in the environment: it is important in rock weathering and in 'acid rain'. In the atmosphere, sulphur is the dominant component of cloud condensation nuclei, and so this element's cycle interacts with the hydrological cycle and with global radiation levels as affected by cloudiness. It is one of the cycles most perturbed by human activity, with natural land–air emissions now equalled by human-produced processes (Fig. 2.13).

The vast majority of the sulphur on the planet is in the rocks of the crust (Table 2.6), but the strongest fluxes exclude that reservoir. The amount in the atmosphere is small but the residence time is short (measured in days) and so there are large fluxes. One of the most important of these is that of dimethyl sulphide (DMS: CH_3SCH_3), which dominates sulphur emissions from the oceans and may even modulate the natural cycle on a global scale: DMS emissions run at 39 Tg/year out

Table 2.6 Sulphur reservoirs (in Tg, where T = 10^{12})

Reservoir	Quantity
Lithosphere	2.4×10^{10}
Ocean sediments	3.0×10^{8}
Sea water	1.3×10^{9}
Soils + land biota	3.0×10^{5}
Lakes + rivers	300
Marine biota	30
Marine atmosphere	3.2
Continental atmosphere	1.6

Source: R. J. Charlson *et al.* (1992), The sulfur cycle, In S. S. Butcher *et al.* (eds) Table 13-3

of a total of 57 Tg/year from biological sources. By comparison, humans are responsible for 80 Tg/year of SO_2 led off into the atmosphere.

The phosphorus cycle

Phosphorus controls many of the biogeochemical cycles in the biosphere. It is, though, unusual in having no gaseous forms (Table 2.7). It is present in the atmosphere only as a minor constituent of clouds and rain droplets, though input to land and sea from this source can be significant in zones scarce in phosphorus. The circulation of phosphorus through the terrestrial biosphere can be regarded as a closed system: phosphorus from the soils is taken up by the biota and then returns to it via the decay of dead organisms (Fig. 2.14). On land, but outside the biota, phosphorus behaves like sediments (with which it travels) in moving downhill through a variety of sinks and transport phases until it reaches the ocean floor. In the oceans, accumulations of phosphate-rich bottom sediment are brought to the surface by the upwelling of cold

Table 2.7 Reservoirs, fluxes and residence times in the phosphorus cycle

	Reservoir load (Tg)	Flux (Tg)	Residence time
Sediments	4 (15)	4 (9)	1.8×10^{8} years
Land	200 000	88–100	2 000 years
Land biota	3 000	63.5	47 years
Oceanic biota	138	1 040	48 days
Surface ocean	2 710	1 058	256 years
Deep ocean	87 100	60	1 452 years
Atmosphere	0.028	4.5	53 hours

Source: R. Janke (1992) The phosphorus cycle. In S. S. Butcher *et al.* (eds), Table 14-5

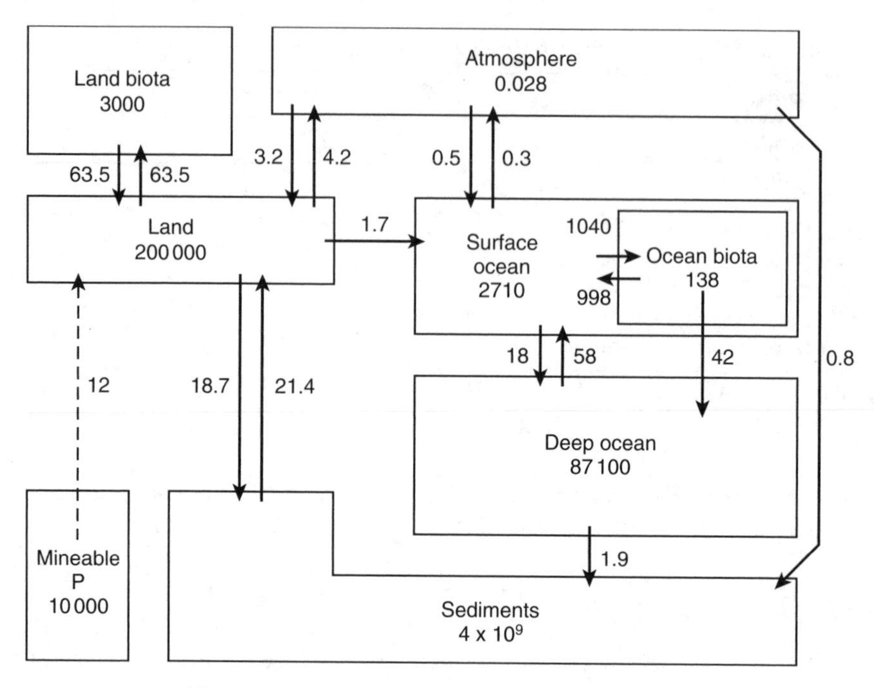

Figure 2.14 The global phosphorus cycle in million metric tonnes. Careful inspection of the values alongside the arrows is necessary to differentiate the magnitudes of the different flows (from Janke, 1992)

water (as in the Humboldt Current off Peru when not capped by the warm water of the El Niño phenomenon). Such accumulations permit very high rates of growth of marine phytoplankton and hence of fish populations.

As with sulphur, the phosphorus cycle is strongly influenced by human behaviour, as in the mining of phosphates and their use as fertilizer on the land, adding to the quantities transported by rivers and added to the sea.

Metals

Our image of metals as the basis for an industrial economy is such that we rarely think of them as parts of living material, although iron deficiency as a cause of anaemia in humans is a familiar enough concept. Nevertheless, the presence of some metals is necessary for life just as a very high concentration may well be toxic to it: copper is a good example.

The main reservoir of metals is invariably in the Earth's crust, from which release comes via weathering of all kinds. When metals are mobilized or deposited, they often change their chemical form and so can be difficult to track. Metallic mercury in sea water, for example, can be changed by micro-organisms to methyl or dimethyl mercury. Neither of these is soluble in water but they are soluble in lipids (fats) and so can concentrate in living organisms. In these forms, mercury is often much more toxic to life than in the metallic form. In many aquatic environments, the movement of metals appears to be tied to particulate organic matter, and so the mobilization processes are often mediated by micro-organisms. These characteristics come together in estuaries, where the physics and chemistry combine to ensure that metals are often sedimented out from the water column and even moved back upstream. Only a small fraction of the metal load of rivers (the major carrier of metals from lithosphere to oceans) ever reaches the ocean.

The use of metals in industry means, however, that additions to the cycles are both frequent and heavy, and may sometimes (see p. 130) exceed the natural concentrations of the metal. An 'environmental problem' almost invariably follows, usually of the type that is labelled 'pollution'.

Human-caused perturbations

In all the major biogeochemical cycles, human activities have produced measurable changes, usually in the period since the Industrial Revolution. The alterations have sometimes been detected more recently due to new processes, or to improvements in the discovery of chemicals in the environment or to increases in concentrations of the concerned component.

In the case of the carbon cycle, the overriding process has been the combustion of fossil fuels. The crustal reserves of carbon which might be burned are estimated to be of the order of 5000–10 000 Pg. In 1860, the global emission from these sources was 0.1 PgC/year (PgC = 10^{15} grammes of carbon), in 1914 it was 0.9 PgC/year, and in the 1980s it was 50 PgC/year. So atmospheric carbon concentration rose by about 1 ppm per year in the 1960–1980 period. Another source of carbon in the atmosphere is the terrestrial biota, which might contribute almost as much as the burning of fossil fuel (estimates vary between 1.8 and 4.7 PgC/year net) but all of which are ameliorated by the 40 per cent which appears to be taken up by the oceans. This latter figure interacts with the N and P (and possibly metals like iron and zinc) cycles since additional supplies of those elements may allow higher productivity of phytoplankton and hence the uptake of more CO_2. The partitioning and flux of carbon is still subject to some uncertainties and is the subject of intensive research because of its key role in scenarios of global climatic change.

The human role in the nitrogen cycle is clearly known in three different aspects. The best known is the emission of NO_x from power plants and motor vehicles. In conditions of good sunlight and relatively still air, atmospheric ozone is produced and this forms a major constituent of smog. Control of the production of nitrogen oxides is the key to lessening the damage from this source. The second aspect is the emission of N_2O from fertilizers, fossil fuels and sewage. Combined with the inert (and hence resident for a long time) chlorofluorocarbons (CFCs), these gases react with stratospheric ozone to produce oxygen. This is especially true in the vortices which form over the poles during winter and so the attenuation of the ozone-rich layers has given rise to 'ozone holes'. International action to phase out the use of CFCs is occurring but the long residence times of CFCs and allied compounds mean that recovery will be very slow. Less O_3 in the stratosphere adds to global warming and allows more ultraviolet light to penetrate and cause damage to tissues of plants and animals, including the increase of the incidence of skin cancer among humans. The third aspect is the fixation of atmospheric nitrogen by industrial processes in order to make fertilizers. Micro-organisms convert some of this to N_2O, whose effects are mentioned above.

The importance of time-scales can be seen in the case of the phosphorus cycle. Mining of phosphate-rich rocks for fertilizer and other industrial uses is the major human-organized addition and it results in the short term in the eutrophication of fresh and offshore waters, often combining with nitrogen in this role. In the longer term, additional phosphorus in the oceans might allow higher rates of photosynthesis by marine plankton so that the uptake of carbon dioxide might be accelerated. This is as yet speculative.

The human effect upon the cycling of metals can be seen well in the case of mercury (Hg). There is now a net transport of Hg to the atmosphere, where 3.3×10^{18} g out of 8.5×10^{18} g is there because of human activity. The transport rate of mercury flowing from the land to the oceans has been increased by a factor of four. These additional loads are not, of course, evenly distributed and so regions of concentration are found, which place considerable stress upon some organisms.

Any overview of the human role in biogeochemical cycles cannot fail to note that these global flows are involved in the following:

- global climatic change (water, C, N, halocarbons such as CFCs),
- acid precipitation (C, N, S are all involved in fossil fuel use),
- food production (C, N, P, S),
- ozone 'holes', where CFCs have long residence times and have contributed to a very rapid depletion of stratospheric ozone, especially in the 1956-1985 period.

The general lesson of this discussion is that global temperature, carbon dioxide levels and methane levels are all positively correlated; in turn these have a negative correlation with the concentrations of sulphate ions. Since there seem to be some positive feedbacks, rates of change can often accelerate due to human actions. Many uncertainties arise, however, from the unexplored nature of many of the relations between the cycles of individual elements. One global model of these systems and their feedback loops has attracted more attention than all the others because of its holistic nature. This is the Gaia hypothesis.

The Gaia hypothesis

The notion that the systems of the planet Earth are all linked has a considerable history. In the 18th century, the geologist James Hutton went so far as to say the Earth was a super-organism and should be studied by physiologists. In 1877, T. H. Huxley (a friend of Charles Darwin) suggested that living organisms were responsible for the disequilibrium of the constituents of the atmosphere. In its present forms, however, the hypothesis owes its importance to the chemist James Lovelock FRS, a gas chromatographer. In the 1960s, he suggested that the planet as a whole had properties that were not discernable by knowing about the component parts and used the term 'Gaia' (the Greek name for the Earth Goddess) as a metaphor for the wholeness of the system.

Versions of the hypothesis

The emergence of the hypothesis in Lovelock's first book (1979) came in the context of the realization of the global scale of some phenomena. Most of the major biogeochemical cycles were seen to be truly world-wide in their linkages, and the ubiquity and speed of electronic communication seemed to add another layer. The views of the whole planet from space, then becoming widely available, seemed to confirm the authenticity of this approach. Although the term was meant as a metaphor, it has been taken up by mystics and by New Age religions. Here we will consider the scientific versions of the metaphor, noting nevertheless that there is a progression from weak to strong versions even within this framework.

The irreducible basis is that Gaia as a total planetary entity has emergent properties not predictable from knowing the behaviour of component parts. At its simplest (i.e. the 'weak' Gaia hypothesis) there is the suggestion that the Earth's biota have an influence over the temperature and composition of the atmosphere. Up the scale is a co-evolutionary model in which biota affect the abiotic world, which in turn exerts evolutionary pressures (in the sense understood by Darwin) upon the living organisms. This leads to a feedback structure in which the living and non-living are linked by negative feedback loops and there is active adaptive control of the abiotic. Beyond this, well towards the 'strong' Gaia hypothesis, is a teleological version. Here, the atmosphere is kept in homeostasis not just by the biota but for the biota; strictly speaking, this is not compatible with Darwinian interpretations of the world. Beyond it, the end of the spectrum is found with the idea that the biota manipulate the abiotic to create conditions that are optimal for life in various forms.

Some versions yield testable hypotheses, other do not. Current controversy often revolves around the notion and meaning of purpose and whether the Darwinian concentration on selection at the level of the individual is too exclusive and whether there could be selection at a group level, which might imply purposive rather than simply adaptive behaviour. Such views argue that although metaphors are not mechanisms, in fact 'mechanism' is also a metaphor and not perhaps an accurate one to describe the potential of life (Schneider and Boston, 1991).

The key intellectual challenge in the Gaia hypothesis is that of dealing with the notion of purpose within a scientific framework. Conventional views do not admit that this is possible within a Darwinian conception of life and its adaptations. The counter-argument is that the word 'teleology' is a misrepresentation of a very close coupling between organisms and their abiotic environment.

Feedbacks

In the current mood of concern about the effects of contamination upon the Earth's systems (and especially upon climate), the Gaia hypotheses have in particular focused attention upon the feedback loops between life and non-life, with an emphasis on the atmosphere rather than the lithosphere as a problem zone. Using the Gaia models as a basis, workers can try to predict whether the linkages between living and non-living are at present negative feedbacks (i.e. self-correcting or homeostatic, in which case any excessive global warming trends will bring about their own stabilization) or positive feedbacks, in which reinforcement of trends will occur and runaway changes are more likely.

Calculations and predictions have been made for geophysical feedbacks such as cloudiness, water vapour transfer and ice/snow cover; for the biogeochemical cycles; for tropospheric chemistry and oceanic chemistry (including the role of DMS mentioned above); the terrestrial biota (including their albedo levels, carbon storage capacity, 'fertilization' by extra carbon dioxide and methane emissions from permafrost and peatlands); and human societies' energy demands. Overall, the gain might be in the region of 0.32–0.98. This implies that the feedback loops are, in total, positive and could mean an increase of global temperature of 10 °C rather than the 2–3 °C normally forecast. Since the climate system is almost certainly not stable beyond increases of 2–5 °C, this is a piece of information that reinforces the views of those who urge prudence in the face of uncertainty.

Much of what we learn about the way the biophysical processes of the world function is derived from sources that we invest with authority. The natural sciences are predominant among these when it comes to the matters described above. But Yearsley (1991) has noted that even with such large amounts of mostly reliable information, there are still difficulties. There are those of uncertainty: in planning of developments, for example, which proposal will lead to less environmental damage? It is often impossible to tell. There is often not enough time or money to collect the evidence needed for certainty. Also, phenomena may be on the margins of observability due to their location (e.g. in the deep oceans) or difficulties of measurement (e.g. in the early days of organochlorine pesticide use). To this can be added the low level of development of theory of ecosystem behaviour, leading to poor predictability. This is set in the context of the complexity of large-scale phenomena, even below the global level. All these features share the characteristic that wholes behave differently from their subsystems and so classic scientific reductionism has some limitations in its ability to explain and control. We have also to acknowledge that natural scientists such as ecologists and physical geographers have not built the systems they are talking about, so there is a residuum of ignorance in every case. Even when there is a lot of information available, other groups (such as corporations or governments) have a large degree of control or influence over these systems. This realization leads us rather neatly to a consideration of the human use of the Earth.

Further reading

There are many books on the nature and social role of science: they range from the ultra-enthusiastic to the downright sceptical. Those by Alan Chalmers (1982) on the nature of science as an intellectual activity and by John Ziman (1980, 1994) on the social relations of the sciences are somewhere in the middle, though they are in favour rather than against. The biography of Charles Darwin by Desmond and Moore (1991) is good reading. The biogeochemical cycles are often described but rarely better than in Butcher *et al.* (1992), since human actions are properly taken into account. The changing physical environments of the Pleistocene and Holocene are discussed in Goudie (1992), Roberts (1989) and Bell and Walker (1992). Human evolution studies are always themselves in a state of evolution but Bilsborough (1992) is a good state-of-the-art account. For sea-level change, see Tooley (1994) for a general account and the more specialized material for Europe in Tooley and Jelgersma (1992). Though now perhaps a little out of date, Chorley's (1969) collection has never been equalled for range and inclusivity. If ice is singled out for further reading then Sugden and Hulton (1994) is a reliable guide. Biomes are described in most biogeography texts but the recent detailed accounts (in which animals are included, which is not always the case) in Archbold (1995) are a rich quarry. The first book-length treatment of the Gaia hypothesis, by J. E. Lovelock, was published in 1979; essays on the implications of it in both intellectual and practical terms are collected together in Schneider and Boston (1991). The uncertainties expressed by Yearsley (1991) are amplified for prediction by Treumann (1991).

Archbold O W 1995 *Ecology of World Vegetation*. Chapman and Hall, London

Bell M, Walker M J C 1992 *Late Quaternary Environmental Change, Physical and Human Perspectives*. Longman, London

Bilsborough A 1992 *Human Evolution*. Blackie, London

Butcher S S, Charlson G H, Orians G H, Wolfe G V 1992 *Global Biogeochemical Cycles*. International Geophysics Series vol. 50, Academic Press, London

Chorley R J (ed.) 1969 *Water, Earth and Man. A Synthesis*.

Desmond A, Moore J 1991 *Darwin*. Michael Joseph, London (published by Penguin in 1992)

Goudie A 1992 *Environmental Change*, 3rd edition. Blackwell, Oxford

Lovelock J E 1979 *Gaia. A New Look at Life on Earth*. OUP, Oxford

Roberts N 1989 *The Holocene. An Environmental History*. Blackwell, Oxford

Schneider S H, Boston P J (eds) 1991 *Scientists on Gaia*. MIT Press, Cambridge, MA

Sugden D, Hulton N 1994 Ice volumes and climatic change. In Roberts N (ed.) *The Changing Global Environment*. Blackwell, Oxford: 150–72

Tooley M J 1994 Sea-level response to climate. In Roberts N (ed.) *The Changing Global Environment*. Blackwell, Oxford: 172–89

Tooley M J, Jelgersma S (eds) 1992 *Impacts of Sea-Level Rise on European Coastal Lowlands*. Blackwell, Oxford

Treumann R A 1991 Global problems, globalization and predictability. *World Futures* **31**: 47–53

Yearsley S 1991 *The Green Case. A Sociology of Environmental Issues, Arguments and Politics*. Harper Collins, London

Ziman J 1980 *Teaching and Learning about Science and Society*. Cambridge University Press

Ziman J 1994 *Prometheus Bound: Science in a Dynamic Steady State*. Cambridge University Press

CHAPTER 3

The human use of the Earth

This chapter deals with resources. These are the materials which human societies perceive as being necessary to their welfare and which they draw from their surroundings. Such acts nearly always result in alteration of the environment, sometimes planned and sometimes accidental. As we saw in Chapter 1, it is often the yield of resources which enables a society to label its environment as a beneficent place.

Resource use

Any immediate consideration of what we expect from resource use will include items such as adequate supplies of basic needs, an absence of threats to our health, materials of symbolic importance and also places in themselves, as in demands for recreation resources. A more systematic ordering might be into environments which are life-supporting, those which deal with wastes (i.e. are sanative), and those where the whole environment itself is the important unit of attention.

Resources for life-support

Resources for life-support must of necessity include materials needed for survival and reproduction of the human organism: a minimal amount of nutrition and clean water, clothing and shelter are universal needs. Equally necessary, though, is the recognition that no humans live at such levels (or below them) by choice. The UNO recognizes that the most basic level of human needs requires more resources than this, and in the HIEs we demand large quantities of fresh water, diverse foods, building materials, transport networks, and services such as education and medicine. The population of Hamburg has a physiological energy consumption of some 5000 TJ/year but the energy expenditure of the community *in toto* is 100 000 TJ/year. So this category includes the use of resources which support particular life-styles or cultures whose future has every appearance of depending upon the continued supplies of these materials. Their use may in some cases be symbolic rather than completely necessary: the killing of eagles to make a Native American chief's headgear was no more necessary than the treatment currently meted out to rhinoceroses to provide aphrodisiacs for Asians, but the imperatives of both have to be understood within a cultural context. When an Indian or a Burmese man has done his life's work and amassed some wealth he may take up a life of meditation and prayer; Westerners are more likely to take a long-haul vacation by plane. The latter uses a lot more resources (especially energy) than the former.

Environments as sanative resources

The use of resources produces wastes. Each human produces both solid and liquid body wastes from their metabolism; the disposal of these is simple where small numbers of people are concerned, but once at the village level of organization and above, it is essential to keep these wastes separate from the people and, especially, from the water supply. At high levels of material use, urban societies wish to dispose of large quantities of solid wastes such as plastics, old TV sets, garden rubbish and discarded clothing, and these too have to be removed from the settlement sites. So in thinking about resource flows, we have to remember the end-phase of the system: how do we avoid the wastes that we produce?

Environments as resources

Sometimes it is not a particular material that is desired but a whole ecosystem or a whole landscape. The wish to protect wildlife from extinction, for example, may result in the designation of large areas of rural land and water as reserves and the management of those environments for the benefit of the wild species. Another demand placed upon whole environments is that of outdoor recreation. Here a satisfactory experience depends upon the right combination of features, such as scenery, environmental features like snow for winter sports or rock outcrops for climbers, and an absence of reminders of more everyday life. In some cultures, there are sacred places where the features of natural and cultural landscape combine to create a numinous place that requires special behaviour within it. Some of the Buddhist monasteries in the vicinity of Kyōtō (Japan) together with their gardens are examples.

Quantities used by human societies

The best available data on the quantities of materials used by humans are given in sources like the World Resources Institute's annual volume (or disk), *World Resources*. Here, some idea of the range of material use and waste production will be given.

Every society starts with the individual and so the annual or indeed lifetime consumption of resources is a key to demands placed upon the systems which supply the resources and deal with the wastes. As life expectancy increases around the world, extra demands are made even in the absence of population growth. In Western industrial economies, for example, a German citizen will in the course of a calculated lifetime of 70 years be responsible for the 'consumption' of 460 tonnes of sand and gravel, 166 t of crude oil, 39 t of steel, 1.4 t of aluminium and 1 t of copper, among other materials. That person will also consume about 190 kg of paper, whereas an inhabitant of neighbouring Poland uses only 31 kg. Abandoned metal and coal mines in the USA cover *c.* 9 million ha, which is roughly the size of Hungary.

The levels of consumption in Western nations are of course very much higher than in lower-income countries: North American per capita consumption of aluminium is 25 times that in India; each Japanese 'uses' nine times as much steel as their neighbours in China. High-income economies use about 10 times as much non-fuel wood as their poorer counterparts, but consumption is rising in both: that of Australia doubled in the 30 years after 1950. Municipal wastes display similar relationships, with annual per capita levels at 864 kg in the USA, 394 kg in Japan and 357 kg in the UK during the 1980s (Smil, 1993). Within these totals, plastics now represent about 8 per cent of the total volume.

The ownership of a motor car over 10 years will enable the vehicle to produce 44 tonnes of CO_2, 325 kg of CO, 47 kg of NO_2, 5 kg of SO_2 and 36 kg of unburnt hydrocarbons. During its life, the car will need 200 m^2 of tarmac and concrete, which in the former West Germany adds up to 3700 km^2 of land surface, 60 per cent more than is devoted to housing. In terms of wastes, the emissions seem to be the equivalent of producing one dead tree and three 'sick' ones. In economic terms, the external costs of the car (pollution, accidents and noise minus taxes on fuel and vehicles) in the West Germany of the 1980s were DM 6000/year, which would have bought 15 000 km of first class rail travel in the 1980s.

Types of resources

It is conventional to divide resources into two main categories: the renewable or flow resources and the non-renewable or stock resources. The former comprise mainly materials of biological origin and water, where the self-reproducing nature of organisms means a continuing supply, and the hydrological cycle keeps water moving through the compartments accessible to humans. The non-renewables are mostly of geological provenance and are identified on the basis that once the coal is burned or the ore refined, then no more is being formed within the time-scales of interest to human societies. Although environments themselves usually belong with the first group, since using them for recreation, for example, can leave them exactly as they were found, the pressure of use can cause irreversible damage. So the classifications are by no means absolute. Heavy culling of a biological stock can reduce it to extinction, just as determined recycling of metals can reduce the need for the mining of fresh rock.

> There are many inequalities in resource use. These are most obvious in the contrasts between the industrialized nations and the LIEs, with use in the newly industrialized countries (NICs) rising quite quickly. The pressure on the resource base to fulfil demands often means stress on the environment, both for the provision of materials and the disposal of wastes. Non-material uses such as recreation and tourism can be non-consumptive but actual practice shows that the levels of manipulation may produce unstable environments with people wanting then to 'move on' to repeat the experience elsewhere.

Renewable resources

If water and biological materials are indeed renewable, then their availability should be assured. We know that this is not the case: there are often regional shortages of water even if the global quantities are sufficient, and many biological resources cannot cope with the demands imposed by rapid population growth. So the concept of **sustainable use** is often applied: a notion which suggests that present use patterns must not undermine the future levels of production of any resource.

Food and agriculture

Humans will eat almost anything that they can get as far as their gullets, from geophagy to cannibalism. Yet consensus about the ideal diet is still lacking. We all need energy-yielding carbohydrates, proteins, vitamins and minerals but in exactly what quantities is difficult to determine since there are so many variables of activity and individual tolerances; also, there are obvious differences between those maintaining their body in adulthood and, for example, children and pregnant women. The average adult person in the world who takes in between 2200 and

3000 kcal/day, with 30–40 per cent of this as animal protein, is likely to be well nourished. That many are inadequately nourished is as clear as the fact that some over-eat (Fig. 3.1).

Time, culture and environment have differentiated food production into a number of systems. Shifting cultivation is one such, being a relatively simple system involving the temporary clearance of wild vegetation (grassland or forest) and its replacement with cultivated plots. More common in the world now is permanent agriculture, where land parcels are devoted only to raising plants and animals. Animals are often an integral part of maintaining the fertility of the soils in these systems, especially where chemical fertilizers are little used. Pastoralism and ranching are systems which rely largely upon animals, whose products are traded for plant materials. Any one of these systems may be supplemented by horticulture, which is the scene of intensive cultivation, especially in backyards. There is also industrial production of food, which uses by-products of, for example, petroleum refining in connection with micro-organisms to produce edible substances.

The success of the world food systems in feeding the growth of world population cannot be denied. The recent past, in particular, has been a period of rising output: in the 20 years to 1989, root crop production increased by 0.8 per cent per annum, cereals by 3 per cent, milk, meat and fish by 2 per cent, and other foods by 2.5 per cent. In absolute terms, this meant an increase in world production of grain from 1.0×10^9 t in 1965 to 1.8×10^9 t in 1989. Within this trend there was some yearly variation, especially in the USA, where effective government policies determine the quantity of land under grains in a given year. In the LDCs most of the growth rates have been higher than in the DCs, though from a smaller base, with India and China being the most effective higher producers. Africa has been the only continent where production per head has in fact declined: wars, poor distribution systems and ineffective government policies have added to the uncertainties of rainfall (Table 3.1). The food security of nations is therefore very variable: it can be measured in terms of the capability of a nation to produce enough cereals for domestic consumption (Fig. 3.2).

Unlike water, there is a global market in agricultural produce, with large cash crop and export trades. Several nations earn most of their foreign exchange this way: Burundi gets 93 per cent of its exchange from coffee exports, Sudan 65 per cent from cotton. Overall, Africa devotes about 13 per cent of its cropland to exports and most LDCs have a slightly higher proportion. The trade is to some extent dominated by North American exports of grain, which provide the world with its buffer against harvest losses elsewhere, with $100–120 \times 10^6$ t/year being a typical exported amount, accounting for 87 per cent of world grain exports. In fact, 53 per cent of the world's cereal food aid comes from the USA, with the EU in second place at 20 per cent. The major

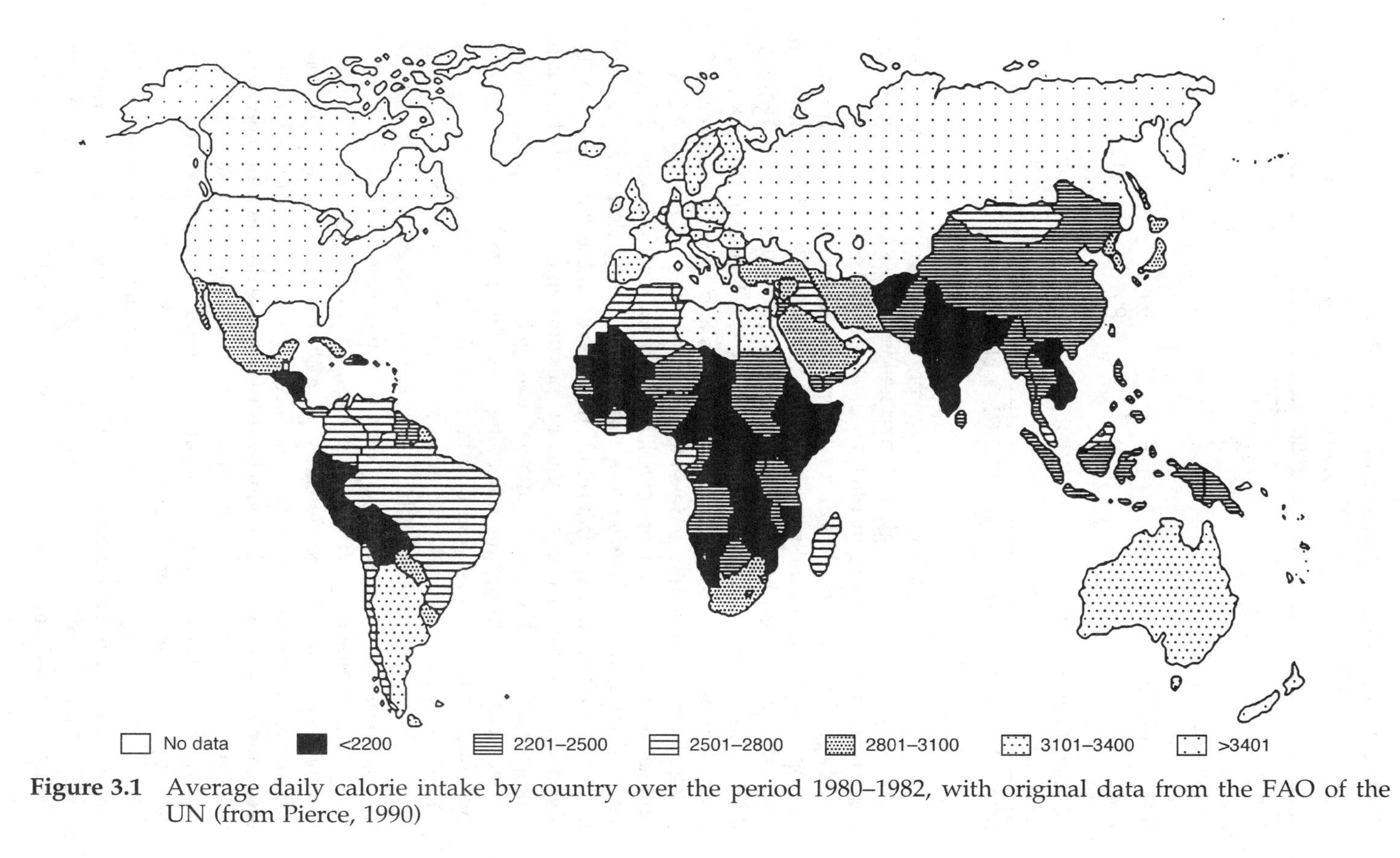

Figure 3.1 Average daily calorie intake by country over the period 1980–1982, with original data from the FAO of the UN (from Pierce, 1990)

Table 3.1 Indices of food production, 1979–81 = 100

	Total 1978–80	Total 1988–90	Per capita 1978–80	Per capita 1988–90
WORLD	98	122	100	104
Africa	98	125	101	96
North & Central America	97	104	98	92
South America	97	126	99	105
Asia	97	139	99	117
Europe	99	108	99	105
Former USSR	105	121	106	112
Oceania	103	109	104	95

Source: WRI (1993) *World Resources 1992–93*, Table 18.1

recipients of cereal aid have been Egypt, Mozambique and Tunisia in Africa, together with Bangladesh and Pakistan in Asia, each receiving more than 400 000 tonnes in 1989.

This production surge has also taken place at a time when the quantity of cropland per head is falling. The world amount (at 0.28 ha per capita in 1990) suggests a projected decline from the 1971–1975 period figure of 0.39 ha per capita to 0.25 ha per capita by 2000. In African LDCs the equivalent numbers are 0.62 to 0.32 per capita and for Southeast Asia 0.35 to 0.20 ha per capita. In the developing nations, the fall is largely due to population growth and land problems (e.g. soil erosion, salinization of irrigated land), and in DCs such factors as over-production and industrial land uses will reduce the figure. Of themselves, these data need not be alarming, since production per unit area per unit time can clearly increase, though such changes may involve external costs. The global state, therefore, appears to be one of keeping pace with population growth and even of passing the threshold of minimal self-sufficiency. But a universal transition to food security and more varied diets has yet to be brought about. There are regional problems of nutrition, where some people are getting too little and others far too much (Brown, 1994).

In the face of such problems, many solutions have been sought. There are those who affirm the necessity of technical developments in food production, and those who are convinced that the problem is basically economic and can be tackled only by increasing the demand for food in the towns of LDCs so that better-off urbanites 'pull' more food out of the rural sector. Slower rates of population growth are a social factor which may back up any other approach. Socio-political developments such as land reform are seen by yet others as the key to the success of any other developments. On the one hand, the virtues of indigenous farming systems and 'bottom-up' development are extolled and, on the other, large-scale technology-based early warning systems for drought or other extremes,

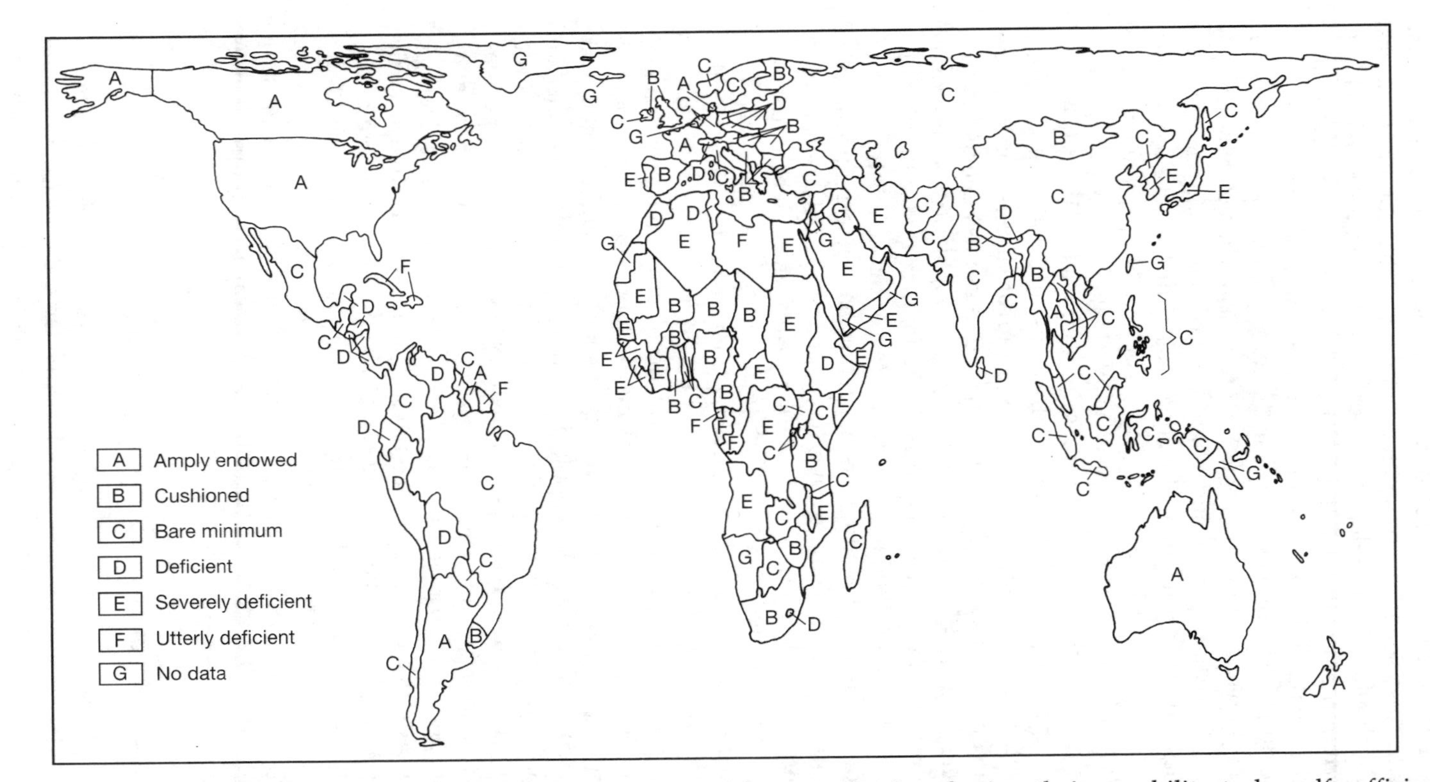

Figure 3.2 An attempt to plot the food security of the world's countries by plotting their capability to be self-sufficient in basic cereals. The assumption is that the shortfalls currently made good by trade and aid might suddenly persist (from Kidrom and Segal, 1991)

like the FAO's GIEWS (Global Information and Early Warning System) scheme are also sought. Those with a primarily global concern think that the DCs should consume less in order to make surpluses available to poorer countries. The lesson of this paragraph is that problems are as likely to be social and political as they are scientific and technological.

Agricultural development is by far the most popular of approaches to nutrition problems, since it does not overtly mean changing the existing patterns of political power and may be perceived as an element of a much-desired 'modernization'. In effect, agricultural development has meant an attempt to transfer the technology of DC agriculture to the Southern nations, using a battery of HYV (High Yield Variety) cereals (especially rice, maize and wheat), irrigation, chemical fertilizers and biocides, and intensive water control. In terms of absolute productivity, the so-called 'Green Revolution' has been a great success but it has not eradicated all the pockets of poor nutrition, because its benefits have not been evenly shared. Hence the more recent popularity of improving indigenous systems, which may benefit the poorest rather than making the rich richer. Even though its benefits were often concentrated in the hands of the better-off landowners, irrigation has underpinned a lot of rises in production: in India during the period 1950–1990, cereal output rose from 55×10^6 t/year to 194×10^6 t/year, with over 50 per cent coming from irrigated areas. In fact, Asia housed 56 per cent of the world increase in irrigated area, rising from 166×10^6 ha in 1970 to 213×10^6 ha in 1982, with 297×10^6 ha forecast for 2000.

From time to time, famine occurs in developing countries. Though symptomatic of long-standing problems of food supply, its cause is usually perceived as some 'external' agent such as drought or civil war, and international aid can be supplied. Of more general importance are those LDCs which exhibit chronic under-nutrition or malnutrition and especially those where intake is less than 2500 kcal per capita per day. Some 82 nations fall into this category, most of them in Africa, along with the poor areas of South Asia and Latin America. For example, Ethiopia, Kampuchea and the Maldives all fell below 1800 kcal in the mid-1980s. The UN estimates suggest that 730 million people are getting insufficient calories and that in total some 950 million people (i.e. 20 per cent of the world population) live off deficient diets. In many ways, however, chronic deficiencies of nutrition are less environmental than social in cause; the local relationships of political power are often as important as the level of soil nitrogen.

Forests and trees outside the tropical lowlands

The present uses and degradations of the lowland tropical forests are discussed (p. 121) as a special case of biodiversity. But bearing in mind that organic evolution has tree growth whenever natural conditions have permitted and that forests are both sources of useful resources and, often, reservoirs of produced agriculturally fertile land, consideration needs to be given to them.

To equate trees with forests is to forget that they grow in a variety of other habitats as well. Open woodland (i.e. woodland lacking a closed canopy of branches), small woodlots and coppices and even lone pines have their place in the spectrum of wood-producing places alongside the major areas of closed (i.e. having a more or less unbroken canopy) forest.

The world today contains *c.* 2800×10^6 ha of closed forests, which is 21 per cent of the land area. The addition of 4500×10^6 ha of open forest and woodland in areas of shifting cultivation brings the total to 34 per cent of the land area, though this latter figure is prone to considerable error. Closed and open woodland together are approximately three times the world's cropland area. The tropics contain 43 per cent of the closed forests, the temperate zones 57 per cent; the developed nations are responsible for 90 per cent of the coniferous woodlands, whereas the LDCs have 75 per cent of the broadleaved forest. About half of the closed forests are in fact found in four countries: Brazil, the former Soviet Union, Canada and the USA (Mather and Chapman, 1995).

The uses of trees, woodlands and forests are manyfold. The greatest single use is still the provision of wood fuel for domestic use in the poorer nations: in 1989 the harvest was estimated to be 1.7×10^6 m^3 for that purpose (Fig. 3.3). Total industrial wood harvested amounted to 16×10^9 m^3. In peasant economies, the uses of wood are for fuel, construction, fencing, tools and a myriad of other purposes; in industrial nations, construction and furniture are major uses but the predominant demand is for paper and paper products of all kinds. The weight of these demands has led to the point where deforestation is proceeding at 10–20 times the rate of reforestation: in the last 10 000 years perhaps 33 per cent of the broadleaved forests have been destroyed at human hands, and 25 per cent of the savanna woodlands and subtropical deciduous forests. Until quite recently only 6 per cent of the tropical moist forests had been felled. Unless intensive production can be maintained from the current area, the world-wide status of forests as a renewable resource is threatened.

Apart from their yields of wood, forests have other values for human societies. Unless they have no understorey and ground-layer vegetation, they can often be used to pasture domestic animals; they hold the soil and thus act as a protective cover for watersheds (a quality recognized by regulation in Tokugawa Japan, for example), and are reservoirs for wildlife. All the qualities of the forest come together in the popularity of the habitat for recreation of various kinds. Finally, forests may have regional climatic roles: in the Amazon Basin, for example, much solar heat is used in evaporating the moisture produced by evapo-

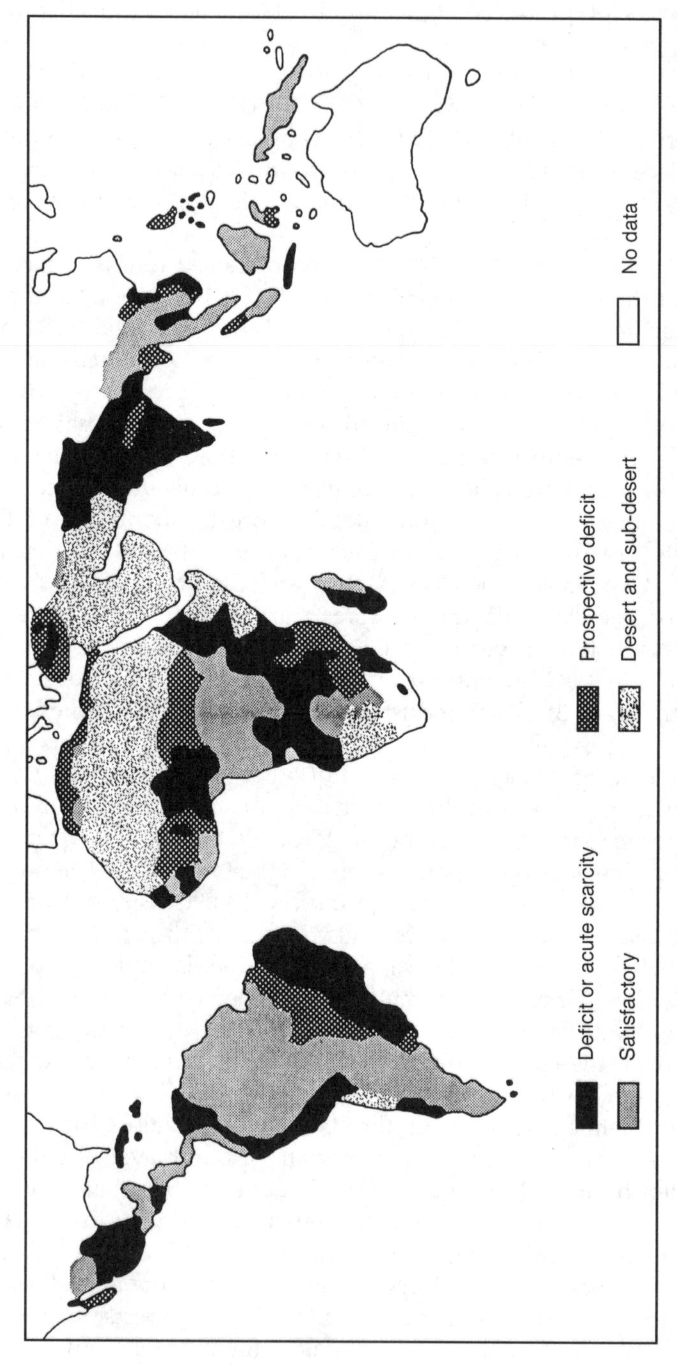

Figure 3.3 Areas of the world with fuelwood deficit in the 1980s. The preponderance of semi-arid and seasonally arid regions, together with mountains, can easily be seen. The HIEs escape the problem since they can afford oil (from Williams, 1990)

transpiration from the trees; without them the region would probably be much hotter. Globally, the forests are one of the sinks for carbon. Without their role in sequestering atmospheric carbon dioxide, for instance, the concentration of that gas would be rising even faster than at present, probably with global climatic consequences. Forests, then, are an integral part of the human life-support system.

Water

The account of the hydrological cycle (p. 69) revealed that the proportion of the planet's water that is available as a human resource is small: about 0.3 per cent of the total fresh water. As a renewable resource (i.e. excluding groundwater at a depth that is supplemented only on a geological time-scale), this amounts to about 41 million km^3 or, in 1990, 7690 m^3 per capita, which amounts to 1870 litres per capita per day. Such an average is exceeded by North Americans, for instance, who have access to 2098 litres per capita per day. The use of water may be totally non-consumptive in the sense that it is not altered by uses such as power generation, flotation or recreation. On the other hand, it may be withdrawn from its original source and only a part returned to the source in some form. The difference is called 'consumptive' use. So data presented as 'withdrawal' include those uses where some of the water is quickly returned to the river or other flow. Table 3.2 gives some of these data on a global scale.

These data tell us, among other things, that water withdrawal and consumption has increased fourfold in the 20th century in the world as a

Table 3.2 World annual water use

	Total withdrawal (km^3) 1980	**Per cent of total renewable resources**	**Total consumption (km^3) 1980**	**Change 1900–1980 (per cent)**
WORLD	3 320	8	1 950	2 741
Africa	168	3	128	126
North & Central America	663	10	224	594
South America	111	1	71	96
Asia	1 910	15	1 380	1 496
Europe	435	15	127	397
Former USSR	353	8	NA	NA
Oceania	29	1	15	28

Sources: S. L. Kulshreshtha (1993) *World Water Resources and Regional Vulnerability: Impact of Future Changes*. Laxenburg: IIASA, RR-93-10; WRI (1992) *World Resources 1992-93*, OUP, Table 22.1, and various sources

whole, with large absolute and relative increases in North America, a large absolute increase in Asia and a relatively large growth in Australia and Europe. Africa has been below world average levels. They obscure, however, considerable variation between countries (Fig. 3.4). Four main types of use can be identified: agriculture (mostly in irrigation); industry (a myriad of uses including food processing and the cooling of power plants), municipal (which includes domestic use) and stored in reservoirs. As Table 3.3 shows, agriculture is by far the greatest use and thus poses the greatest problem because of the linkages to population growth and possible climatic change.

In absolute numbers, the diversion of the hydrological cycle does not look very great. The highest ratio of supply to use is about 18 per cent (in Asia), with a world average of just over 8 per cent. But of course,

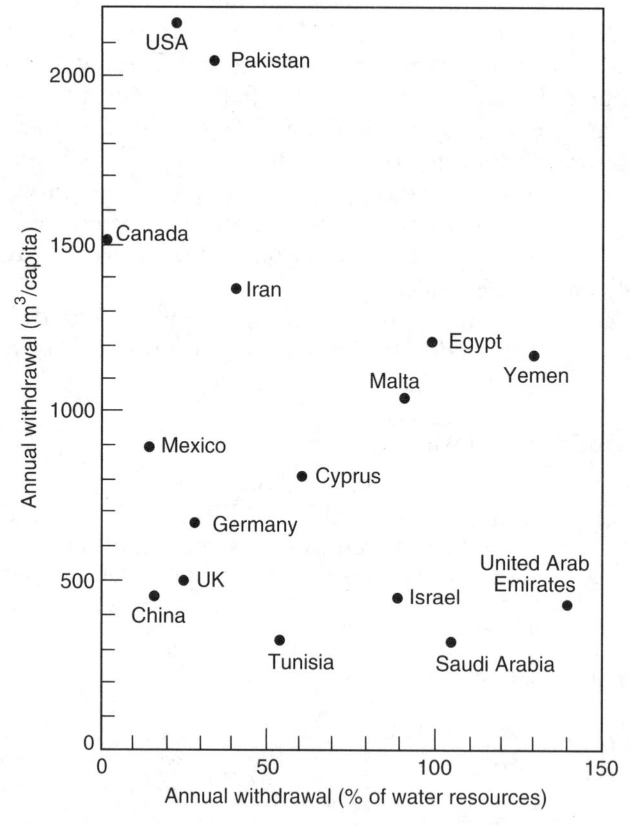

Figure 3.4 A chart of the absolute quantity of water withdrawn against annual withdrawal as a proportion of total water resources. Thus to be in the upper right-hand corner is to be in the worst position and in the lower left is to be relatively 'affluent' in water (from Smil, 1993)

Table 3.3 Uses of water (km^3)

Type of use	Withdrawal 1950	Withdrawal 1980	Consumption 1950	Consumption 1980
Agriculture	1 130	2 290	859	1 730
Industry	178	710	14	62
Municipal	52	200	14	41
Reservoir	6	120	6	120
TOTAL	1 370	3 320	894	1 950

Sources: compiled from various sources

these numbers conceal a great deal of local and regional variation and in addition say nothing about water quality, which may be seriously affected by rapid industrialization. Thus, climatic fluctuation, local topography and geology, rates of local population growth and urbanization, rising demands for clean water and for energy, all mean that some regions are viewed in terms of exposure to water stress or scarcity within the next few decades (Gleick, 1993). This may be especially so where a nation already utilizes a high proportion of its available fresh water. Examples (1990 data) include the UK at 82 per cent (the highest in Europe, with Belgium second at 73 per cent), Egypt at 111 per cent, Libya at 168 per cent, and Israel at 96 per cent. The Gulf oil states are all well in excess of 100 per cent but they can expend energy on desalination without much concern. So the nations which currently experience *water stress* and *water scarcity* (defined using criteria of supply per capita and the use–availability ratio) are the UK and Peru in the first category, and core nations of the Middle East together with Libya in the second. Using the best available assumptions about population growth, climatic change and industrial development, the more severe category might by 2025 bring in most of Europe, North Africa, the African Sahel zone and South Africa, as well as the Indian subcontinent.

Too much and too little water can both cause considerable dislocations in human affairs and are often labelled 'natural hazards'. Floods and droughts are relative terms but both describe conditions outside the range of values of water presence (or absence) to which a society has adjusted. Thus the 1970s and early 1980s were an especially bad time for droughts in both the USSR and the Sahel zone of Africa. In the latter, the peak year of 1984 saw 30–35 million people affected, of whom 10 million were displaced. An average year in Southeast Asia sees the destruction by floods of 4×10^6 ha of cropland. Adjustment to these hazards can be of four types. First, an attempt can be made to modify the event, such as tapping extra water supplies in the case of drought, or building flood-control dams on the upper stretches of rivers.

Second, damage susceptibility can be modified, such as land-use zoning to prevent investment on flood plains or adaptation of drought-tolerant varieties of crops. Third, the burden of loss can be lessened by the provision of adequate insurance, and lastly there is the stoical hope that the hazard will not recur.

Water resources do not appear to be a global problem of the first order; there is untapped water and there are many possibilities of reuse of water if it is carefully managed. But that is not to ignore the fact that three-fifths of the population of the LIEs have no access to safe drinking water, let alone extra supplies for other purposes. In between them and a better availability lies a requirement for such matters as the collection of better data, which are sparse and inaccurate in the LIEs. There is also a need for the setting up of the proper laws that will move towards economic and equitable water use and recognize that water quality and presence depend upon external influences such as land use, industrial effluents, recreation and wildlife conservation. Institutions are often important: the drainage basin is a useful unit but one which may transgress political boundaries, including international frontiers.

The oceans

Since the seas cover 71 per cent of the globe's surface, hopes of them being a cornucopia of resources surface from time to time. Since the salinity of the oceans averages 35‰, and is kept topped up by influx from the land, then minerals sometimes enter that framework of anticipation. However, apart from salt itself and some extraction of magnesium and boron, sea water is not currently much valued for its mineral content. Indeed, it is valued more for its usefulness without minerals, as desalinated (i.e. fresh) water. Many oil-rich nations and some islands without many indigenous resources resort to one or more of the high-tech processes such as flash distillation or reverse osmosis to separate the water from the minerals. Some of the Gulf states are highly dependent upon desalinated water for all kinds of use and so were highly threatened by oil slicks moving towards the intakes during the Gulf War of 1990–1991.

The living resources of the sea are without doubt the major focus of resource users. These are most common near the interfaces with the land areas and over the continental shelves, for in the open oceans the lack of mineral nutrients usually retards primary productivity by phytoplankton. Exceptions occur where cold upwellings bring mineral nutrients from the ocean floor to the surface, as with the Humboldt Current off Peru, where capping by warmer water in some years gives rise to the El Niño phenomenon (see p. 62). The total carbon fixed in the oceans (20–60 $\times 10^6$ g/year) is about the same as that on the land; the offshore productivity is of the order of 50–170 g/m^2/year (and in

Table 3.4 Biological productivity in ocean zones (reprinted with permission from 'Photosynthesis and fish production in the sea', *Science* **166**, 72–6 (Ryther, 1969). © 1969 American Association for the Advancement of Science)

	Per cent area	**NPP (gcarbon/m^2/year)**	**Fish production (10^6 t/year)**
Oceans	90.00	50.00	1.60
Continental shelves	9.90	100.00	120.00
Upwellings	0.10	300.00	120.00

upwelling zones 1800-4000 g/m^2/year) and the open oceans no more than 100 g/m^2/year, which is equivalent to a terrestrial desert (Table 3.4).

Since oceanic food chains start with very small organisms, those which are actually harvestable by humans may be very high up those chains, with all that is thereby implied for energy loss from the ecosystem. An NPP of 900 kcal/m^2/year of phytoplankton in the North Sea, for instance, becomes 0.6 kcal/m^2/year of adult carnivorous fish such as cod. So fish, which are the main renewable resource of the seas, may relatively easily be over-exploited, especially since they swim in shoals or lie on the ocean bed (Cushing, 1975). The global fish catch in 1950 was of the order of 20×10^6 t, and in the late 1980s nearing 100×10^6 t/year; the United Nations' Food and Agriculture Organisation (FAO) has suggested that 100 million t/year is probably the limit of sustainability. Of the current catch, about 35 per cent is converted to fish meal, which is used as animal food. It is estimated that a staggering 27 million tonnes (about 8 per cent of the total catch) is simply discarded after being caught. Fish are not the only catch: many other marine animals (molluscs, cephalopods, crustacea and mammals) and plants are also used; likewise, fishing in either artisanal or industrialized form is not the only method of exploitation, for aquaculture is widely used (and amounts to about 11 per cent of the total landings), especially in the brackish-water shore zone of Asia, often on land reclaimed from mangrove.

The cultural ecology of most fisheries revolves around their ability to sustain constant harvesting. In artisanal fisheries (as contrasted with industrial methods), there is rarely sufficient impact upon the populations to prevent them from reproducing and growing to adult size. On the other hand, a map of the collapse of fisheries is also a map of the introduction of industrial technology (Fig. 3.5), such as the steam-powered trawler and net winch in the late 19th century and, after World War II, nylon nets and electronic aids to the location of shoals of fish

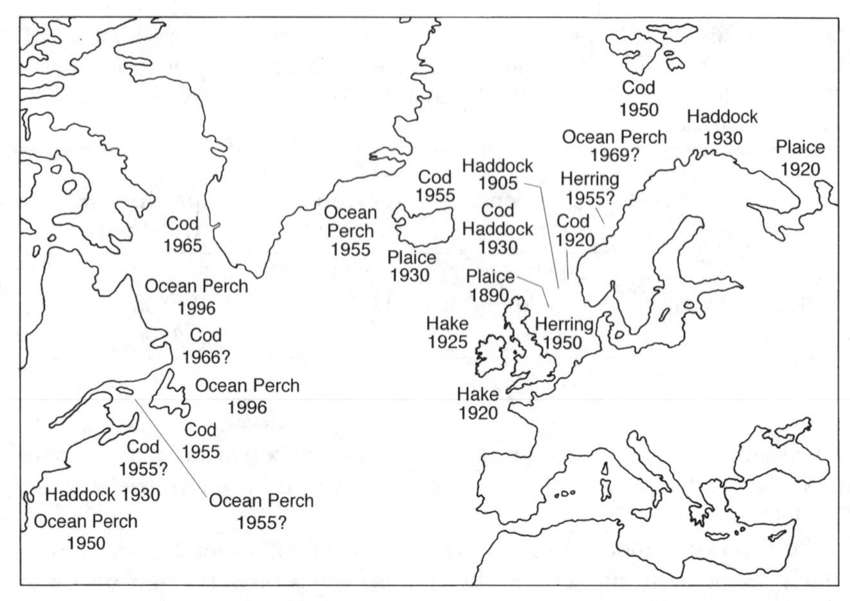

Figure 3.5 The dates of significant decline in the fish stocks (and hence the catches) in the North Atlantic. Note the concentrations of dates in the inter-war period when the effects of the steam trawler had worked through fishing fleets; and in the post-war period with new technologies such as nylon nets and echo-sounding (from Pickering and Owen, 1994)

(Cushing, 1988). This reminds us of the relation of fishing to energy use: if we measure energy input per unit of protein landed (and wet fish averages 9 per cent protein) then for artisanal fisheries with boats under 5 tonnes, the relation is of 37 kcal of protein per 1 kcal of fossil fuel input; for industrial fisheries with boats over 5 tonnes, 14 kcal per kcal. Equivalent figures of land-based food production are 20–44 kcal for feedlot beef, 2–4 kcal for vegetables, and 2–4 kcal for grain. As an example of the key role of culture, the example of whales must be considered. There are now perhaps 200–1100 blue whales left, whereas a century ago there were 250 000. Similar declines can be charted for all major whale groups. Nowadays, there is no need for killing whales, since their oil is not needed for lighting or for lubricants and their flesh is eaten in only a few cultures. Yet the international moratorium on whaling agreed in 1982 proved incapable of lasting for more than 10 years, after which minke whales were again the target of catches by Japan, Iceland and Norway: partly to keep up employment but mainly, we might suspect, for the same reasons as others eat thick slices of beef or drive powerful motor cars.

The moral seems to be that all fisheries can be managed as renewable resources into any foreseeable future but that restraint in catch has to be exercised, otherwise fewer and fewer fish come to maturity and so the population declines. Three consequences may follow: one is that there is always a search for new stocks of hitherto unexploited animals (including squid and krill in recent years, for example); a second is more intensive production under controlled conditions of aquaculture; the third is the unplanned and querulous decline of a way of life which although dangerous had the sanctions of time and tradition.

Here we have seen once again the importance of living organisms: they are self-renewing and also adaptable via breeding programmes and genetic engineering. Water is 100 per cent essential: it is always renewable in quantity but the quality desired is often so high. For many purposes it needs to be as free as possible of both organisms and minerals. Most of it is in the oceans where it is not in fact a very great resource except near the land masses, where, as might be expected, it is easily possible to reduce its utility through contamination.

Environments as resources

Two very different matters are discussed here. The first is the highly practical matter of the supply of biological materials to human societies. A variety of species have always been used by humans but the onset of industrialization has narrowed the spectrum in the interests of uniformity. The same processes have changed many habitats and made many species locally extinct. So there is concern about the pool of genetic material that will be available in the future. The second derives in part from industrialization as well. Greater wealth allows people to take holidays or shorter periods of recreation in places they consider beautiful. It also allows them the luxury of feeling detached from nature and to desire its preservation in relatively wild places. Both these cultural traits come together in movements to protect what is often seen as the natural world, where environments are considered as having value as totalities.

Biodiversity

Cutting across these questions of supply of renewable materials is a concern for the future of the reservoir from which they have come. In this case, we are talking about the variety of living forms which are the foundation of so many systems of renewable resources. The term **biodiversity** is used to mean (a) genetic diversity within species, e.g. the different breeds of domestic cattle on a continent; (b) the variety of species in a given area, e.g. the number of different moss species within a swamp forest; and (c) the number of different ecosystems within a biogeographic province, e.g. the swamps, woodlands, savannas and grasslands which traverse parts of East and central Africa (Groombridge, 1992).

The advent of genetic manipulation as a new element in biotechnology has meant some concentration upon genetic diversity as a focus in biodiversity research and policy-making. It is estimated that <1 per cent of genetic material is expressed in the form and function of the organism, so that the search for manipulable genes is always going to be costly. Nevertheless, the need is strong, for the exigencies of modern cultivation have greatly reduced the genetic base of agricultural production. Wheat production in Canada (one of the world's few areas with a large surplus), for instance, relies for 75 per cent of its output on four varieties, with over one-half from a single variety. Four varieties of potato account for 72 per cent of the US crop and that nation's pea production rests on only two varieties. US soybeans all came from six plants from the same place in Asia; more or less all the coffee trees in Brazil are the descendants of one plant. Of the 145 recognized breeds of cattle in Europe and the Mediterranean, 115 are likely to become extinct.

Most of the wider public attention has been given to species' extinction, often because charismatic animals have been involved. The brink-of-extinction status of the giant panda, the Californian condor or the Indian tiger are examples. Perhaps 5–50 million species of plants, animals, fungi and micro-organisms exist, of which *c.* 1.4 million have been described by science. At present, about 4 per cent per decade of these species are becoming extinct, which is estimated to be the fastest rate since the end of the Cretaceous: in absolute numbers an estimated 25 000 plant species and >1000 vertebrate species and subspecies are threatened with extinction. Though small areas of great species richness occur (like islands such as the Galapagos, where isolation has meant the formation of many species not found elsewhere, known as *endemics*), the tropical lowland forests which cover about 7 per cent of the land surface contain 70–90 per cent of known species.

Concern about loss of ecosystem types is bound up with changes in land cover and land use due to economic development. It is often associated with a transition from wild land inhabited by wild species to tamed land with domesticated species, and is measured by any of the data for the extension of agricultural land, the diminution of forest and grassland area since the 19th century, or the extension of cities and transport networks in the last 100 years. Soil erosion, too, takes the whole ecosystem (including its mineral nutrients) with it on its downward journey. This latter set of processes is at the heart of loss of biodiversity. Although the industrialization of agriculture is responsible for the loss of many indigenous varieties, the scale of the enterprise means that new varieties can be bred, either by genetic manipulation or by the more traditional techniques of scientific plant and animal breeding. But no industry stands ready to replace plant and animal species lost by over-cropping or made extinct because of the intensification of land use or the extension of cropland to feed growing human populations. In all cases, too, global extinction means that the genetic material is lost for ever. Extirpation on a lesser scale is less of a loss, though the risk of eventual total extinction is inevitably increased.

Reversal of these trends towards biodiversity loss centres around two processes:

1. *Ex situ* conservation, in which the genetic material is preserved in artificial conditions. This may be in living form in botanical gardens and zoos, where efforts centre around captive breeding programmes, with an ultimate aim of releasing specimens back into the wild if conditions for their survival have improved. The other mode is in gene and seed banks, usually associated with research institutes (Sandlund *et al.*, 1992);
2. *In situ* perpetuation, with two main types:
   (a) Conservation by protection, with specified areas of land or water set aside from 'normal' economic development processes and managed in a biocentric fashion, with the long-term welfare of the conserved species as the primary aim of management, though secondary aims (recreation, tourism, scientific research) may also be accommodated. Curious extremes of management may have to be tolerated, such as the de-horning of rhinoceroses to make them less attractive to poachers.
   (b) Perpetuation by sustainable use. It is recognized that far more people will identify with a conservation endeavour if there is a direct benefit; in this instance, elephants are managed for an ivory yield, maintaining a reasonable supply at moderate prices, rather than trying to prevent the poaching of totally protected elephants. Sustainable use does not of itself guarantee that any income will be evenly distributed.

In the last two cases, the concept of a minimum viable area is important. Populations of plants and animals below a certain size do not contain enough genetic variety to prevent inbreeding and a consequent loss of fertility. Hence, much research has normally to be done to establish what area must be conserved in order to keep a viable population in a self-reproducing state. The problem will often occur in zoos in an especially acute way.

The transformation of the tropical moist forests (TMFs) to logging, mining and grassland has been the scene of one of the biggest debates about biodiversity loss (Park, 1992). In the late 1980s, the conversion rate was something like 27 ha/minute, which equals *c.* 110 000 km^2/year and a total loss of the biome within 85 years (Fig. 3.6). The right to exploit these forests is contested: should it be dominated by local people, by big business, the governments, or by the international community in the interests of global climatic considerations such as the sequestration of CO_2? The diversity of possible uses of these forests *qua* forests is very great, encompassing the following examples:

- plant food crops (200-3000 species);
- wild genes which can be bred into extant crop plants for better yield, or resistance to viruses;
- timber (1989 export value $6000 million);
- rattans (>90 per cent of species endangered);
- medicinal plants and animals (90 species in world-wide use, annual trade $350–500 million);
- ornamental plants and animals (orchids, feathers, skins: export trade $3000 million/year);
- meat and eggs, mostly locally consumed;
- working animals (16 000 Indian elephants thus used);
- sport hunting ($1600–2300 million/year turnover);
- captive animals for pets and display (annual turnover $1300-2300 million);
- animal domestication: *c.* 200 breeds of domestic animals from tropical forests; selective breeding material for some pigs and peccaries.

What is common to all these processes is that little of the wealth gained therefrom is being reinvested in the forests, which are turned into a non-renewable resource in the search for rapid riches or for an overflow region to relieve densely populated areas. Turning such a set of values around is a formidable task, and one that will need to contain the following themes:

(a) a broader constituency for sustainable use, to include those engaged in production as well as those in conservation;
(b) the engagement of local people and their indigenous knowledge, to give them a real stake in the use of biological resources to provide productive employment, better health and improved nutrition; equally, management structures need to be consonant with the local culture;
(c) using the information already to hand and expanding it where necessary: ants cycle most of the energy in the terrestrial habitats of the tropics, but in the 1980s there were only eight entomologists in the world with the competence to identify the relevant species;

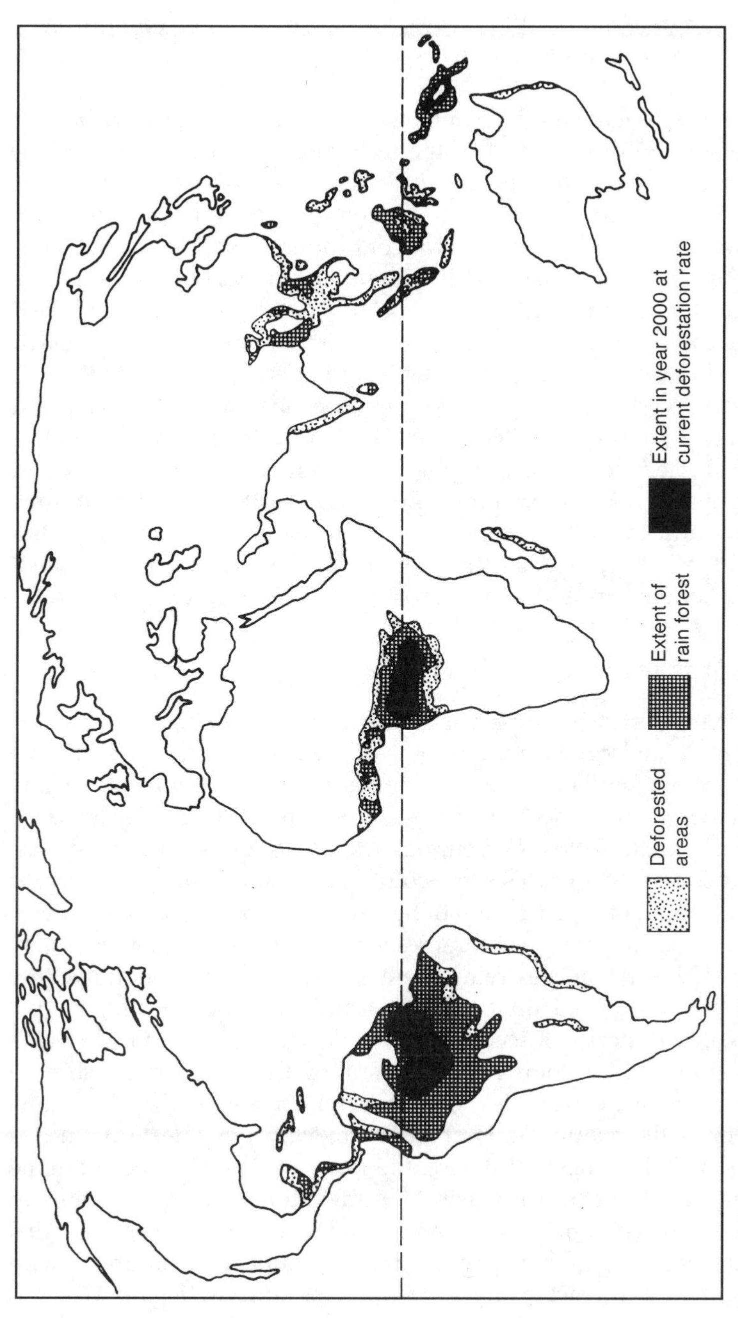

Figure 3.6 The progress of deforestation of closed tropical forests (i.e. those with complete canopies). The 'extent' shading is the extent earlier this century; 'deforested' areas are those now lacking forest; and the black areas extrapolate from present rates of loss those places where forest will be left in a few years time (from Pickering and Owen, 1994)

(d) devising new institutions that can focus directly upon biodiversity: natural systems are integrated entities, government departments often a fragmentation. What ways, for instance, can be found to help government and business to work towards the same ends?

Recent developments in biotechnology have raised the level of interest in biodiversity: will the advent of the technology needed to manipulate DNA outmode the necessity for *in situ* conservation? Although of considerable interest in crop production, population genetics and evolutionary biology, the potential for contribution to applied genetics in conservation seems relatively small. Too little is known about the nature of genetic information to predict the behaviour of genetically modified organisms if released into the wild, just as with commercially engineered organisms. In everyday terms, the ownership of engineered organisms is nearly always vested in large companies from industrial nations and it seems likely that they would benefit the most from any productive processes; this argument also applies to the discovery of new uses of wild organisms (Mannion and Bowlby, 1992). At present, the exploitation and marketing capacity is often likely to be lodged with a First World company and so the profits (e.g. most of those associated with safari holidays in Africa) are repatriated to the already rich.

Landscapes as resources

In many cultures, and especially those of the West, a cultivated appreciation of landscape is part of the world-view. Originally inspired by paintings and now influenced by many other media and the whole swathe of the 'heritage' movement, it finds outlets in demands for outdoor recreation of various kinds. In a curious way, it seems as if the sand-and-sea holiday is also part of this same cultural demand, although we might hesitate to call it cultivated by anything except the profit motive of those engaged in its supply. So the natural resources called upon are very varied (Simmons, 1991). At one extreme there is the kind of ecotourism that motivates people to walk long distances and camp under very simple conditions in the hope of seeing a particular species of animal, like a mountain gorilla. The comfortable variant of this is a cruise ship to Antarctica. The other extreme is the mass search for sun, sea, sand or snow, which involves the manipulation of environments a great deal (Mathieson and Wall, 1982). The coasts of the Mediterranean or the Caribbean are the best examples of the replacement of semi-natural shorelines with an urbanized set of conditions; the French and Italian sectors of the Alps illustrate the dumping of winter sports upon a mountain environment with little thought for its impact in any terms upon the surroundings.

In such circumstances the concept of renewability is difficult to analyse, for it must relate to both the environment and the users.

Clearly, recreational use of a beach or a forest can leave it untrammelled and just as fit for the next batch of visitors as for the last. Equally clearly, it can be altered by such use in a minor fashion or devastated entirely. Scavenger animals which emerge to beg food from visitors are ubiquitous: various species of rodents, birds and monkeys are instances of this category. In some contrast, a careless recreationist's cigarette end may result in a fire which leaves very little forest at all. The experience of the users is a much more plastic quality but there is obviously one mind-set which dislikes the presence of fellow humans in any numbers so that there is a continual search for recreational experience made exclusive by money or by difficult physical conditions. At the other end of this spectrum is the desire for a similar experience in the company of the same kind and numbers of people, which keeps the hotel industries of Honolulu and of Majorca in business. Renewability is, so to speak, annual in a kind of pre-agricultural cosmogony.

Environments as wholes sound a splendid idea but they are often more difficult to manage than a single system. One problem is that managers are unlikely to have sufficient predictive knowledge of the complexities they wish to control. Unforeseen effects are therefore likely to be great. 'Integrated management' is a fine slogan but in practice may not be much more than that; this is especially true where biological populations fluctuate naturally, as in the seas for example. The larger the unit, too, the more chance that it will transgress existing political boundaries and that different authorities will not agree to common programmes of management.

Non-renewable resources

Apart from materials ejected into space by modern technology, the Earth is a closed system. Therefore, when we talk of non-renewable resources we mean primarily inorganic materials together with hydrocarbon energy sources which are so transformed by their use that they are not immediately employable again by human societies. But in one form or another they are still present on the planet.

Major categories of resources

There appear to be three main categories of resources:

- those which are 'consumed' by use, such as coal, oil and natural gas; their complex molecular structure is broken down into much simpler components;

- theoretically recoverable materials such as minerals, which are technologically capable of being recovered after use; cost is the major factor in determining the degree of reuse;
- recyclable substances such as metals and glass which can be reused without an enormous amount of reprocessing.

To these must be added materials which are not being created anew in the way living organisms create complexity out of simple chemical components, for example, or the manner in which solar energy translocates water molecules in the hydrological cycle.

Practicalities of use

In practical terms, materials such as metals, ceramics and plastics may become so dispersed by use that they cannot be sieved out from the repositories into which they are led: plastic bags and wrappings are an example. Some substances may be physically impossible to recover since they are used in a very dilute form and become dispersed in water or air: the lead in aerosol form used as an additive in petrol is an example. At the other extreme they may be so concentrated that they are too toxic to handle: some of the wastes from nuclear power generation fall into this category. Lastly, the materials may be sequestered for such a long time that for all practical purposes they are lost to reuse: the steel frame of a large building locks up great quantities of materials.

Overall characteristics

General characteristics of non-renewable resources, in summary, are that they are usually products of the lithosphere, they usually need complex processing before use (with linkages to energy consumption and the production of wastes), they enter world trade and so are moved around the globe and have been much more important quantitatively since the 19th century, and they become expended since so much of their use is a 'once-through' process. This raises the question of the optimal depletion rate: should it emphasize the perceived needs of future generations and thus conserve the material as much as possible, or will we do the best for our descendants by using as much as we wish in order to turn it into knowledge of how to do without it? The complex field of resource economics, for example, is much concerned with this last question.

How much of it is there?

In such analyses, a fundamental question about any non-renewable resource is always, 'how much of it is there?' This is not a simple question, if only for the reason that exploration of the Earth's mineral resources is not complete. Indeed, the amounts available in practice at a given time depend upon five factors:

- The availability of technological knowledge and equipment and their location in the right places and amounts.
- Levels of demand, which encompass many constantly changing variables such as population growth, affluence, tastes, government policies and the availability of alternatives.
- Costs of production and processing: these reflect the nature of the material and its location, the state of the art of production as reflected in its costs, including those of energy but also capital, the rate of interest on loans, taxation, and the risks of being nationalized or terrorized.
- End-price: this will reflect not only the factors above, but also pricing policies of the producers and government subsidies or taxes.
- The attraction and availability of substitutes, including the use of recycled products as against virgin materials.

Hence the resource is scarcely a fixed physical quantity (though this must exist) but a rather fluid economic and social construction. A common variable is price: as the price of a material increases it becomes more worth while for prospecting to take place, or for better methods of recovery to be developed. In this way, the recovery of crude oil from rock strata has risen from *c.* 25 per cent in the 1940s to *c.* 60 per cent in recent years.

Land resources

Each year, there are land gains and losses, with portions of the surface of the Earth becoming a resource in the sense of a useful surface, or losing that status. Coastal erosion and deposition are the most obvious categories, but landslides and soil erosion are also significant. Some of these changes are the result of natural processes, as when cliffs of soft material are exposed to high-energy seas; others result from human activity, as when coastal structures provide traps for silt and sand and thus build up ground above tide-levels. Occasionally, more spectacular losses occur, as when a volcano emits lava over former forests or cropland; the equivalent gains are made when a nation like the Netherlands dykes off large areas of coastal mud flat and salt marsh for conversion to pastures and crops. If land is in short supply, then the response of many societies with a choice is analogous to agricultural expansion. Option one is to intensify the use, often leading to multistorey buildings in sought-after areas of cities; option two is to extend outwards by 'reclamation', which may mean many things but usually signifies the bringing into the economic framework land where benefits were formerly negligible (e.g. industrial wasteland), or not normally quantifiable (e.g. coastal marshes).

One exacerbation of land shortages is by degradation of the land surface to the point where it has little or no monetary value. For example, dumps of toxic waste can have no other function since they are dangerous and sterile; land prone to subsidence (due to fluid withdrawal or mining) may have a few uses but if it is unstable then it may just be left; and even in planned land-use systems, a planning blight may occur in which land awaiting a change in function often grows nothing but weedy vegetation and discarded hypodermic needles. So although not a classic example of a non-renewable resource, land approaches the category of a material that is not being made by natural processes at anything like the same rate at which human societies are transforming it.

Mineral resources

Mineral resources are typical non-renewable resources in the sense that the deposits from which they are taken are formed on geological time-scales of a totally different order from that of the human scales of their use. Possibly 90 per cent of the human population now depends upon minerals not simply for industrial life-styles but just for survival (Blunden, 1991).

Production

About 100 non-fuel minerals are traded, and these contribute about 1 per cent of world GNP; of these there are 20 metals of considerable importance and 18 non-metals of equivalent significance, including aggregates, asbestos, clay, diamonds, fluorspar, graphite, phosphate, salt, limestone, silica and gemstones (Vanecek, 1994). Some are needed only in small quantities, e.g. steel hardeners such as tungsten. Some others rely for their importance upon rarity, as is the case with gemstones. Further, the use may seem humble but be very important: consider, for example, the use of metals in all phases of the industrial food system, from machinery through to processing and packing. The end-point of this scale of values is the designation of some minerals as strategic minerals and the stockpiling of them against the kind of politically induced shortage on world markets that would impede the manufacture and use of military hardware.

Consumption

In the past 80 years or so, consumption of minerals has risen by a factor of 12 and we can say with confidence that since about 1950 the world has consumed more minerals than in the whole previous history of humanity (Fig. 3.7). Most of this consumption has been located in North America and Western Europe (with Japan catching up fast) to the extent that the US economy demands 20 t of new minerals per capita per year. The LDCs consume only about 10 per cent of the total but their growth rates are higher at present.

World trade in minerals in the past 20 years has exceeded GNP growth by a factor of more than two, so Western Europe, Japan and the USA have come to be dependent upon imports; if fuel minerals are included then 40 per cent of Japan's total imports are of minerals. Such trade patterns can mean that some LDCs with large deposits can come to rely on their export for foreign currency: copper comprises 95 per cent of Zambia's exports, and in Zaire, 67 per cent comes from iron. With some concentrations, OPEC-type cartels are always possible and indeed

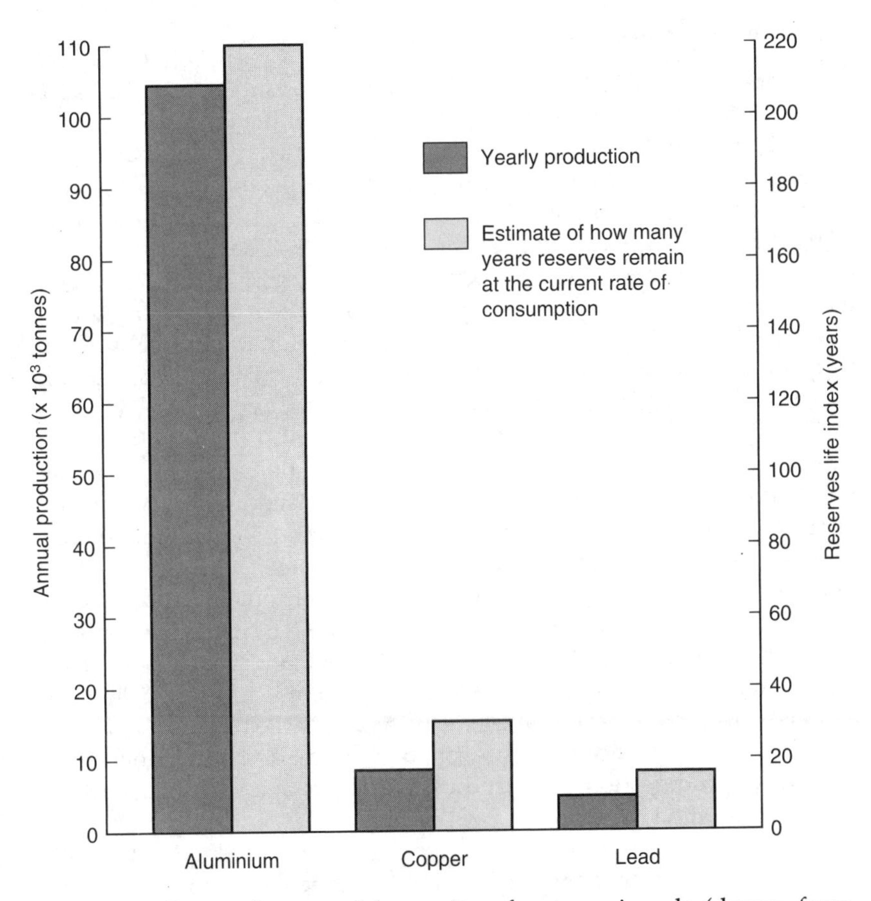

Figure 3.7 Rates of use and longevity of some minerals (drawn from data in *World Resources 1994–95*)

likely given that LDCs usually see the terms of trade as being unfavourably stacked against them. Such factors encourage recycling in the DCs and so we note that 50 per cent of British Steel's output is from scrap and that in a nation like the USA 35 per cent of lead, and 20–30 per cent of copper, nickel, antimony, mercury, silver and platinum, is reused. The trends suggest more trade and less self-sufficiency in the consuming nations. The LDCs, Japan and the former Centrally Planned Economies are set to increase their consumption and in general this will lead to more large-scale production at the lowest possible cost except where environmental considerations are seriously taken into the reckoning. Exploration of less conventional sources will increase and there will be conflict over the exploitation of the resources of the ocean beds outside the EEZs (Excusive Economic Zones).

Environmental considerations

The world over, the effects of mining and processing of minerals upon the environment cannot be ignored. In the case of the land, $2–3 \times 10^{12}$ t/year of rock and soil is estimated to be moved in this cause and projections of current rates suggest that by 2000, some 24×10^6 ha will be affected, which is about 0.2 per cent of the land surface of the globe. Within these totals, experience in the USA points to some 60 per cent of this disturbance being due to extraction, with most of the rest being used for the disposal of wastes and a mere 3 per cent is land subsidence due to underground operations. For an individual mineral such as copper, the production in the USA of 5.5×10^6 t of copper ore concentrates means the mining of 245×10^6 t of copper ore, and the leaving of 240×10^6 t of tailings. The output of 1.6×10^6 t of blister copper results in 27×10^6 t of solid waste in the form of slag. The production of heaps and holes and poisoned land was very much a feature of the 100 years following the Industrial Revolution, though common enough before that on a smaller scale. Our attitudes now require attention to be given to the reclamation of mineral land, and one estimate has given the world-wide figure of 40–60 per cent for such reuse. In the DCs large holes are always in demand near cities for the burial of urban and industrial wastes, and access to energy in those nations makes possible the levelling of tips and their conversion to all kinds of other uses: housing, forestry, agriculture and recreation are all possible depending upon the kind of treatments available and effective in a given location.

Minerals and the atmosphere interact partly through the use of energy in large quantities. Smelting of metals is a particular feature of mineral processing and one of the products is usually sulphur, a noted contributor to acid rain. A large copper smelter, for example, can emit

7400 t/day of sulphur if there is no treatment of the waste gases. Downwind from such plants there is usually a plume of affected vegetation, with very few living things in the areas of highest fallout; it is very similar to the vegetation pattern around sulphurous hot springs. Humans are also badly affected in such places: neither their bodies nor their possessions react well to high levels of sulphur, for example.

Rivers may also be affected by mineral extraction, when the plant manager is allowed to discharge waste materials into streams or when runoff from waste heaps enters the surface flow. Salt from the potash mines in Alsace is one of the noted pollutants of the Rhine, and rivers in the Pennine districts of England still bear the effects of lead-rich discharges in the 19th century. Some minerals may build up in sediments in low-energy environments such as estuaries and thence via bioaccumulation become lethal to a wide variety of organisms. If rivers flood, then toxic metals may be spread over farmland; in the Philippines some 130 000 ha of irrigated land is said to be affected by mine tailings that have entered the water distribution system.

The effects of deep-sea mining on ecosystems are unknown. If oceanic mining (Fig. 3.8) of mineral nodules (e.g. manganese and nickel) is pursued, then what will be the impacts upon the sparse life of these regions? The effects on offshore marine ecosystems are, by contrast, rather well known; the stirring up of great quantities of seabed silts usually has a positive effect in releasing nutrients into biological systems that are limited by lack of phosphorus and nitrogen, for example. On the other hand, the silts may blanket and kill organisms, or at the very least cause levels of turbidity that will inhibit photosynthesis.

It is no surprise, therefore, that in DCs there is usually strong government intervention in the mineral extraction process to ensure some degree of amelioration of environmental impact; many LDC populations would like the same consideration extended to them but the strength of their governments *vis-à-vis* the need for foreign exchange and the power of TNCs often results in second-class treatment.

Futures

Concern over future supplies of all kinds of resources has sometimes invested the term 'non-renewable' with a kind of negative air, as if a normative judgement was being made that somehow it was wrong to use these materials. Nevertheless, current civilizations are built upon them in the most literal of senses, and the history of their use is a very long one indeed. At the same time, the growth in population, the rise of material expectations and the extra energy needed to extract poorer

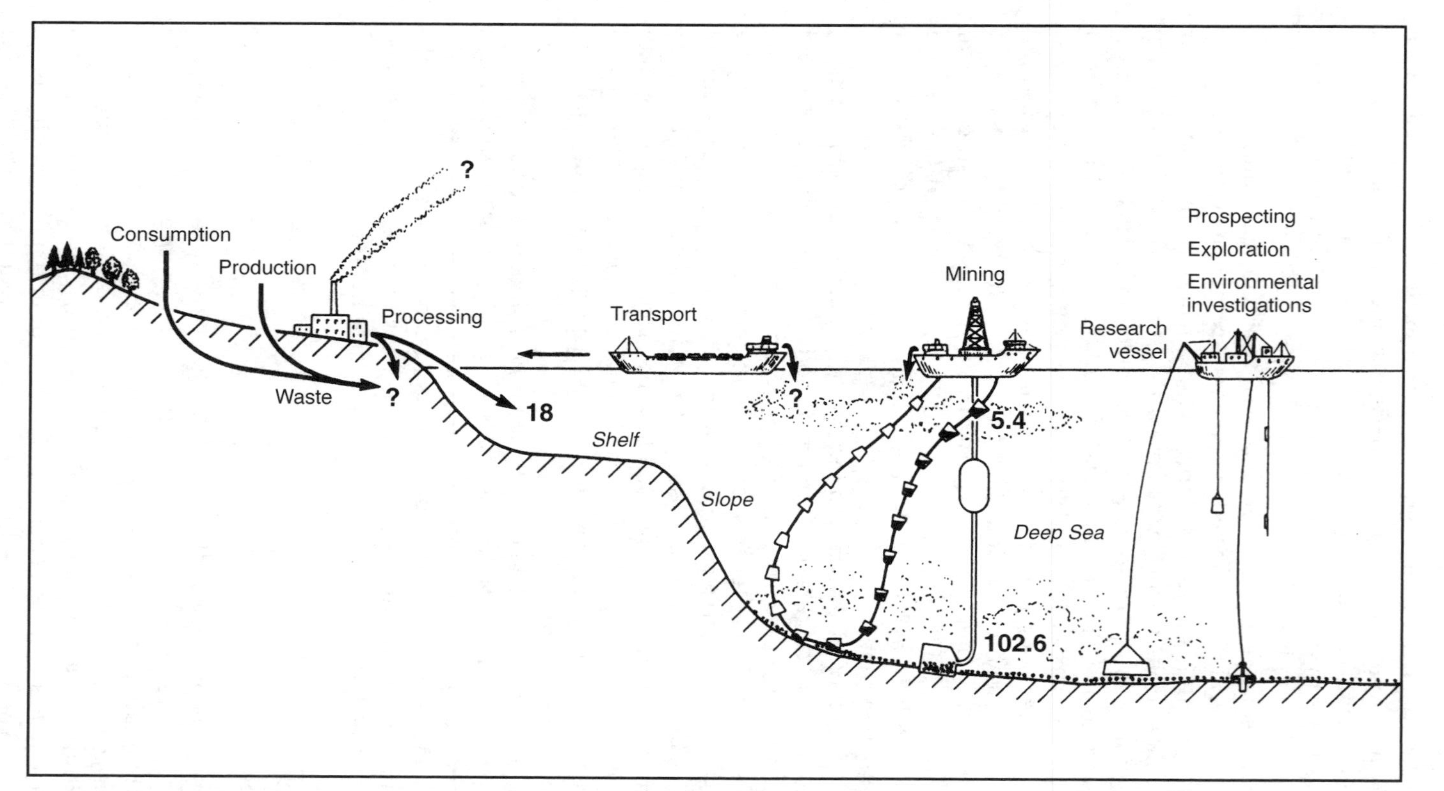

Figure 3.8 A scenario for the environmental contamination (in million tonnes) and disturbance resulting from the mining of perhaps 10 vessels-full of manganese nodules from the ocean bed (from Simmons, 1991)

grades of minerals all make a cornucopian future unlikely. Given that in most cases it is less energy-consumptive to recycle, and less costly overall to reuse, then many societies will find it in the interests of the population to move towards reuse as a first stage. Whether Western societies have the capability to go to the next stage after that is another matter, for it involves thermodynamic and materials thrift brought about by longer lifetimes for many goods. Are we capable of adapting (except under the most pressing of conditions) to societies in which the lifetime of vehicles, for example, is 50 years, or where a pair of shoes bought at age 17 lasts the rest of one's life? Further, are all societies prepared to give up using non-renewable metals, other minerals and energy in preparation for, and conduct of, warfare?

The classic instances of substitution should be found here. We cannot do without water or some kind of food. But a culture does not have to be built on steel (none were until the 19th century) and so another material ought to be feasible should supplies fail. In fact, the price and availability of energy is often likely to be the key factor in substitution decisions since the lower the quality of virgin ore, for instance, the more energy required per unit of output for the smelting processes. Of even greater interest is the question of non-renewable energy sources such as the fossil fuels: should they be eked out or be transformed into knowledge of how to do without them?

Wastes and their flows

In the course of using resources, human societies produce wastes which are perceived as harmful to the body of society and which must therefore be disposed of. In many cases 'the environment' (i.e. the atmosphere, the seas, rivers and underground) is a place where the adage of 'out of sight, out of mind' applies. In some instances these are genuinely toxic to humans and other living organisms (e.g. high concentrations of sulphur compounds in gases); in others they are evaluated as undesirable, like pairs of old shoes. Generally speaking, the richer the society, the more easily are materials perceived as wastes; poorer people are apt to regard such *disjecta* as sources of raw materials.

Like resource consumption, waste production has grown in scale as human numbers have increased and as their capacity to use more materials and energy has grown. At low concentrations, wastes can be ignored by people because they are not offensive and offer no threats to health or livelihood; in such circumstances we may talk of environmental *contamination* by humans since the flows of the natural world have substances added that would not otherwise be there or would be present in lower concentrations. At higher concentrations, wastes may cause damage to human health, to ecological systems and to the built environment, and be offensive culturally; at this level they are generally labelled **pollution**. In addition, science and technology can produce substances not present in the natural world and whose side-effects in the environment are not always predictable. Unforeseeable interactions between contaminants can occur in the process, called **synergism**.

Several classifications of pollutants are possible: here we will organize their description according to the environmental compartment into which they are mostly emitted.

Pollution of the land

Since land is where humans live, it is no surprise that concentrations of contaminants are high here, though strenuous attempts are made in most societies in the West to dispose of them in less 'visible' environments. Some emissions are, of course, not detectable by the unaided senses: radioactivity is probably the most important of these. Above-ground testing of nuclear weapons has left a legacy of bioaccumulated radionuclides in the lichens and herbivores of the tundra. Accidents like the Chernobyl partial meltdown of 1986 have also contributed to pollution thousands of kilometres away from the actual site, as well as sterilizing the immediate surroundings for many purposes. Nuclear power stations have a limited life (of the order of 25–40 years) and after that the structures are heavily contaminated with radioactivity. They must be disposed of either by being dismantled and the parts treated like other nuclear wastes or by being entombed in concrete on site. Monitoring will be absolutely vital in these places, which will be quite plentiful elements of the land-use pattern of any nation becoming heavily involved in electricity production by this means.

A recent addition to the repertoire of invisibles has been the possibility of deleterious effects on human health of the electromagnetic emissions from high-voltage powerlines upon those living close by. As with some of the effects of low-level radiation, definite relationships are hard to prove. No such difficulty exists with noise and vibration, however, where precise scales of damage to hearing from elevated noise levels are known, and the overall effects of vibration (e.g. from heavy vehicles) on structures and people are closely documented.

Many solid wastes are tipped onto land that can be spared from other purposes. This happens especially at redundant mines and quarries, but is also found at other industrial sites and is particularly of household and institutional ('municipal') solid wastes, the composition of which reflects in both quantity and quality the contributing society. Garbage tips ought to be watertight at the base and each day's debris must be filled over with soil. Failure of the first condition means that toxic wastes can leak into surface or groundwater; failure of the second means that a habitat is provided for rats, and fermentation bacteria that will generate methane, which can catch fire. Controlled production of methane is a useful source of energy but this requires even more careful management of the rubbish tip. Overall, disposal sites for solid wastes are undesirable neighbours, though their resources are often sought after by anybody who can make a living from unofficial or sanctioned recycling and reuse of discarded materials. They may eventually provide a land surface suitable for playing fields or afforestation, for example.

A last general category is that of derelict land produced by industries that have often vanished. The land may be morphologically difficult to use (e.g. full of cavities and mounds), or may be toxic, with high concentrations of metals such as lead or copper. Reclamation is usually possible and such land can be restored to uses of many kinds, though the costs to society of doing so are high if the exploiters of the original resource have not been made to pay for restoration. Most advanced countries now impose in advance restoration conditions and costs on the makers of holes and heaps, though some poorer governments are unable or unwilling to do so.

Productive land may contain all kinds of residuals which although not immediately toxic or offensive, nevertheless form a pool from which they can be taken up into, for instance, food chains. Substances such as organochlorine pesticides (e.g. those of the DDT group when stored) break down very slowly in soils and very little at all in animal metabolism or in fats. When they do break down, their metabolites may in some cases be more toxic to life than the original substance. The result has been a series of lethal and sublethal effects upon wildlife in agricultural areas since animals vary in their ability to excrete organochlorines and the substances may thus build up in their fatty tissues. Levels may become lethal and kill a proportion of the population; sublethally they may depress the level of reproductive hormones or in the case of some birds induce thinning of the eggshell so that few eggs hatch successfully. This phenomenon has been noted especially in predatory birds such as sparrowhawks and peregrine falcons, whose populations have recovered where this class of pesticides has been replaced with less persistent chemicals. Declines in common songbird populations from this cause were the trigger for Rachel Carson's book *Silent Spring* (1963), which was catalytic in raising environmental consciousness among Western societies. Industry may also contribute materials such as lead. Concentrations of lead are high in urban areas and near roads, especially where it can build up in slow-growing parts of plants like fruits. There is a direct pathway to human

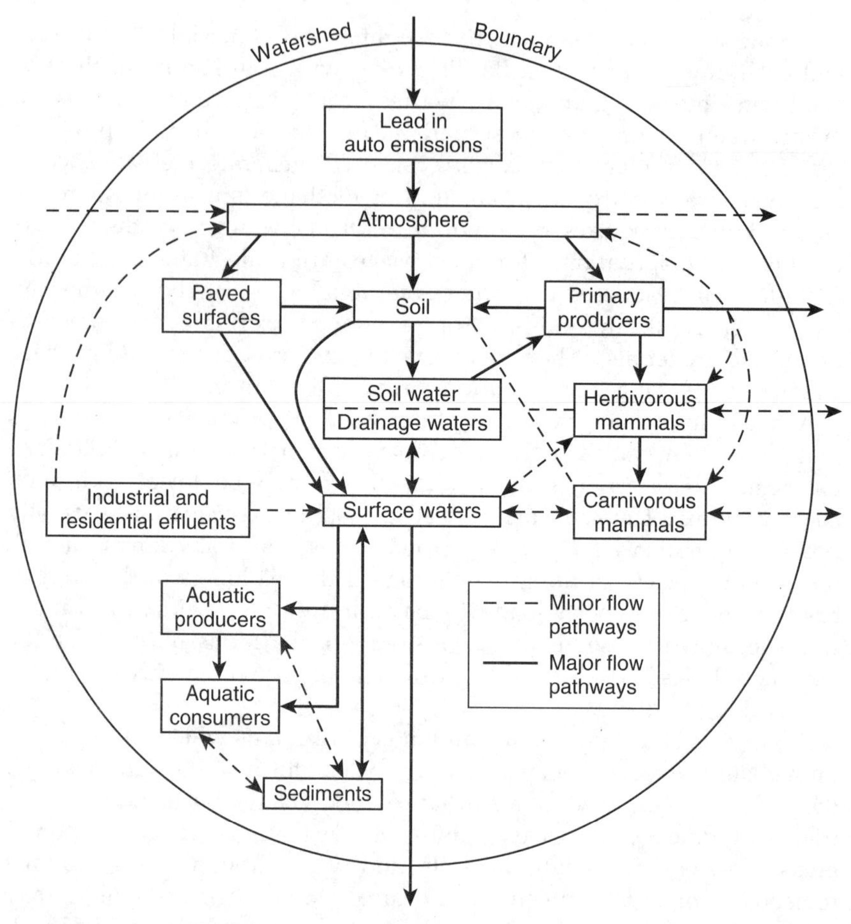

Figure 3.9 A flow model of the movement of lead within a local watershed from its emission into the atmosphere in aerosol form as an additive in vehicle fuel. No specific compartments for humans are designated (from Burgess, 1978)

ingestion, and above certain levels lead can act as a poison of the central nervous system, although it is rarely lethal. Figure 3.9 shows that the flows of lead can be complex – as complex as any local ecosystem with plants and animals, in fact.

We would all like the land where we live to be only a temporary resting place for wastes but enough has been said to show that, alas, its biochemical nature, combined with human habits, ensures that a great many substances stay around long enough to cause changes in the metabolism of individual animals (including our own species), some of which are bound to have negative effects on well-being.

Fresh water

Rivers are perceived to be especially good places to dump wastes, since the flow of water carries them away to be diluted or to be somebody else's problem. So it is no surprise to find that rivers and lakes are often so changed that their original biota are no longer present. The Hangpu River near Shanghai, for example, has one volume of untreated waste for every four to six volumes of water, and 54 out of the 78 major rivers of China are said to be seriously polluted. One of the most pervasive of water contamination processes is that of human-caused nutrient enrichment, or **eutrophication**. Rivers vary in their natural concentrations of elements such as nitrogen and phosphorus but low levels are common since the flora takes them up. So runoff from agricultural land or from sewage that contains high N and P levels is likely to lead to biological changes. The most common are 'blooms' of algae, whose populations have been hitherto limited by the low levels of nutrients like N and P. Released by the contaminating supplies, the algae grow rapidly, but as they die, bacteria also burgeon and take up much of the oxygen in the water. If the water is warm the oxygen levels are depressed anyway, so the fish die quickly. In the USA, for example, farm runoff affects 64 per cent of all rivers and 57 per cent of lakes. In Europe, the levels of NO_3 have been increasing in many rivers; the WHO recommended limit is 11.3 mg/l but one-third of all UK rivers exceeded this in 1978–1988. Sewage may be subjected to varying levels of treatment before it is released into rivers or lakes. The index is the level of coliform bacteria present in water: in India, for example, the Yanuna River before coming into New Delhi has had 7500 coliforms per 100 ml water, and below the city, 24 million coliforms per 100 ml water. The use of chlorine usually kills most of the bacteria, which cause disease if the water is used for drinking and cooking without being boiled or otherwise disinfected.

If the runoff is contaminated by elements or compounds in solution then it is likely that the groundwater will reflect such flows. For NO_3, the US recommended limit in wells is 10 mg/l but an estimated 8200 wells exceed this limit; the effects of high levels of NO_3 on adults are unknown but babies are affected by 'Blue Baby Syndrome' at 745 mg/l. Pesticides are also found now in many aquifers (e.g. aldicarb in 40 states in the USA); in LDCs, many more would probably be detected and at high concentrations. Groundwater also receives leached nutrients and compounds from cesspits and leaky landfills: as many as 7300 landfills in Western Europe are estimated to need immediate action to contain contaminants of various kinds.

Ever since about 1800, power plants have been increasing the amount of sulphur emitted into the air, and much of this has fallen out in acidic form. In lakes and rivers, the effect is to lower the pH, with consequent impacts upon the flora and fauna of all kinds (some lakes show changes in diatom flora as early as 1850) but especially upon commercial and sporting fish populations. The exact course of the acidification is not predictable, since the ecology of the entire watershed exerts influences upon runoff (the soil acidity level is an obvious variable) but many fresh waters in industrial nations formerly at pH >6.0 are now in the 4.8–5.5 range. The changes are especially noticeable in Scandinavia, the northeast USA and southeast Canada, and Scotland. In southern Norway, for example, 1700 lakes have lost their fish populations. The consequences of acidification of these waters are far-reaching: for instance, the decomposition rates of lignin and cellulose are reduced because of lower levels of bacterial activity, so plant debris begins to accumulate on lake bottoms and becomes decomposed only by anaerobic bacteria so that methane and H_2S are given off. The biota of river substrates is especially vulnerable, so that snails, clams and crayfish disappear. The lower pH accelerates the rates of leaching of some metals into rivers, whence they may be found in drinking water: pH levels of 4.5 have been found in some wells, along with detectable concentrations of lead, mercury, cadmium, aluminium and cobalt. Reversal of these trends has to begin at source and it is possible to scavenge out much of the sulphur in power station wastes, though not without cost. In Europe, most nations have committed themselves to a 30 per cent reduction in sulphur emissions by the end of the next decade, with an increase of about 6 per cent in the price of electricity. Central Europe has the greatest task since lignite (brown coal) is such an important energy source and it is very rich in sulphur. Lake acidification can be reversed by liming, though this is expensive and is available only in the richer nations. Sweden, for example, planned to lime 3000 lakes in the first 10 years of a programme during the 1970s and 1980s.

Probably one of the longest-established contaminations of river water is that of sediment, which must have been changing water quality ever since agriculture and pastoralism began. Yields to runoff are especially heavy in areas of intensive agriculture, of deforestation (notably where slopes are steep), and of rapid urbanization or industrial construction. In a country like the USA, sediment is the primary pollutant of 47 per cent of the rivers, measured as a proportion of the length of rivers classified as polluted in some manner. Heavy sediment loads change the ecology of rivers very radically. The increased total volume increases the likelihood of flooding; the finer inorganic material blankets and kills much animal life (especially fish), and inhibits photosynthesis. Ubiquitous though sedimentation is, it is very difficult to control since it stems from the whole pattern of land use.

Any overall view is difficult to establish and the situation is changing rapidly in many nations, for better in some and worse in others. A gross generalization might be that point sources of contamination are being subjected to clean-up and improvement but the non-point sources (especially nitrogen and sediment) are not, and at that scale there is a challenge to the technology and institutions of most countries. In many industrialized nations, regulatory mechanisms based on legislation are used to try and reduce such flows of wastes; indeed the lead content of the Rhine, to quote one example, fell from 24 μg/l in 1975 to 8 μg/l in 1983. The mechanism of achieving such falls is variable from nation to nation and often has an ideological content. The stronger the socialist bias in government then the more likely it is that regulations will be used; at the other end of the spectrum, free-marketeers favour systems of licences to pollute which can be bought but whose cost in general makes it cheaper for a firm or individual not to pollute in the first place. Both these processes depend on the presence of an independent monitoring organization, paid for from public funds, which can provide scientific data, and on pressures from the general public, which in turn needs access to objective data. Such data are often kept secret however: many socialist administrations fear the political consequences of the truth and many capitalists are in thrall to the idea of commercial secrecy.

The oceans

There are six or seven main kinds of wastes that reach pollutant levels in the seas. For most of us, the most obvious ones in terms of personal experience and media attention are sewage and oil. Sewage is the main example of an oxygen-depleting waste affecting the sea, but others in this class include other organic materials (e.g. the effluent from paper plants), and eutrophicants such as phosphates and nitrates. Sewage is either fed raw into the sea by means of offshore pipelines of varying lengths, or dumped as sewage sludge, which is the dewatered solid residue after treatment. Since 1000 people will give rise to 25 t/year of solids, large areas of offshore waters can be affected. The major effect of these substances is to provide a substrate for bacteria, which in the course of their growth and reproduction take up some or all of the dissolved oxygen in the sea water. If all the oxygen is metabolized then anaerobic bacteria may flourish, releasing gases such as methane, hydrogen sulphide and ammonia. Fish usually escape to other areas, but any that remain get their gills choked by plankton and there is a heavy rain of plankton onto the benthos which consequently smothers most of the living organisms there. High levels of nutrients also make for 'red tides', which are blooms of dinoflagellates whose excretions are often toxic to filter-feeding organisms such as molluscs. Raw sewage is also a plentiful source of coliform bacteria and parasite eggs which can affect human health. The effect of raw sewage and accompanying solids upon the amenity value of the costal zone is also highly unpleasant.

Oil enters the sea from natural seepages – perhaps 0.5×10^6 t/year of it. To this, however, is added 13.8×10^6 t/year from the human economy, which is not surprising since 50 per cent of the 3.0×10^9 t/year of oil consumed is transported by sea. The most conspicuous and attraction-creating form of pollution is the major spill from a tanker accident, but there is also a continuous flow from minor spills at terminals, tank-cleansing stations, offshore production and effluent from refineries. When a major accident occurs, various measures are deployed: floating booms, slick-lickers and dispersant chemicals are used at sea; on the beach steam hoses and more dispersant can be used on hard surfaces, whereas oil-soaked sand has simply to be scraped up by hand or machine. Both the spill and the clean-up produce environmental effects. Some of these are usually short-lived: plankton, fish and tourist populations are not usually affected for many months but birds, especially those with a low reproductive capacity, may undergo declines which last for years. Fish in the area of a spill may be tainted with an oily flavour and hence unmarketable.

Metals are natural constituents of sea water, to which they make their way from volcanic eruptions, dust, the erosion of rocks, smoke and decayed vegetation. Very many human activities add to the mobilization rate of metals, which in total increase the metal loads of rivers and the atmosphere, as well as occurring in direct discharges of sewage. The following metals are of most concern:

- transition metals (e.g. iron, copper, manganese and cobalt), which are essential for life in low concentrations but toxic at higher levels;
- heavy metals or metalloids (e.g. mercury, lead, tin, selenium and arsenic), which are not used by living organisms and so are toxic to them at low concentrations.

Mercury (Hg) provides a good example: natural inputs from rock weathering and geological degassing amount to *c.* 28 500–153 500 t/year, and humans are responsible for adding about 8000 t/year, some via the atmosphere and some as industrial effluents via runoff into estuaries and inlets. Microbial systems in the seas and estuaries can convert inorganic mercury into methyl mercury, which is (a) readily accumulated by living organisms and (b) much more toxic than the inorganic forms. Thus most sea creatures have some Hg in their tissues, though levels are much higher in areas of high input. The mercury can also accumulate in food chains, so that in Minamata Bay in Japan in the 1950s plankton contained 5 ppm but fish 10–55 ppm. Human deaths (43) and chronic disabilities (700 permanent cases) resulted, since mercury affects the brain functions: the syndrome is now known as Minamata disease. Standards have now been adopted in Japan which limit the amount of Hg in seafoods for sale: these are mostly between 0.5 and 1.5 μg of Hg per gram of fish or other seafood.

The class of entirely synthetic chemicals known as halogenated hydrocarbons appear to be permanent additions to the biosphere. They withstand oxidation and bacterial action and are thus able to undergo bioaccumulation if conditions are right. So unlike both, they are substances unknown in nature: totally humanly produced. The two main groups are pesticides and PCBs. The pesticides include the well-known DDT group as well as aldrin and its relatives, lindane and toxaphene; PCBs (polychlorinated biphenyls) are industrial chemicals renowned for their stability. Most of both sets are under careful regulation in the industrialized world but in the LDCs they are still widely employed: DDT against malaria for example. Disposing of used PCBs is often done by incineration at sea (due to be phased out in the North Sea by 1995), which produces carbon dioxide, hydrochloric acid and metals, though some traces of dioxin are said to have been detected. As on land, bioaccumulation can occur, with progressively higher concentrations of halogenated hydrocarbons up the levels of a food chain. The concentration factor between sea water and an organism is often of the order of 40 000–700 000 and so metabolic effects have been recognized. In phytoplankton, for instance, NPP rates are reduced at quite low aqueous concentrations of DDT. In sea-birds, eggshell thinning is common in areas of DDT concentration in sea water. Effects upon human health are hard to document with certainty. PCBs caused some human poisoning at Yusho in Japan in 1968, producing all the symptoms of acne, but other cases are debatable. Control of DDT levels in foodstuffs has meant that at times (especially in the late 1960s) human milk would not have been passed for sale in the supermarkets of North America. But we cannot make confident predictions about the halogenated hydrocarbons still present in the world's oceanic ecosystems: dilution may render them harmless to life or bioaccumulation may mean they are a kind of chemical time bomb.

Radioactivity is present under natural conditions. Average levels would produce sea water with an activity of 12.6 Bq/litre, sands with 200–400 Bq/kg and muds with 700–1000 Bq/kg. The additions made to these levels by human activity come from military testing in the atmosphere (now stopped), planned releases from nuclear power and reprocessing plants, and the dumping of solid radioactive wastes. Some radionuclides remain in sea water (e.g. caesium-137), whereas others are sequestered onto sands and muds: plutonium-240 is an example of that tendency. No known effect of radionuclides on wild populations of plants and animals has been proven under planned conditions of discharge from civil installations. That bioaccumulation occurs is acknowledged and emission limits are set which aim to keep body burden levels well below the international standards for various radionuclides. Potential pathways to humans are also closely scrutinized: heavy fish and shellfish eaters around the Irish Sea are closely monitored and lobsters are reckoned to be a critical pathway for the discharges from the Sellafield reprocessing plant, for example. The adsorption of plutonium onto sediments means that particles can be blown onshore from beaches at low tides, forming another pathway. In spite of a great deal of monitoring and of close study, the fear of childhood leukaemia clusters around nuclear plants still persists, and until these are shown epidemiologically to be unrelated, or a causal link is discovered, then emissions of radioactive waste to the sea (as elsewhere) will generate much public suspicion.

Solid wastes have always been dumped in the sea, and strand-lines have borne evidence to the fact they have not always been borne away. Today, seabed dredging creates very high burdens of particulate matter, which can blanket bottom fauna but release large quantities of nutrients at the same time. The effects are therefore complex but undeniable in the short term; some idea of the intensity may be gauged from the fact that $30–35 \times 10^6$ t/year of sand and gravel are dredged out of the North Sea alone. The east coast of North America has similar impacts. The newest member to the family of added solids is plastics, as a walk along most beaches in the world will now reveal. Most of the non-degradable bottles come from shipping, but industry is the source of the many small pellets (3–4 mm in diameter) of partially broken-down plastics which are found in most oceans both near and far from land. Another material that is largely 20th century in origin is munitions which have been dumped at sea because they are obsolescent, or have sunk along with the ships carrying them and are too dangerous to move.

Last in this catalogue of contaminants of the sea is the addition of heat, mostly in the form of a plume of hot water from power stations, probably 12–15 °C above the ambient sea temperature. In cool waters this mixes with the receiving water, resulting in rises of perhaps 0.5–17 °C of the sea water. In warmer waters, living organisms may be nearer to the upper tolerance levels for temperature and so the impact on life is greater: around a North Sea or North Atlantic outfall, less than 1 ha may be affected biologically, whereas in the subtropics up to 40 ha has been recorded. Life may also be affected by chlorine and leached metals in the outflow water.

One general comment that may be made upon all these contaminants is the vulnerability of estuaries from almost all of them. Because estuaries are shallow, because they accumulate sediments, and because the tides may produce low net outfall rates of water (along with its burdens of contaminants), they are prime sites for pollution. But under more natural conditions they also have a high biological productivity, which shows in their role as breeding grounds for many commercial fish populations. The estuaries have undergone more change by human hands than any other part of the world's seas.

Contamination of the atmosphere

The atmosphere is a complex system whose dynamics are only partially understood and for which accurate prediction is often impossible, beyond the level of short-term weather forecasts. So the pathways and concentrations of contaminants are often difficult to trace and to predict – even more so than in the oceans. Most contamination of the gaseous envelope of the Earth does in fact come back to affect us in a number of ways: one measure suggests that in DCs air pollution causes damage to the tune of 1–2 per cent of gross domestic product (GDP).

Particulates (more properly called suspended particulate matter or SPM) consist mainly of very small pieces of carbon, hydrocarbons or sulphur compounds with a diameter between 0.1 and 25.0 μm. These are given off by a number of natural processes such as windflow over deserts and the oceans, volcanoes, forest fires, and soil erosion. Human economies add to these in the form of fuel combustion, fires, ploughing and the creation of other unvegetated land, and industrial emissions. The natural processes add some 1320×10^6 t/year to the atmosphere, and estimates of the human-led emissions vary in the range 60–300×10^6 t/year, but at all events a much smaller source than those of nature. The majority of SPMs fall out at relatively short distances downwind from the point or area of emission and so with the human-induced sources there is a directly observable link between deleterious effects and their immediate cause. Thus damage to all kinds of materials, to visibility and to human health can be traced directly to the kind of emissions which typically produce concentrations of 20–100 $\mu g/m^3$ in cities and under 10 $\mu g/m^3$ in rural areas. Calcutta in the 1970s reached 360 $\mu g/m^3$, whereas Brussels was only 18 $\mu g/m^3$; the WHO recommended limit is 60 $\mu g/m^3$. Because of the obvious nature of the problems created (in the London smog of 1952 the levels of SPMs were 6000 $\mu g/m^3$ and 4700 excess deaths occurred) and the existence of technology to reduce the levels of emissions, there has been a decline in levels in most DCs in the last two decades, usually brought about by national legislation accompanied by an enforcement agency.

A great cocktail of gases is emitted from the surface of the planet into its atmosphere, some from natural sources and some from human activities. As with a number of environmental contaminants, there are natural flows of many of these substances (carbon dioxide, taken up by plants during photosynthesis and emitted by animals during respiration, is an obvious case) and some of these are added to by human activities. In the case of gases, the production and use of fuels (especially the fossil fuels) stands out as a polluter at all scales. There are, however, some unexpected sources: domestic cattle for example emit large quantities of methane and this contributes to the possibility of global warming. Naturally enough, our concern is focused on two major hazards: those of human health and of unpredictable global climatic change. But a number of substances produce other effects which we ought not to overlook. Some may reduce the level of amenity for instance: photochemical smog in which ozone and peroxyacyl nitrates (PAN/s) are major constituents reduces visibility considerably, with a brown haze of varying density spreading characteristically across cities that lie in basins such as Los Angeles, Tokyo and Mexico City. Such smogs may often produce no more than irritation of mucous membranes for the majority of people but thus contribute to the lowering of the quality of life. Involved in such perceptions of quality, too, are areas of forests which are suffering die-back and death from two airborne sources: sulphur dioxide and nitrogen oxides. The sulphur, which rains out as part of the phenomenon of acid precipitation, is accompanied in many areas by high concentrations of nitrogen oxides, mostly from car exhausts. Together, these chemicals kill or inhibit the growth of trees, especially in temperate industrial nations (Germany is especially badly affected), and reduce the economic yield of the forests as well as removing them as a source of recreation and aesthetic pleasure.

When fuels are burned for any purpose, waste heat is emitted. This may be led off directly into the atmosphere or via cooling systems such as ponds or towers. The major effects are local in scale; downwind from large power stations, for example, it is possible to detect increased incidences of cumulus cloud and of fog formation. Agglomerations like cities give off more heat since they are concentrations of energy use and their structures often absorb solar radiation during the day and give it off at night. So night-time 'heat islands' are found during still air periods. These are commonly 1–2 °C above the surrounding rural areas but the difference may be as high as 5–10 °C. All the time, however, a city is radiating heat into the atmosphere and the flux may reach high levels: in London the outgoing heat flux has been measured in some districts at 100 W/m^2 and up to 234 W/m^2 in one area; the average solar input figure for London is 106 W/m^2. In New York, outgoing fluxes of 630 W/m^2 have been recorded. The extra heat of cities and the increased roughness of their surfaces seem to produce increased precipitation and thunderstorm frequency downwind, and the energy parks sometimes proposed for industrial areas might well do the same. The data for cities have led to some speculation as to whether heat might be a factor in any global-scale warming. One set of calculations suggests that human-organized emissions are at present only 0.01 per cent of solar flux at the Earth's surface, representing an energy consumption from all sources of 7.4 TW/year. If global heat fluxes were to be affected then energy consumption would have to be many orders of magnitude higher, so we can forget this for a while although reminding ourselves that any heat from fossil fuels and nuclear sources is additional to, and does not replace, that from the Sun.

The atmosphere only provides a transport mechanism for radioactive wastes; there is no analogue of the bottom silts of the seas to take up the materials. Its importance comes in its role of distributing the radioactive cloud from nuclear accidents, such as those at Windscale (now Sellafield) in 1957 and Chernobyl in 1986. The latter was about 100 times worse than the former and the initial cloud contained 286 million curies of isotopes with a half-life of >1 day. The cloud drifted over Europe for 7–10 days and exposed the inhabitants of 20 countries to health-threatening levels of radiation from the fallout. by 1995, the child thyroid cancer rate in the region of Belarus, north of Chernobyl, had risen 200-fold.

The ecology of wastes in the world is very diverse but our attitudes are basically very simple. They are dominated in the richer countries by the desire to be rid of them: to the next county, to the sea or the air, to another country. The costs are high in terms of paying for the disposal and in terms of impact upon ecosystems and upon the health of poorer people. The latter in general regard wastes as mines of recoverable materials and in a few DCs the embedded energy is being recovered as heat. What seems to be needed is an economics which will look at the whole resource process and not simply the 'back-end', so that all of it mimics more closely a natural ecosystem in which one organism's wastes are food for another and the whole is closely recycled.

Which pollutants matter and which can be ignored as unpleasant but unthreatening to life or ecosystem integrity? The erection of priorities may well depend on where and who you are. The LIEs are not often worried about the 'greenhouse effect' (they contribute little of the carbon dioxide, for example) unless their national territory is mostly low islands. Indeed they may resent being called on to provide a sink for HIE wastes, as does Brazil, for instance. The HIEs worry most about short-term localized effects like ozone holes since cancer rates are so closely linked to the extra UV radiation. But they also see that longer-term instabilities would endanger stable governance and capitalist business sees the same processes as endangering its profits.

Connections

Time and time again, accounts of resources and wastes mention their diversity. Equally often, words such as 'world-wide' and 'globalization' occur. Although a task of some complexity, we have to consider how all these various processes may be linked together and how they are in turn connected to the great global cycles upon which they impinge.

Energy flows

One great global process linking humanity and the universe is the flow of energy. We encounter the cosmic scale of this in the form of incoming solar radiation and we add our mite to it in the form of heat produced by the combustion of hydrocarbon fuels. Some of our other cultural practices can also be characterized by their energy flows (Table 3.5).

It seems from these data as if humanity's efforts are still small-scale compared with most of those of nature, but two things should be noted:

- Infrequent natural geophysical events are lower down the scale than some human-dominated processes, even though we perceive them to be very damaging. This has to do with, among other things, their spatially concentrated character.
- The relation of fossil fuel use (0.3 cal/m^2/day) to the truly renewable resource of net primary productivity by plants (7.8 cal/m^2/day) is not so distant as to make it insignificant, especially when we recall that about half of the NPP is in the oceans.

This latter idea has been extended into the calculation that out of a global NPP of 2245 Pg/year (where 1 Pg = 10^{15} g), 60 Pg is under human influence of various kinds. Most of the impact falls on the land surface, so the proportion is more like 58 Pg out of 150 Pg, which is equivalent to *c.* 39 per cent. This puts humans in a more dominating position than that conveyed by Table 3.5, which reminds us that there are geophysical processes beside which we stand very small indeed. But since energy as plant matter is central to our lives and since it is the

Table 3.5 Global mean energy flows for various processes

Process or event	Energy flow (cal/m^2/day)
Solar energy to Earth	7 000
Solar energy absorbed by Earth	4 900
Primary production by plants	7.8000
Animal respiration	0.6500
Forest fires	0.3000
Fossil fuel	0.1100
War (non-nuclear)	0.0500
Floods	0.0400
Earthquakes	0.0010
Volcanoes	0.0005

Source: J. F. Alexander, 'A global systems energy model', in R. A. Fazzolare & C. B. Smith (eds) (1979) *Changing Energy Use Futures*. New York and Oxford: Pergamon Press, vol. 1, 443–56

emissions from hydrocarbon use that currently scare us most, the role of energy as a linking medium between us and most facets of our environment needs very little extra emphasis (Smil, 1991).

Telecommunications

What makes the ecology of the world a humanized ecology (Smil, 1993) is a basis of information acquisition, retrieval and transmission. These are fundamental to any culture. Central to today's industrial and post-industrial economies is the use of electronics in the shape of the digital computer, television and the other myriad forms of data manipulation that now form a close network of global proportions. Instantaneous transmission is perhaps the outstanding feature and even the remotest groups in desert and forest can see the latest clothing fashions in Milan, even if they lack a neighbourhood boutique that will stock them. Revolutionaries and missionaries alike want to capture the TV station, since the screen is the validating medium for information, as was once print or pulpit. The result is an implosion of cultural diversity combined with an explosion of demand and the results for resource use, given population growth as an element of any equation, will be immense. Forecasting the exact pattern, however, is for the time being best left to the imagination.

An implicit question that runs through all resource considerations is that of sustainability. How long will the resources last, if non-renewable? How much can we take, if they are renewable? The discussion of these questions is at the heart of the debate about the nature of the concept of sustainability and its translation into operational terms, a discussion which will have to be amplified at a later stage in the book. But for now, let us note that any concepts which are grounded in a notion of a past Golden Age, or in ideas of stasis with a return to some kind of equilibrium, are non-starters. The future is, as it always has been, open-ended to change unless a very conscious decision is taken to prevent it (as happened in Tokugawa Japan) and such a step is probably not now possible given the permeability of any culture to novel ideas via telecommunications.

Further reading

The conflict between those who think that resource scarcity (and environmental degradation) is inevitable and the optimists is still vigorous. Two proponents of each case have their views encapsulated in Myers and Simon (1994). The optimistic view was set out at some length in Simon and Kahn (1981). Less polemic books include the largely economic approach of Rees (1990), and the more environmentally linked Simmons (1991). A new approach to environmental economics, which emphasizes the ecological linkages, is Barbier (1993). The works of Vaclav Smil, especially the 1993 volume, link together the processes of the planet and the human use of the Earth in an especially satisfying way and are vigorously written. The Open University texts cited below are part of a series of four books from 1991, all published by Hodder and Stoughton for the Open University, in which a series of authors present lucid and well-illustrated chapters on most of the material discussed here. The issue of biodiversity, which was an important theme of the UN Rio Conference in 1992, is copiously documented in Groombridge (1992), and in Sandlund *et al.* (1992). There is a special issue of the journal *Ambio* on the economics of biodiversity: volume 21(3) 1992. Data on resources are published annually by the World Resources Institute's annual volume *World Resources* (OUP). The environmental consequences of many of the resource uses feature in the WRI's annual volume edited by L. R. Brown and his team, *State of the World*. An overview of resources and their use, of a comprehensive if conventional nature, is in Mather and Chapman (1995). Barrow (1995) presents an environmental perspective on development in LIEs.

Barbier E B (ed.) 1993 *Economics and Ecology: New Frontiers and Sustainable Development*. Chapman and Hall, London

Barrow C J 1995 *Developing the Environment, Problems and Management*. Longman, London

Brown L R 1994 Facing food insecurity. In Brown L R (ed.) *State of the World 1994*. Earthscan, London: 177–97

Groombridge B (ed.) 1992 *Global Biodiversity, Status of the Earth's Living Resources*. Chapman and Hall, London

Mather A S, Chapman K 1995 *Environmental Resources*. Longman, London

Myers N, Simon J L 1994 *Scarcity or Abundance? A Debate on the Environment*. W W Norton, New York and London

Rees J 1990 *Natural Resources. Allocation, Economics and Policy*, 2nd edition. Routledge, London and New York

Sandlund O T, Hindar K, Brown A H D (eds) 1992 *Conservation of Biodiversity for Sustainable Development*. Scandinavian University Press, Oslo

Simmons I G 1991 *Earth, Air and Water*. Edward Arnold, London

Simon J L, Kahn H (eds) 1981 *The Resourceful Earth*. Blackwell, Oxford

CHAPTER 4

A humanized world

In an earlier chapter, we saw how the natural world has been changing during the last 10 000 years. Some of this has been due to relatively gradual shifts such as those of climatic change; some has been sudden, as when volcanoes erupt. During this stretch of time there has also been the influence of human communities upon their surroundings. In order to gain access to natural resources in the quantities we have noted in the last chapter, humans have altered their environments. Basically this has happened in two ways: those changes which have been brought about deliberately, as when a forest is felled to make way for cropland; and those which are accidental by-products, as when the soil from the newly cleared area washes off into a river and exacerbates flooding.

A periodization of 10 Ky of change

During most of this period, the world, including its human economies, was largely solar-powered. As always, the natural ecosystems fixed the radiant energy of the Sun and cascaded it through a series of organisms. Some of these ate plants, others animals; some subsisted off dead organic material; a few might utilize all of these, like the chimpanzee. The systems modified by humans for purposes such as agriculture were also solar-powered: no matter how the flows of energy and matter were diverted in the creation of near-natural, semi-natural or cultural ecosystems, the incidence of solar energy was the limiting factor. Such confines were broken only when all these systems could be subsidized by an extra flow of energy, which came from the fossil hydrocarbons (i.e. stores of photosynthesis from earlier geological eras). This was the ecological basis of the

Industrial Revolution, which was irreversibly launched and well into the fairway by AD 1800.

One practice which links both pre-industrial and industrial eras has been the use of fire in releasing the heat from combustible materials for use in myriad purposes; a further nexus between the economy of nature and that of humans comes with the use of fire in the landscape to manipulate vegetative and animal communities. In the time between the Upper Palaeolithic and the generation of electricity from nuclear fission, therefore, fire has been a basic element, either directly or indirectly, in the relations of humanity and its environment. However, before the 19th century, the great storehouses of concentrated potential energy in the form of coal, oil and natural gas were not totally unknown. Coal was used to warm the soldiers on Hadrian's Wall and natural gas piped through bamboo stems lit streets in a few medieval Chinese cities. But the realization of the potential of the steam engine, coupled perhaps with the rising price of both wood and labour, made the new process of smelting iron with coke the key to a complete revolution in human–environment relationships. Here then is a shift to much older energy (millions of years in fact) and furthermore to energy that is not being formed today, i.e. is a non-renewable resource. The results were first seen in the metals revolution, which gave rise to a technology based on iron and steel, and then the chemicals revolution. To this we may possibly add the plastics revolution coupled with other new materials as well, such as carbon fibre and superconductors, which have given humans the power to make over the natural surroundings on a scale never before contemplated. One more recent change has been in the speed and ubiquity of communications, starting with the steamship and ending, for the moment at least, with satellite television. Another has been the setting up of the factory system for the production of manufactured goods and the consequent attraction of a labour force: hence the explosive growth of cities in nations undergoing industrialization. Industrialization is not confined to industry, but is now interlinked with most forms of economic activity: agriculture is often heavily subsidized by commercial fuels, for example, and the mass tourism industry relies totally on them for cheap transport. Even climbing Everest now needs an industrial base to produce the oxygen cylinders and the high-tech clothing.

In 1942, Enrico Fermi engineered the first controlled chain fission reaction, thus opening the way for both civilian and nuclear power. Beyond the immediate horizon lies the possibility of generating energy from atomic fusion, in which an easily available isotope of hydrogen would be the chief fuel; just appearing over the horizon are the so-called 'alternative' sources of energy which tap the renewable

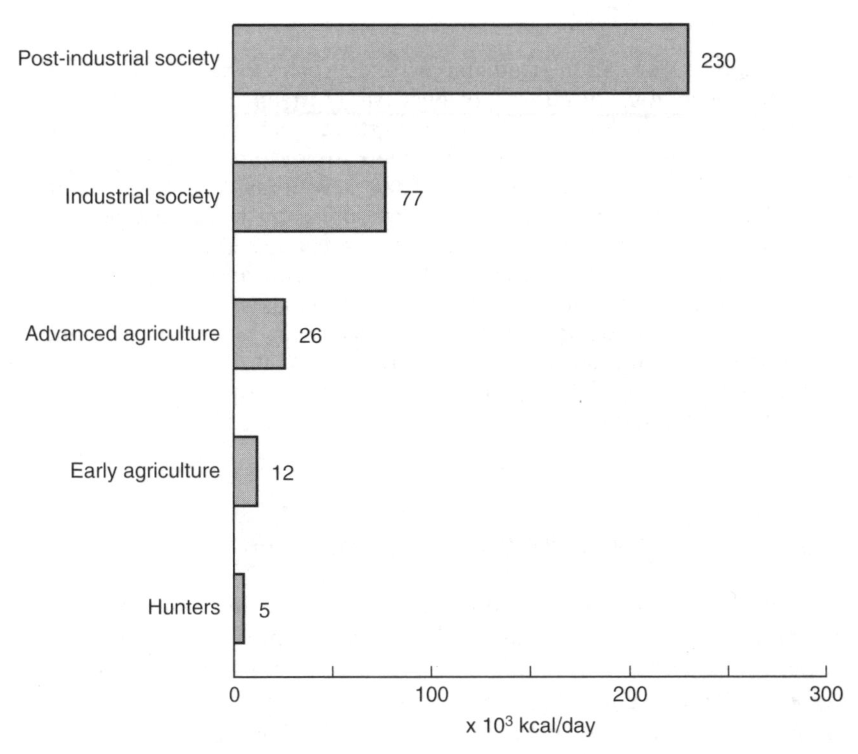

Figure 4.1 The daily energy consumption of different kinds of society. The lowest three are dependent upon solar sources, the upper two depend largely on fossil and nuclear sources, with hydropower being regionally significant. All these stages have occurred in the last 10 000 years

processes of nature such as wind, waves, tides and sunlight. So we can periodize the last 10 000 years in terms of human access to energy (Fig. 4.1) as a way of organizing our knowledge about the capability of societies to change their environments. Putting accurate dates to the periods depicted is scarcely possible since they came on at different rates and intensities at different places at various times. At 10 000 BC, all the members of *Homo sapiens* were hunter–gatherers; now very few indeed could be so described but there are a few. Between *c.* 3000 BC and AD 1800 most of the world's population were agriculturalists but there were substantial hunter–gatherer groups left in tropical and boreal forests and in the Arctic. The industrial impact on agriculture was gradual and is incomplete but most people in the world now have some contact with the industrial sphere; increasing numbers have contact via satellite TV, for example.

Although energy can be used in a historical sense to delimit periods of human economy, this does not necessarily mean that it determines the way people act. Although wood shortages may have played a part in the European transition to fossil fuels, many other societies were not forced to become industrial. They saw it as advantageous in economic terms and they saw the resources as being available to them. The same was probably true of the change from hunting and gathering to agriculture. Except where migration and conquest may have forcefully introduced it, tillage was probably adopted by hunters and gatherers as being a way of life which was perceived as being preferable. Some societies may have taken it up to overcome population pressure upon gathered and hunted resources, but there is no evidence that this was universal.

The human impact on the environment

A small number of humans possessing only simple technologies may garner resources without any lasting impact: shifting cultivators in a tropical forest may leave behind a clearing which in only decades becomes indistinguishable in appearance from the surrounding, unfelled, forest. On the other hand, some cities have covered over their sites with stone, brick or concrete for thousands of years. The following discussion is mostly about long-term (though not necessarily quasi-permanent) changes, but we need to recall that many minor actions have a short-lived environmental impact which may be difficult to trace in the historical and archaeological record: the winnowing of time normally removes the records of all but the major changes.

Hunter–gatherers and their environments

We often have a romantic image of hunter–gatherers. We envisage them living in some primitive but paradisiacal Eden as true children of nature. They lived, we think, simply off the usufruct of nature, taking only what was strictly necessary for survival using a simple material culture. They needed to move often and so family size was kept down, avoiding anything reminiscent of a population explosion. Any such picture needs critical examination to see if we are creating an illusory Golden Age myth.

Ecology and energetics

In this section we review briefly the variety of natural environments that have been occupied by hunter–gatherers during the Holocene and also look at the energy flows by which they were linked to those surroundings and the control of which made it possible for some of them to manipulate the ecological systems in which they found themselves. The number of hunter–gatherers has declined during the centuries as agriculture and now industrialization have taken over. In 10 000 BC, the human population was 100 per cent hunter–gatherers; now perhaps 0.001 per cent follow that way of life and their practices have been heavily influenced by the world of higher energy-intensity living around them.

The interaction with nature of the pre-19th century hunter–gatherer group then involved the collection of such plant and animal material as was seasonally available within the territory of a particular band. The seasonal availability of foods together with the kinship and other social arrangements of the people might make it desirable or even essential for there to be a yearly movement round a territory where they had rights of access to the resources. In areas where resources are not scarce then hunter–gatherers who maintain relatively low population levels may simply live off the usufruct of nature, taking what they need without any necessity to worry that sufficient will not be there in future. Thus the !Kung bushmen of the Kalahari, studied in the 1960s, were dependent upon the nut of the mongongo tree as a staple food even though their total dietary spectrum was very wide. They had no need to plant mongongo trees, or ensure their pollination or any such manipulations: in season, enough nuts could be gathered to feed everybody with a decent diet of calories and protein. Their neighbours, the G/wi bushmen (Fig. 4.2), relied on the *tsama* melon for part of their diet and its abundance was variable: thus the G/wi had to travel further in a bad melon year to get their required quantity.

But there were also environments where the people felt that nature would not always provide unless some more positive action were taken. For some folk this was mainly of a non-material kind in the shape of practices designed to propitiate the gods into maintaining food supplies; for others it was a case of the gods helping those who helped themselves and so management of plant and animal populations was practised. In the case of plants, this might mean seasonally diverting some water over a stand of wild grasses, for example, to ensure their seed crop was heavy, or perhaps transplanting some wild yams into an easily accessible location. Animal populations might be protected from the kill in some areas so that depleted numbers might expand again, or there might be a prohibition on the killing of gravid females. In semi-arid environments like part of Australia, canals might be dug in and near swamps to keep plenty of the right habitat for eels.

On a world scale, there was a broad gradient from high latitudes where hunting of animals was dominant, through mid-latitudes where

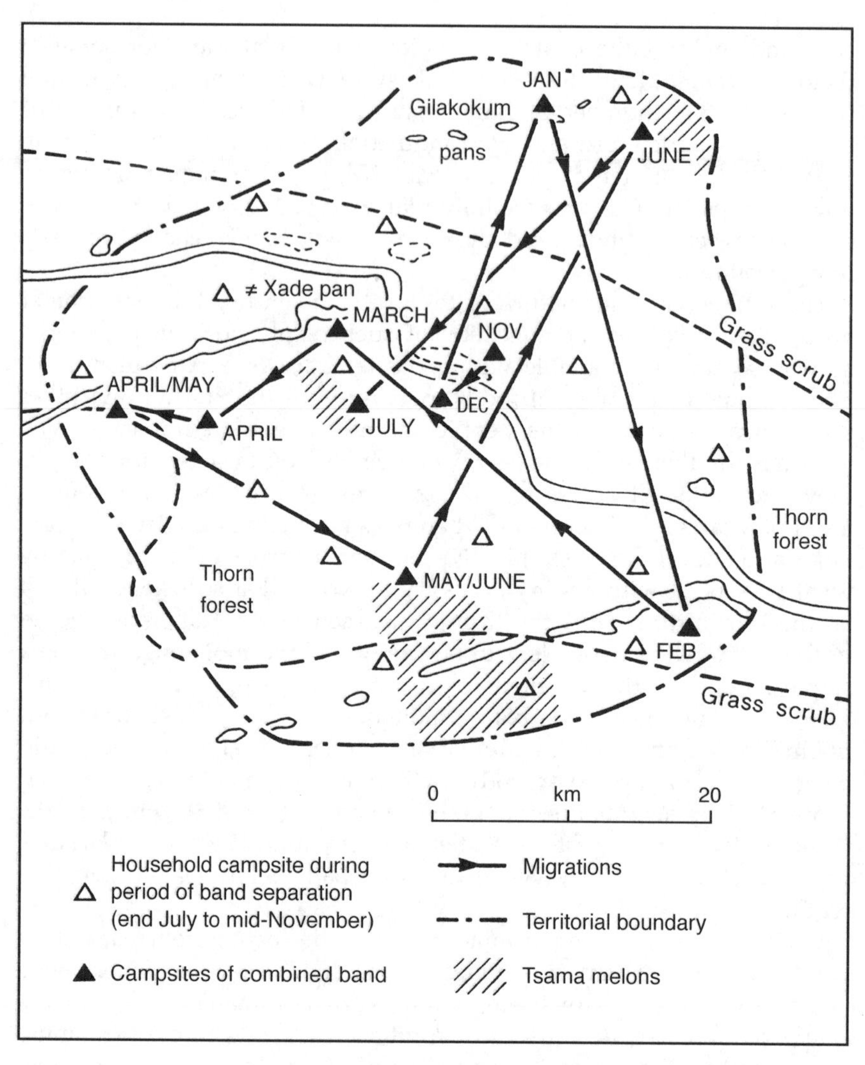

Figure 4.2 The yearly movement of a band of G/wi bushmen in the 1960s at a time of abundance of the *tsama* melon. The group splits up during the year: a common practice with many hunter–gatherers (from Simmons, 1989)

fish were very often important, to the tropics, where the gathering of plant food was paramount. But in all groups, hunting seemed to provide at least 20 per cent of the diet. Plants must therefore have been key foods in all except extreme environments like the Arctic. Hunting and gathering seems to have been founded upon the consumption of as much plant food as necessary and as much meat and fish as possible.

Ecological change produced by hunter–gatherers

Not all hunter–gatherers produced ecological change but those who did contrived to do so in both temporary and permanent ways. Impermanent alterations often centred on the impact of a hunting group upon an animal population: this could be severe but yet not so damaging to the reproductive capacity of the beasts that their numbers could not recover. Buffalo hunting on the High Plains of the USA before the advent of Europeans is one example. Numerous archaeological excavations have shown that herds of buffalo were driven over cliffs, into box-canyons or into dune slacks and then slaughtered wholesale. Yet long-term diminution of the animal's availability was in general avoided by concentrating upon herds of males and so letting the females carry their young unmolested. Analogous practices were carried out further north, where boreal forest Amerindian people developed traditions of 'resting' particular areas within their territories so that elk and beaver populations, for example, might recover from heavy hunting. The occasional use of fire as a hunting aid (i.e. to drive out animals from cover rather than to try to change the vegetation) might well not change the ecology for more than a season or two.

Of more interest and significance, however, are the cases where hunter–gatherers have affected their environments permanently. In these instances we often find that fire is one of the most important tools and the case of the Aboriginal inhabitants of Australia exemplifies this. In the interior, fire was regularly used as a hunting aid since it flushed many animals out of the bush and from underground, with the result that the vegetation types encountered by Europeans were in fact fire-adapted ecosystems produced by human agency. In the north, women regularly fired the vegetation containing a particular cycad tree: it then produced more fruit and produced them more or less simultaneously, which is always an advantage to people on the move. In North America, the burning habits of the native populations of the forest–grassland edge (again as a hunting aid, including keeping down the quantity of undershrubs which prevented a hunter getting a clear sight-line for his arrows) kept a mosaic of open woodland and grassy glades: when the Indians were extirpated, the forest rapidly reclaimed the land. We have an instance of permanent change from the uplands of England and Wales during the period of the last hunter–gatherers, the Later Mesolithic of *c.* 7000–3500 BC. The present-day moorlands such as the Pennines, Dartmoor and the North York Moors have yielded palaeoecological evidence to show that in those times such uplands were largely forested. Yet among the woodlands there were clearings which seem to have been maintained by fire and their frequency in time and space is such that natural causes are unlikely. In some places, these openings disappeared when agriculture started, which suggests that

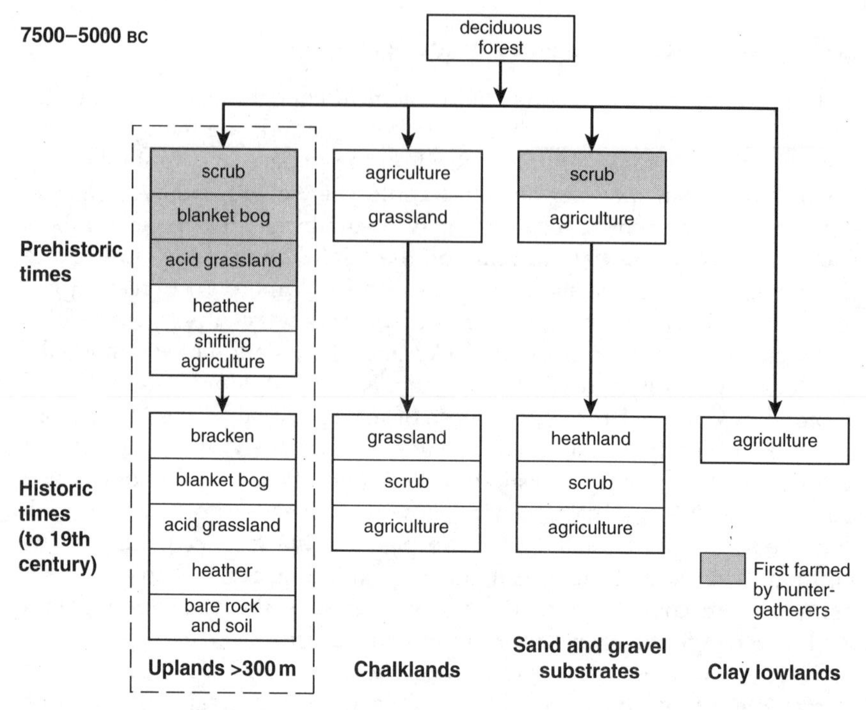

Figure 4.3 A diagram plotting the course of the differentiation of land uses out of a largely forested matrix, between Mesolithic times and the 19th century. The variety produced in the uplands was greater than at lower altitudes, though this has been reversed in the 20th century (not plotted here). Settlement is not dealt with at all in this diagram (from Simmons, 1982)

burning was an integral part of the hunter–gatherers' way of life. But on other sites, the removal of the trees allowed the soils to become waterlogged (since the trees acted as large water-pumps) and peat grew. In favourable places, this grew to depths of 2–3 m and formed a blanket over the land, in which condition (though often now eroding) it can still be seen today: an example of a current landscape element formed by hunter–gatherers (Fig. 4.3).

A further set of examples of more permanent change can often be seen when hunter–gatherers came into contact with agricultural or even industrial populations. The boreal forest again provides good examples, for many areas were almost entirely depleted of fur-bearing animals by indigenous trappers who sold their catch to the Hudson's Bay Company and similar agencies. The insatiable market for beaver, for example, wiped out many populations in spite of the company's efforts to install rotational trapping schemes; had fashions not changed, the beaver might have disappeared from most of North America, as it did from many areas to the south of present-day Canada.

Adapting

One mediation between hunter–gatherer life with nature is through technology, which has developed a number of forms during the millennia. At the end of the Pleistocene such groups had access only to organic materials and to stone, but they quickly took up metal whenever it became available to them. In recent times, relict hunter groups have rarely been reluctant to absorb the products of the industrial revolution when these came their way. The traditional technology centred around wooden tools for grubbing up plants, baskets or slings in which to bring plant materials back to camp, together with hunting aids such as the spear, bow and arrow, blow-pipe, slingstick, woven nets, and poisons. Metal tools and points, the rifle, the outboard motor and the snowmobile slipped into this repertoire with ease at various times, though none has prevented the virtual demise of this way of life.

Agriculture

The success of agriculture has been so great that the human population grew from perhaps 170 million in AD 1 (approximately the present population of Indonesia), to 957 million in AD 1800, which is less than today's total for China. The energy relations of agriculture move towards a more concentrated production per unit area/time than in food collecting. Hence, more people can be fed off a smaller area. A lot of human energy input is required into manifold tasks and subsistence agriculture is often an unremitting existence. Once above a purely subsistence level, however, it often produces surpluses that permit the differentiation of a wide variety of human occupations.

Ecology and Energetics

Compared with hunting–gathering, agriculture represents an intensification of use. This is fundamentally achieved by using culturally selected crop plants and animals. The product is a domesticated plant or animal that would not survive in the wild, for its life-cycle has been so tailored to meet human demands for food, fibre or skins that it has lost its evolutionary niche and would be unable to compete with wild species. Like any food system, agriculture has to provide a positive energy balance; unlike hunter–gatherers, agriculturalists have to expend a lot of 'front-end' energy in tasks such as sowing, ploughing, fencing and herding before any crop can be taken. Yet at its most productive, agriculture provided magnificent surpluses: enough to feed those who built the Pyramids of Egypt for example.

Origins

The beginnings of agriculture are still only partly known. A number of places in the world were important in the emergence of different agricultural systems but these do not encompass the whole spectrum of domesticates (Fig. 4.4). In the hill-lands of western Asia we know that cereal-growing around permanent villages grew up in the millenium either side of 7000 BC, and that these people kept domesticated animals as well. Perhaps about 4000 BC in the same region, nomadic pastoralism based on herds of domesticated animals became fully fledged. In Southeast Asia, rice was domesticated around the same time as cereals such as wheat and barley in western Asia; millet arose from northern China in the same era. In the 4000–2000 BC period, New World agriculture developed on the basis of maize, potato, beans and squashes. From these origins, the various types of agricultural system spread and unfolded into most parts of the world, often replacing hunting and gathering on the way, though unable to dispossess it from the more marginal places such as the very dry, the very cold, and the remotest tropical forests.

Global Impacts

If agriculture represents a concentration of both energy input by humans as well as increased intensity of yield, then the scope for enhanced environmental impact is correspondingly higher. In agriculture, conversion of natural systems is deliberately undertaken on a much larger scale (both spatially and ecologically) than with hunting (Fig. 4.5). One of the greatest changes wrought by humans has been to clear woodlands permanently in order to grow crops. Examples abound: the deciduous forests of Europe yielding to the plough in prehistoric and medieval times; those of North America disappearing along with their indigenous inhabitants after the European colonization. Analogous stories can be told from Asia, the former USSR, South America and Oceania, and the process is still continuing. Clearing a forest has strong ecological consequences: soils erode more easily and more water is shed more quickly so that floods, valley aggradation and nutrient dumping are all common below the deforested area. To combat loss of soil (especially in areas of seasonally intense rainfall), many societies constructed terraces. In essence, these convert a sloping field into a flat one and so both soil and water are held up for further use. Water is often seasonally scarce and so small reservoirs were built to store it; at

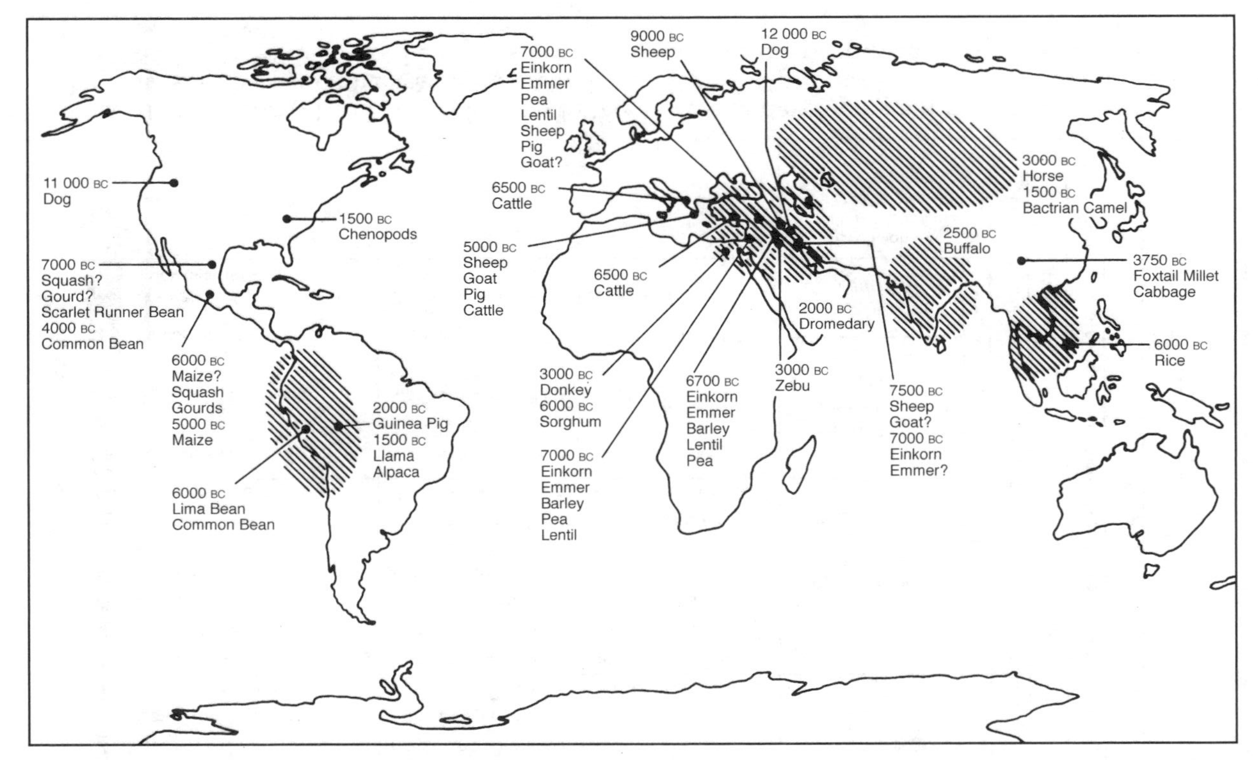

Figure 4.4 Although the domestication of plants and animals has been a continuous process, certain places have been important in the early phases. The key locations are indicated on this map: it suggests for example that mountain areas have been very significant (from Simmons, 1996)

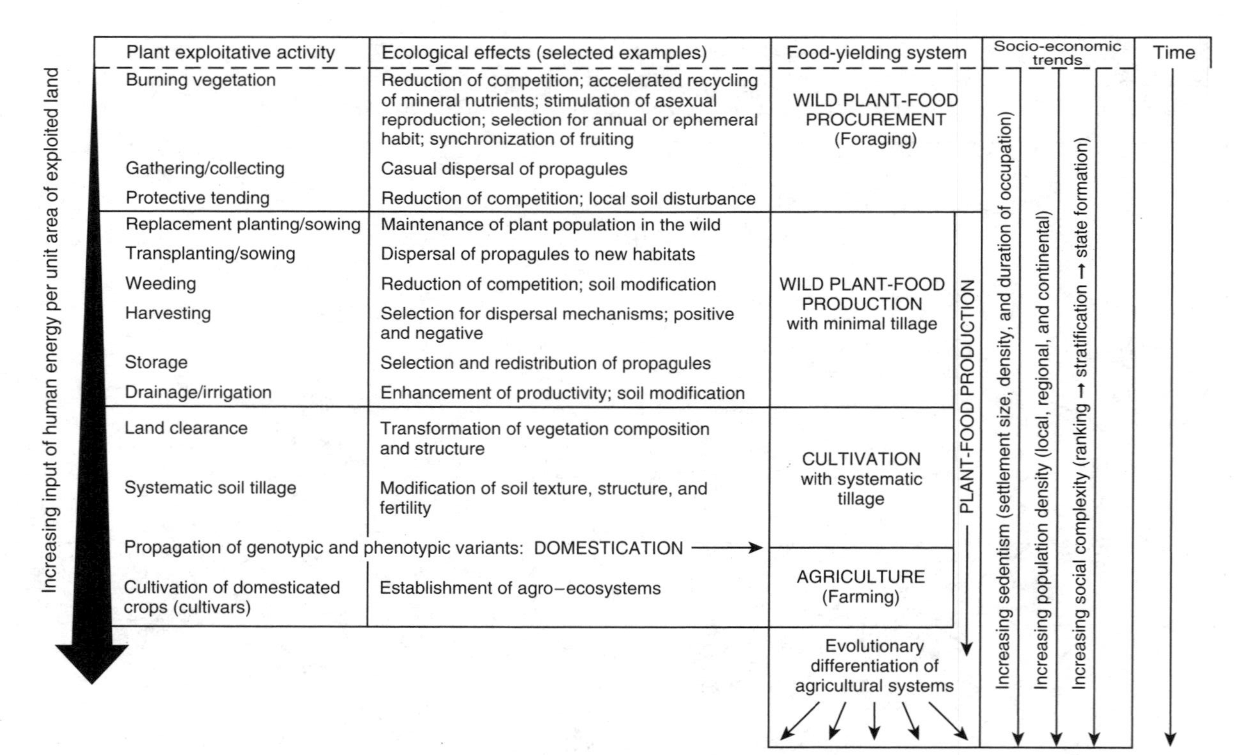

Figure 4.5 Some of the ecological and cultural stages in the process of developing agriculture in its early phases. Though the diagram gives prominence to the material–ecological changes, the social contexts implied by the right-hand vertical lines must not be overlooked: the culture of the people involved must have been crucial (from Harris *et al.*, 1989)

the extreme here are the flights of irrigated fields that we associate with *padi* rice. Terracing here converts slopes and flatlands alike into aquariums in which not only rice but animal food in the form of fish and shrimps grow well; multi-cropping of the cereal is often possible.

All agricultural systems must mimic the natural condition in circulating nutrients and so humans must fertilize their crops with animal manure, night soil, silt, soot, marl or whatever else is to hand. Such systems, like irrigation, imply complex forms of social organization as well: in Bali today the temples are the main nodes for the organization of the water distribution. In ancient Mesopotamia and Egypt it seems likely that authoritarian rule was necessary to keep the irrigation systems in working order.

Less intensive forms of cropping produce their own levels of impact. Shifting agriculture allows the return of the original type of ecosystem but often the actual species composition is changed; the more frequent the rotation, the greater the change. Nomadic pastoralism also changes the composition of vegetation. Domestic animals are selective in their feeding habits and so certain species are preferentially eaten and above some threshold level of consumption will not regenerate, leaving the site to thorny or toxic plants or perhaps simply to bare soil, i.e. the essential ecology of desertification. Many high mountain pastures in the Andes, for example, are as much the result of centuries of grazing as of climate and soil. The same can be said of the grassy areas of the moors of Britain: the species composition reflects the density of sheep, with the wiry *Nardus* predominating where grazing intensity is high.

Other pre-industrial impacts

Agriculture (including pastoralism) is central to societies because it feeds people. Thus most human groups at some time make attempts to extend their area, either to feed more people or to feed some of them better. But pre-industrial societies also reached into their environments for a variety of other purposes: for materials to use in construction, in farming or in cities, and as fuel. They also sought pleasure in parks and gardens, and like most humans before and since, engaged in warfare which was sometimes at the expense of the environment as well as of themselves.

Extending the agricultural area often meant land transformation, usually called 'reclamation' in English. Areas already relatively flat (Fig. 4.6) have always been popular, and converting salt-marshes and other estuarine areas has a long history, as has draining wetlands. In the former, the ability to build a good sea-wall has always been crucial, and in the latter, the windmill was an important piece of technology in using the energy of the atmosphere to lift water out of one channel and into another in order to speed it to the sea.

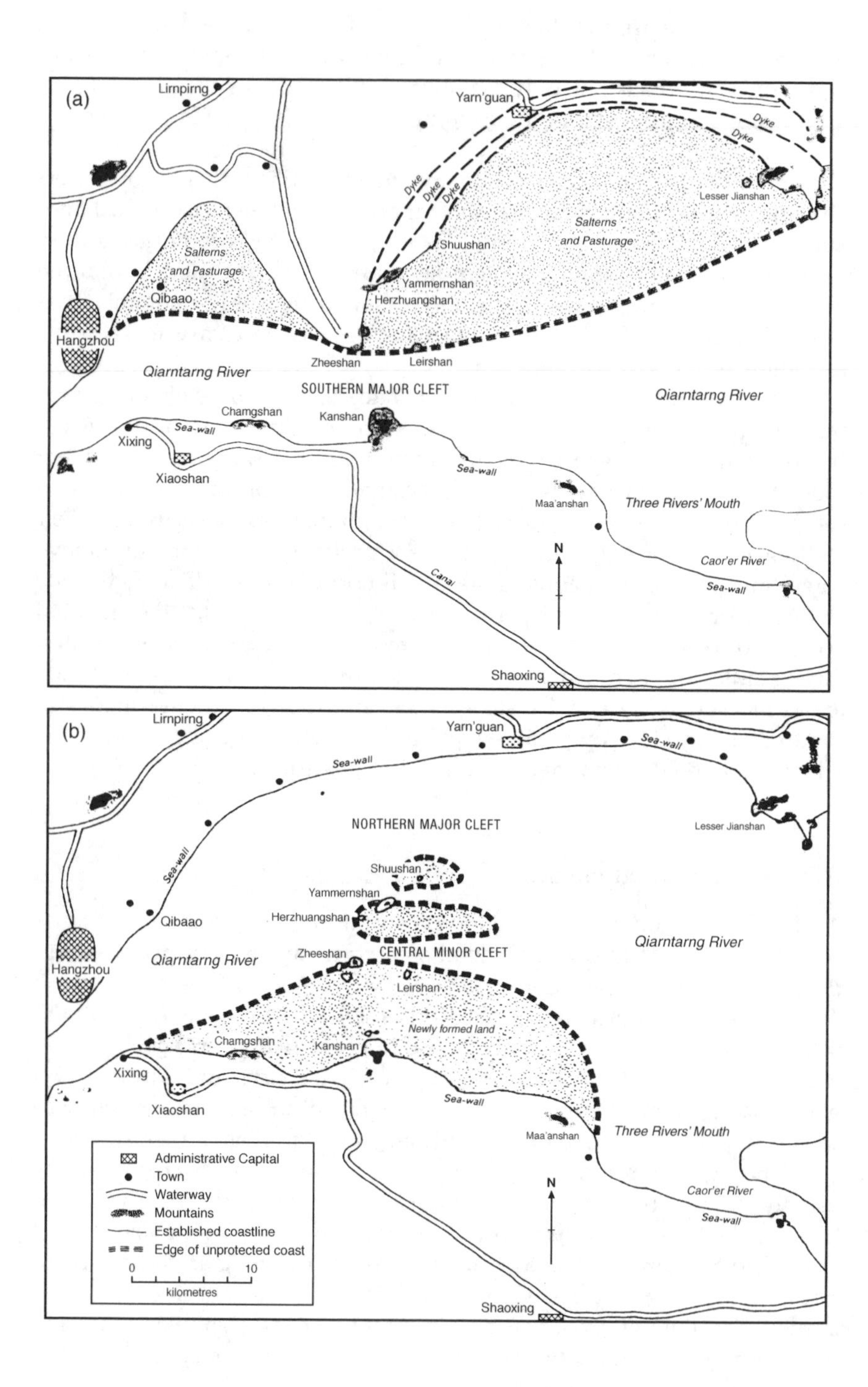
(a)
Lirnpirng
Yarn'guan
Dyke
Dyke
Dyke
Dyke
Dyke
Lesser Jianshan
Salterns
and Pasturage
Shuushan
Salterns
and Pasturage
Qibaao
Yammernshan
Herzhuangshan
Hangzhou
Zheeshan
Leirshan
Qiarntarng River
SOUTHERN MAJOR CLEFT
Qiarntarng River
Chamgshan
Kanshan
Sea-wall
Xixing
Xiaoshan
Sea-wall
Maa'anshan
Three Rivers' Mouth
N
Canal
Caor'er River
Sea-wall
Shaoxing
(b)
Lirnpirng
Yarn'guan
Sea-wall
Sea-wall
NORTHERN MAJOR CLEFT
Lesser Jianshan
Sea-wall
Shuushan
Yammernshan
Herzhuangshan
Qibaao
Qiarntarng River
Hangzhou
Qiarntarng River
Zheeshan
CENTRAL MINOR CLEFT
Leirshan
Newly formed land
Chamgshan
Kanshan
Xixing
Xiaoshan
Sea-wall
Maa'anshan
Three Rivers' Mouth
Administrative Capital
Town
Waterway
Mountains
Established coastline
Edge of unprotected coast
0
10
kilometres
N
Caor'er River
Sea-wall
Shaoxing

Many other wild ecosystems have also been reclaimed. Heaths and moors in Europe, for example, were targets in the medieval period: the Cistercian monks were especially active in such regions because they were supposed to live away from the temptations of the world and what better place than a desolate moorland valley? Forests have always been useful for their products as well as being land banks: wild foods to supplement basic diets are usually present and, above all, there is wood. Most pre-industrial societies depend upon wood for all their fuel requirements (domestic and industrial), for much construction (shipbuilding, scaffolding, the frames of buildings) and for a host of small things such as tools, and animal fodder. Woods can be managed to supply these materials: for example, coppicing and pollarding provide a sustainable supply of poles (for fencing or for charcoal-making), and shredding is a good source of animal fodder. Deciduous trees growing in semi-open conditions may develop crooked branches which are essential for the hull timbers in shipbuilding; contrariwise, dense stands of conifers produce the best masts.

The mention of industry reminds us that this was not exclusively a feature of the 19th century and after: there were workshops, quarries, mines and mills in many parts of the world by 1800. Many needed power and this was provided by the diversion of stream courses into water-mills; others needed heat for smelting, for example, and so woods might be managed for a continuous supply of charcoal, like those of the Sussex Weald of England in the 15th century. Wastes begin to be a problem: the tanning industry has always produced noxious effluents and in pre-industrial Holland special drains (*stinkerds*) were constructed for the removal of tanning wastes.

Where agriculture produced good surpluses then a leisured class might be sustained. This might devote itself to religion, to learning, to building monuments to itself, to acquiring an empire or even simply to pleasure for much of the time. Thus we have environmental manipulation for non-material purposes, as in gardens. Even though most gardens produce useful things to eat, and often herbs and medicines, there is always an accent on pleasure. In many cultures, then, combinations of shade trees, flowers, water and grass (and moss in Japan) are found (Fig. 4.7) which reflect particular values of that

Figure 4.6 A pair of reconstructions of reclamation attempts in Harngzhou Bay in China (a) during the 12th century, with small-scale dykings on the northern side of the estuary, and (b) during the early 18th century when larger-scale attempts to bring more intertidal land under control were made. In both, however, note the presence of sea-walls to protect existing land (from Elvin and Su Ninghu, 1995)

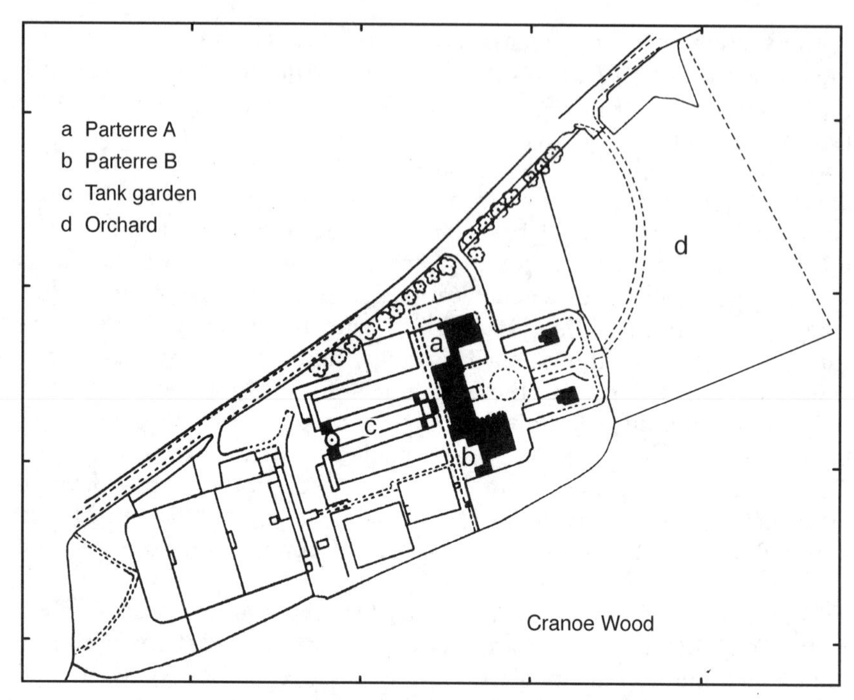

Figure 4.7 An English garden plan with formal elements such as flower gardens (parterres), a rectangular pool with flower beds either side (tank garden) and a productive orchard forming the approach to the house (from Tooley and Tooley, 1982)

society: the desert origins of Islam, for example, are shown in the primacy of water in their gardens. Other values may be reflected in the Gardens of Love in the courtly era of European history. Killing of wild or semi-wild animals for pleasure appealed in most cultures and to ensure the day's success, walled parks were often constructed (as far apart as England and China) to confine the quarry: deer were very popular, as were wild boar. The opposite might also apply, with edicts for forest preservation (under the Buddhist rulers of India, for example) and prohibitions on killing certain species, such as elephants in India in the 3rd century BC or kingfishers in China in 1107.

Killing of fellow humans has always been popular, and we may note in passing that the environment has sometimes suffered as well: at the battle of Pylos all the vegetation of one island was burned off so that the Athenians could see the movements of the Spartans; similar depredations were made on Scottish forests during clan conflicts. The Romans sowed the fields around Carthage with salt and wells were often poisoned. Most of these changes were temporary, however, and at Massalia, Plutarch put the positive case: so many Teutons were killed that the soil 'grew so rich and became so full to its depths of the putrefied matter that sank into it that it produced an exceeding great harvest in after years.'

Adaptations

Behind these human-led changes in the natural scene were the lineaments of human culture which made it possible to transform natural ecosystems. Technology is the major mediator between humans and their surroundings and during this period most societies were not hugely different in their access to tools, as they would become after the Industrial Revolution. The possession of iron, knowledge of metals generally, the virtues of the plough, awareness of navigation techniques and the recording of information on paper were all common to most human groups, with most societies lacking one or two of these inventions. Just as interesting, perhaps, is that the deployment of such technologies was given sanction by widely differing cultural ideologies. Thus the Chinese transformed the surface, vegetation, soils and water regimes of much of south and central China during the ascendancy of a Taoism which preached a kind of environmental quietism. In the end, the result was ecologically not very different from the Christianized cultures of Europe, where the Benedictine motto of '*labore est orare*' was symbolic of a divine approval of clearing forests and draining marshes. Nor should we forget that agricultural economies provided enough surplus energy for tangible accomplishments such as the Egyptian Pyramids or Chartres Cathedral, or less physical achievements like the *haiku* of Bashō Matsuo (1644–1694) or the music of J. S. Bach (though Bach owned shares in a coal mine and so might be said to have one foot in industrialism, just as the Archbishop of Salzburg, W. A. Mozart's early patron, was very rich from salt revenues.)

So the outcome of 12 000 years of human occupation is much as we might expect, given that in those years the human population grew from 4 million to 957 million. Where it was possible to produce edible materials, then they were produced in ever-increasing quantity to feed many people adequately and a few very well. At the margins, the original hunters were allowed to survive or a new breed of hunters-for-pleasure took over. Often at the junction of both, cities grew up to regulate the exchange of goods. The result by 1800 was an Earth of which large parts were physically humanized. The imminent erection of an efficient net of rapid human communication, importing more information into many cultural ecologies, was to carry on and virtually complete the process of humanizing the Earth's ecosystems.

Industrialization

Industrialization is the mode under which most of us now live in the West and the material benefits of which most of the rest of the world aspire. Since it delivers high material standards (of nutrition and health, for example) and allows wide margins for leisure and cultural development (Australian TV soap-operas, for example), its basis and likely sustainability are always under scrutiny.

Ecology and energetics

The essence of the changes which came to full flower in the 19th century and which have borne their fruits ever since is access to the stores of concentrated energy in the Earth's crust. Coal, oil and natural gas differ from the power sources of hunter–gatherers and agriculturalists in two main ways: they are much more concentrated and they are non-renewable. Because they are all so rich in energy per unit volume, a relatively small investment of effort in getting at them will bring a manyfold yield. Their use has been to generate steam to power machines, to fuel machines directly, to make possible the chemicals and plastics industries, and to underwrite an increase in the world's population from perhaps 600 million in 1700 to the 5300 million of the present.

From 1800, therefore, it has been possible for humanity to transform the natural world (and the previously existing humanized world as well) in ways not previously possible. The new sources of energy can be directed via a versatile and ever-inventive technology at manipulating the ecosystems of land and sea, and the effectiveness of this process is greatly increased by the increased amount of, and reliability of, knowledge conferred by science. In effect, science is fossil-fuel energy converted into information. The magnitudes of energy now available are summarized in Table 4.1; what is also true is that access to energy is now one way of defining wealth. Those countries which industrialized in the 19th century have become the kernel of the rich nations of the world and all have access to large quantities of energy. Those which are still largely primary producers have access to smaller amounts and are less rich in material goods, including energy access.

The ecology of industrialization means the addition of this new energy to the solar flow so that there is a subsidy which increases the intensity of output from systems that were previously solar-powered. Agricultural productivity per hectare or per person-hour can be greatly increased: hence the ability to feed most of the extra population since 1800. Equally, the technology now available means that we can get at almost any part of the globe and alter it: the top of Everest, the South Pole and the floors of the deep oceans bear evidence of the presence of humans or machines. Dredging of the ocean floors, the hunting of

Table 4.1 Cumulative rates of energy use in the 19th and 20th centuries

Year	World population (millions)	World industrial energy use rate (TW)	Per capita use (W)	Cumulative energy use since 1850 (TW-years)
1870	1 300	0.2	153	3
1890	1 500	0.5	333	10
1910	1 700	1.1	647	25
1930	2 000	2.0	1 000	55
1950	2 500	2.9	1 160	100
1970	3 600	7.1	1 972	200
1986	5 000	8.6	1 720	328

1 TW = 10^{12} watts = 31.5×10^{15} J/year.
Source: I. G. Simmons (1991) *Earth, Air and Water*, London: Arnold, p. 38

whales in remote waters, the placing of restaurants at the tops of mountains, the conversion of Pacific islands into quarries or airbases: all these are now possible. Given, too, that our wastes are often in gaseous or aerosol form, we can affect the whole planet and its atmosphere. All this is due to our command of the fossil fuels and concern about future energy supplies is therefore never surprising (Flavin and Lenssen, 1991).

Origins and spread of industry

A phenomenon as complex as modern industry is difficult to define, let alone assign it a birthplace. For the moment we will take as our criterion the idea of an economic activity powered by fossil fuels plus electricity from whatever source. For our spatial starting point we will take England of the 18th and early 19th century as being the best candidate for the world cradle of industrialism, and the changeover from charcoal to coal products (notably coke) plus the use of the steam engine in the smelting of iron ore as the key process. A charcoal bloomery might extract 15 per cent of the iron from the ore but a blast furnace will recover 94 per cent. This technological development released the whole gamut of both qualitative and quantitative transformations that led to a fully formed industrialism in England by 1840.

From Great Britain, the new ways spread quickly to form what are now the core nations of the developed world. By 1870 there were distinct industrial regions in France and Belgium, Russia, Germany, the USA and Japan, all with the characteristic of being based on coal as a fuel, even if it had to be imported or even if, as in the case of the USA, wood and water were the founding fuels of industrialization even in the 19th century. Along with industrialism came rising population levels: by 1880 the population of Great Britain, for example, had trebled in less

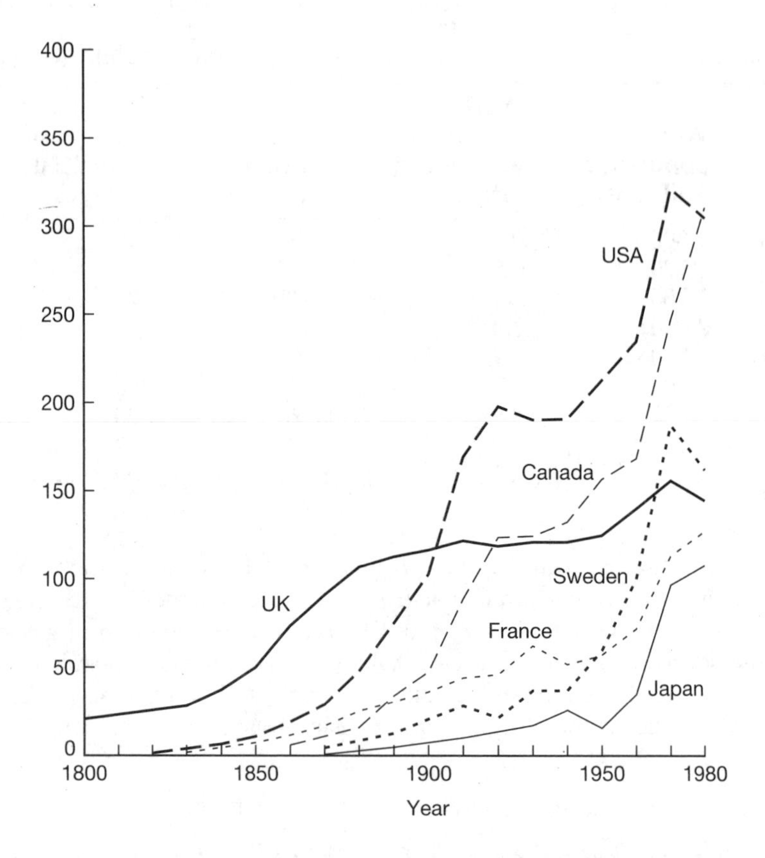

Figure 4.8 Per capita energy consumption in industrial economies from 1800 to 1980. The units are Gigajoules (1 GJ = 10^9 J). Note the gap which opens up between North America and Western Europe in the first half of the 20th century and the rise of Japanese consumption after 1950 (from Smil, 1991)

than a century and this experience was repeated elsewhere. Unprecedented, too, was the way in which these extra people were concentrated into towns and in emerging conurbations like the West Midlands of England and the Ruhr District of Germany. Once set in motion, industrial growth was for a while very rapid nearly everywhere: in the UK, energy use per person rose from 1.7 tce (tonnes of coal equivalent) per capita in 1850 to 4.0 tce per capita in 1919. Figure 4.8 shows the rise in energy consumption through this period in the context of its even greater rise during the later 20th century.

Outside these core areas, the older ways persisted as the ground of economic occupations: the opening of the Stockton & Darlington Railway to passenger traffic in 1825 or the launching of Brunel's screw-driven *Great Britain* in 1843 probably went unremarked in Ulan Bator or in Fiji but in time the echo of the steam whistle and the outer ripples from the docks in Bristol reached even such far-flung places. For the railway, the steamship and the telegraph were to be the means by which industrialism was to be fed (in terms of primary products) and spread (in terms of trade, helped by gunboats where necessary), and its concomitant ideas transmitted all round the world.

Global impacts in the early 20th century

To organize this material, we will employ the concept of core and periphery. We can use it at two scales: that of the impact of an industrial plant upon its local and regional environment, and that of the developed world upon the rest of the globe's ecosystems. Figure 4.9

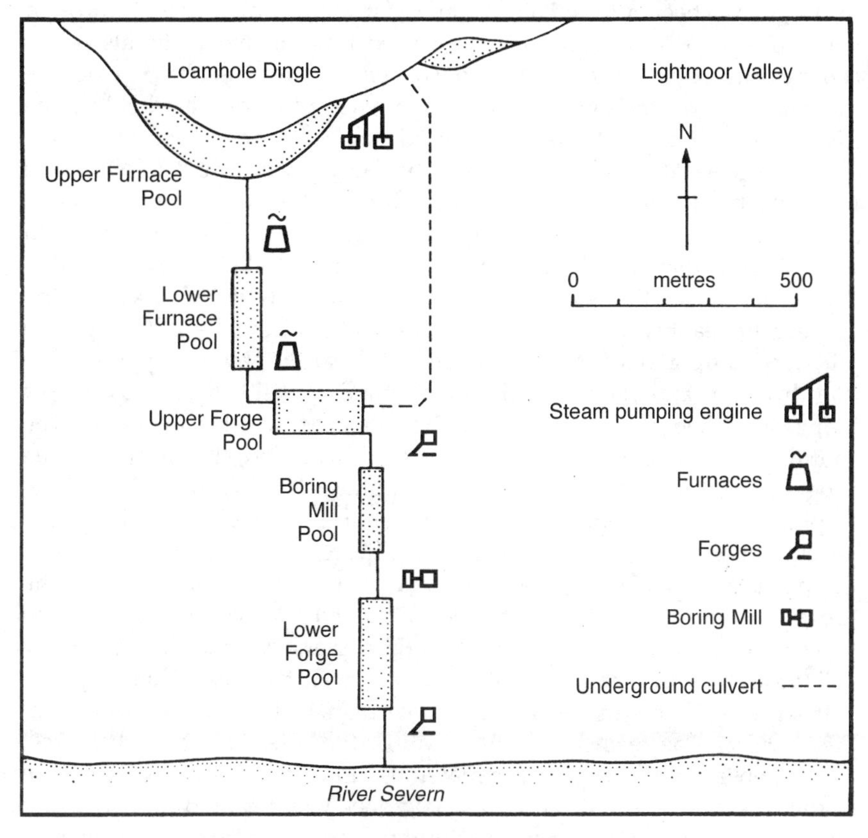

Figure 4.9 A diagrammatic version of the land and water transformations at Ironbridge (Shropshire, England) used in the supply of water for steam pumps used in iron smelting and forging. All this took place before AD 1800 (from Ironbridge Gorge Museum Trust, 1979)

reminds us that some of the developments superseded even older industrial installations.

Starting at the local level, we can imagine a turn-of-the-century industrial plant such as a large coal mine or perhaps a coke works or steel mill. The local environmental impacts are dominated firstly by the change in land use, with a large area being now covered with buildings, heaps of materials, roadways and yard surfaces, and transport features such as railways. When under construction there would have been a great deal of bare soil which would have shed considerable quantities of silt to the runoff, but when completed the plant would show all the hydrological features of urbanization, especially the ability to shed water very quickly after rainfall, since the new surfaces absorb little or no water. A heavily built-up zone is therefore likely to make the local watercourses more 'peaky' in their response to precipitation. The quality of the water is also likely to have been affected since some of it will have percolated through waste tips, and some will have been used for washing materials or for cooling purposes. Its load of both suspended and dissolved matter is therefore higher and some of the substances now carried are likely to be toxic to life. Water that has now carried domestic and industrial sewage and sullage is also different in quality; largely untreated, it will have raised the biochemical oxygen demand (BOD) of the water by virtue of the action of bacteria upon the organic components of the sewage. So downstream from the plant, the watercourse is not likely to have a great deal of life in it until the effluents have become well diluted. Likewise the air near the plant will be contaminated and fallout from the effluent plume will depend upon distance from the source. Near the chimney, particulates will drop out and thus be a disamenity and a health hazard to people living near their work: in the UK the moneyed will try to live on the southwest fringe of a large industrial area. Downwind, the effects become progressively less, but sulphur compounds will rain out as dilute sulphuric acid and destroy building stones as well as acidify the vegetation on wet uplands, for example. This reaching-out of the effects of the plant is paralleled on the input side by the tentacular stretch of the growing conurbations for water: local upland valleys are likely to be submerged by reservoir construction. Each plant is linked more widely still through the demands it creates and the products of its technology.

If we extend this model to the whole planet, the developed areas of the world can be regarded as the plant or the conurbation, and the rest as the zones of (a) outreach for materials, and (b) sinks for wastes. One of the chief demands of the industrial zones was for food and so many temperate grasslands were either ploughed up to provide cereals such as wheat (e.g. the North American prairies) or converted to ranching to provide meat, especially after the invention of refrigeration, as in Australia and Argentina. Not all the foodstuffs were staples: tropical products such as tea and coffee were in great demand and so large areas

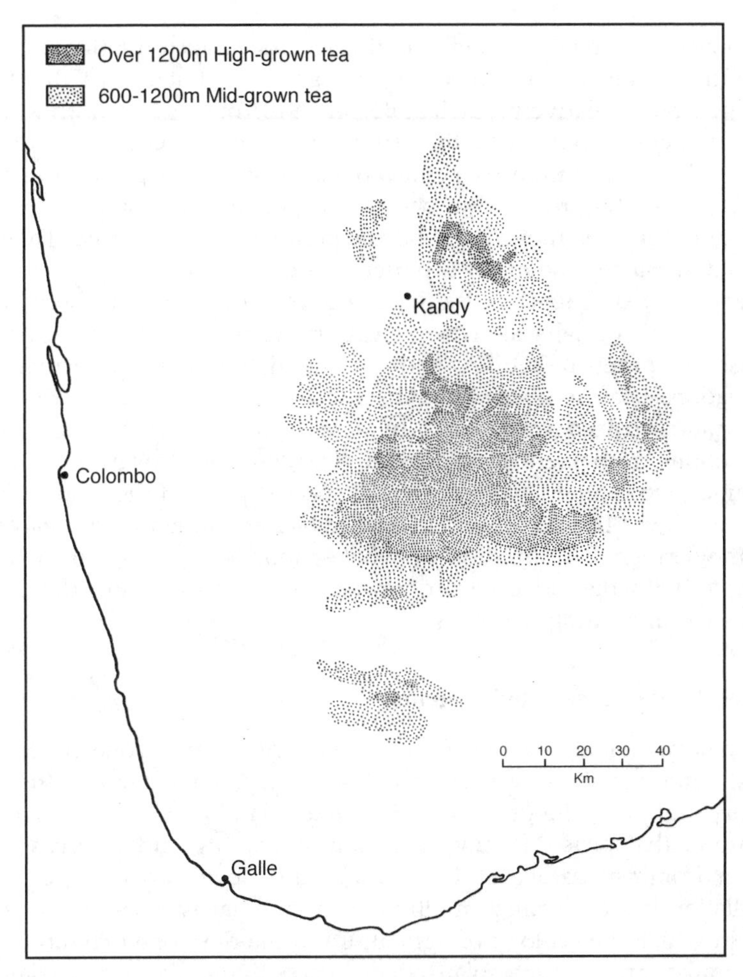

Figure 4.10 Tea plantations above 600 m in Ceylon (now Sri Lanka) in the late 19th century. Before the plantations there was forest or open woodland and grassland (from Forrest, 1967)

of forest and bush were transformed into plantations (Fig. 4.10). Colonial governments tended to think of such lands as 'empty', though they may well have supported seasonal populations of pastoralists or hunter–gatherers. As motor vehicles took hold, the demand for rubber increased greatly and so tropical forests in Malaya, for instance, were converted to rubber plantations, with sales of the original timber an additional source of profit. Many boreal forests began to have their ecology changed by the large-scale harvesting of trees for the pulp and construction markets during the later 19th century as well. Other agricultural systems now found that growing for export was possible and so some changed their ecologies drastically by importing irrigation techniques. We tend to think of the less developed world as major suppliers of minerals resources as well as crops, but in fact this trade was generally relatively small-scale: in 1913, only six countries were major mineral suppliers to the industrial nations, compared with the 1960s, when major industrial states obtained about 30 per cent of their minerals from the tropics. But cheap transport by steamship meant that foodstuffs could be moved around the periphery itself: Burma, Thailand and Indonesia became major exporters of rice after 1870.

Another great impact of the industrialized regions upon the planet's resources was in the seas, where steam trawlers facilitated the catching of fish as never before. A map of the dates when particular fish populations became uneconomic to utilize is a map of (a) proximity to early industrialized areas and (b) dates starting in the 1880s. Some populations recovered if left alone (e.g. in the North Sea during wartime) but in others the commercial species' place in the food webs of the oceans was taken by other species, which might not be of interest to the trawlermen. The hunting of whales, too, was getting much more effective, with the aid of steam-powered mother ships and the use of explosives in heading harpoons.

The ecology of associated systems

As we have seen above, the effects of industry go far beyond the factory gates. Although the city itself is not solely a phenomenon of the 19th century and after, the growth of cities into conurbations is very much a feature of that phase. In environmental terms, large urban areas affect the environment rather in the manner described above for a single installation but on a much multiplied scale; what is more subtle is the way in which the ecology of agriculture in the developed countries has been transformed from a solar-based process to one that is underlain by large quantities of fossil fuel.

Earlier this century, this metamorphosis had not proceeded to today's extent (there was no sliced, wrapped bread) but we can seen the bringing on of the basic changes. In essence, this consists of the application in indirect form of fossil fuel to the food and fibre production sequence. Some of this takes place on the farm or other site of production: the use of powered machinery, for example. Steam engines were the first signs of this, in threshing, and also in ploughing in North America. In both cases petrol and diesel tractors proved more flexible in use, though they replaced the horse quite slowly in Europe (Fig. 4.11). The tractor and other machines not only consume fuel on site but represent a great deal of embedded energy used in their manufacture; this is also true for chemical fertilizers, which began to replace manure in the pre-1918 period. The extension of cultivation into grasslands is facilitated by late-19th century inventions such as barbed

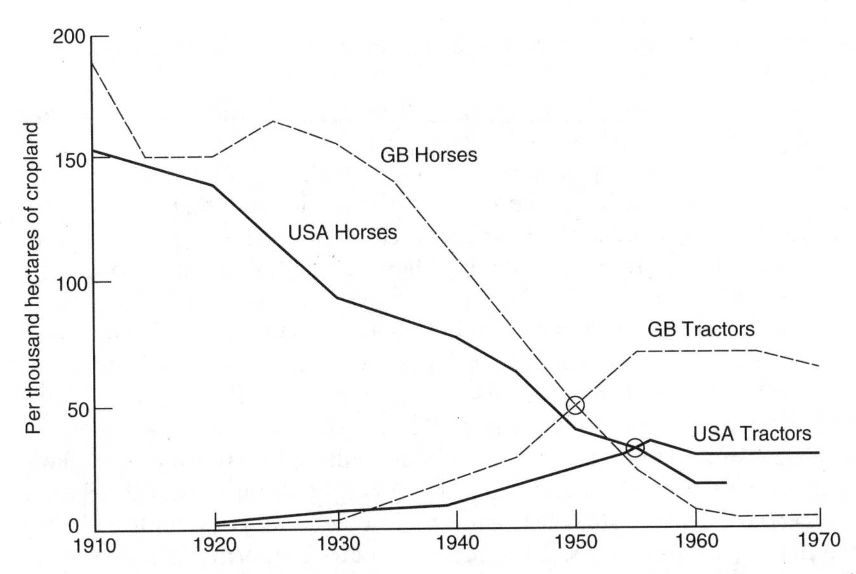

Figure 4.11 The numbers of horses and tractors per thousand hectares of cropland as an index of industrialization in agriculture. The cross-over points are *c.* 1950 for Great Britain and *c.* 1955 for the USA. However, since tractors are much more efficient, their importance would have been felt much earlier, though bear in mind, for instance, that the German army in 1944 was still dependent upon the horse for much of its supply functions (from Simmons, 1989)

wire, and the winter of the temperate zones could to some extent be subverted by the early appearance of salad crops and spring flowers grown in metal-framed glasshouses. All this embedded energy is usually referred to as 'upstream' energy; 'downstream' from the farm, yet more energy is required for transport, storage (especially if the temperature and humidity are controlled to prevent spoilage) and processing. Thus wheat may be stored for some time at a controlled humidity, achievable only by using a set of air pumps, and then it is milled using steam power and baked in an electric oven. This intensity of energy use was common early in this century but was low compared with that of today. The overall lesson is clear, however: the energy balance of the food system begins to move away from that of the surplus which must be produced in subsistence societies if they are to survive. Instead, a negative balance can be tolerated because fossil fuels can be mined to subsidize the whole process from agricultural college to table (Briggs and Courtney, 1985). With the exception of Europe, however, the productivity of agricultural land did not rise a great deal before 1950: more output per head was achieved by land-use conversion from forests and grasslands to cropland. Since 1950, though, technological change has been the driving force in the increases of productivity per unit area of cropland.

The greater access to energy resources made available in the 19th century had its impact upon pleasure as well. In Great Britain, one environmental consequence could (and still can) be seen on the moorlands of the drier uplands of northern England and of Scotland. Before the 1840s, it had been the practice of the owners of shooting rights to kill red grouse by walking them up with dogs and firing over the dogs at the rapidly retreating birds. But this changed: First, there was the advent of the breech-loading shotgun, which could be reloaded very quickly and so demanded more targets per hour than its muzzle-loading precursor. There was also new money, and the railways meant that the financier could leave his desk in the City at 4 p.m. and be on the Aberdeenshire moors next day at 10 a.m., willing to pay for a weekend's sport. But to go to these lengths to kill a couple of birds was clearly not acceptable. So the system was evolved of burning the moors to enhance the density of grouse: since the breeding pairs are territorial, eat mostly tender shoots of heather and also require bushier heather for nesting sites, then it made sense to burn the vegetation to the advantage of this fire-tolerant low shrub. Virtually a 100 per cent monoculture could thus be provided, in patches of different heights and bushiness. The denser grouse could then, come August 12th, be driven across the guns to be shot in huge numbers (over 1000 birds in one day by one man is recorded for the late 19th century) by guns sheltering behind butts, each sportsman having one or two loaders. All potential and actual predators were killed by keepers and people kept away during the breeding and shooting seasons. The firing of the moor at perhaps 15-year intervals meant that a great deal of its scarce nitrogen budget was lost in smoke, and soil erosion is always faster on such moors. Thus the long-term ecological processes were deleterious, just as the economics and social prestige were advantageous.

The railways and steamships initiated the era of mass travel for pleasure, of which today's inheritor is the package holiday to Spain or the Seychelles. The steam railway fostered the burgeoning of great resorts in Europe, North America and Japan to cater for the day trip for immersion in the newly declared health of sea water. The steam packet even made the British marginally less insular, since cross-channel travel was not now so subject to wind ('Fog in Channel: Continent isolated' ran, it is said, an English newspaper headline in the early 1900s) and so the middle classes began to move out and about. One target for this migration was the Alps, very different from Britain in both summer and winter; in the latter season exotic sports like skiing could be sure of snow. Then, under the aegis of operators like Thomas Cook, really exotic places such as Egypt might be safely ventured to.

So the ecology and energy balances of both necessity and pleasure had begun to change markedly by the time of World War I, though their full course to our present situation had yet to be imagined, let alone experienced.

Today's industrial impact

The industrial nations today experience such a plethora of connectivities that it is impossible to encompass them all in a short account. In the industrial areas themselves many environmental changes spring from the development of different types of installation and their associated housing, transport and other urban features. Both planned and accidental emissions occur. Such is the power of industry that it exerts its effects at a distance as well, calling in resources and hence provoking environmental change many thousands of kilometres away. The same may well be true of its products and by-products. Even at smaller spatial scales, the impacts of energy transformations, such as that of coal to electricity, may well be profound (Fig. 4.12).

One common theme of today is the spatial concentration of industrial processes of all kinds, producing environmental change on scales never before achieved. The presence in the West and Japan of spatially extensive urban–industrial zones (e.g. the Tokyo–Hiroshima belt or the Boston–Washington DC corridor) means that within them there is probably no truly natural feature left. The development of industrial plant, housing, transportation and leisure facilities has left some less intensively used land and water no doubt, but such areas are either carefully managed or subjected to the by-products of all that surround them. So there is probably no part of the land and offshore waters of such areas that could truly be called natural if we except the occasional rare bird blown in by a storm, but even it would have to feed from a human-managed environment. Very often, such regions concentrate processes (e.g. concreting or tarmacadaming of soils, or the production of sewage) which occur elsewhere but at lower intensities. An isolated farm can probably discharge its sewage into the local river without too much effect, but a city of a million people is a different matter.

Another common theme is the emplacement in the environment of compounds unknown in nature. The success of science and technology has allowed the synthesis of many substances (of which the plastics are obvious examples) that were not in existence before the 20th century. Most of these materials circulate separately from the environment but some are led into it as wastes. Here, in general, they find no natural breakdown pathways and so they may accumulate. In the case of many plastics, the results are obvious: in the instance of the organochlorine pesticides of the

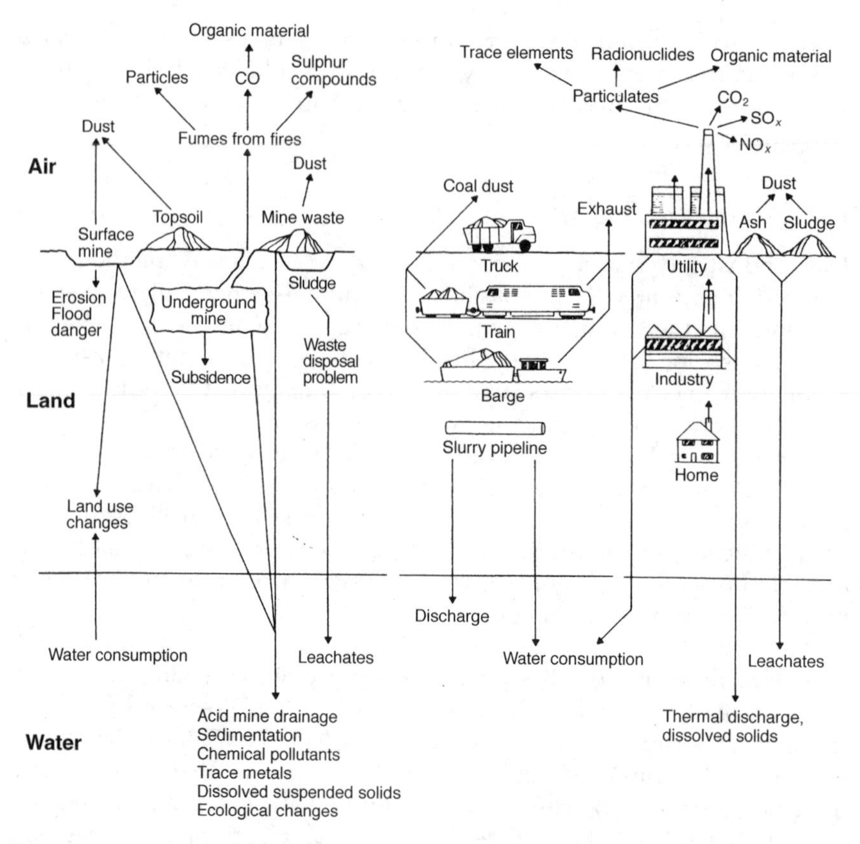

Figure 4.12 The sequence of environmental linkages of coal from its extraction to use. The processes shown are mostly those of output. The land-use changes needed to get and transform the energy source are not shown; neither are manipulations needed to ensure the water supply for the cooling needed at the power station (from Simmons, 1989)

DDT generation, much less so (Fig. 4.13). But bioaccumulation of DDT, aldrin and other similar compounds produced concentrations in non-target organisms that were many hundreds of thousands of times the application rates (Ware, 1983). Some herbicides carried with them contaminants from the manufacturing process, such as TCDD (dioxin). While the herbicides were themselves not toxic to humans, dioxins are teratogenic to human foetuses, as was shown during and after the Vietnam War.

Another version of this same process is the introduction of natural substances into environments where they are not normally present, as happens in coastal oil spills. The action of waves and bacteria will break down the oil but it is often a slow process, in the course of which mortality to plants, animals and micro-organisms is very high. At one point in the

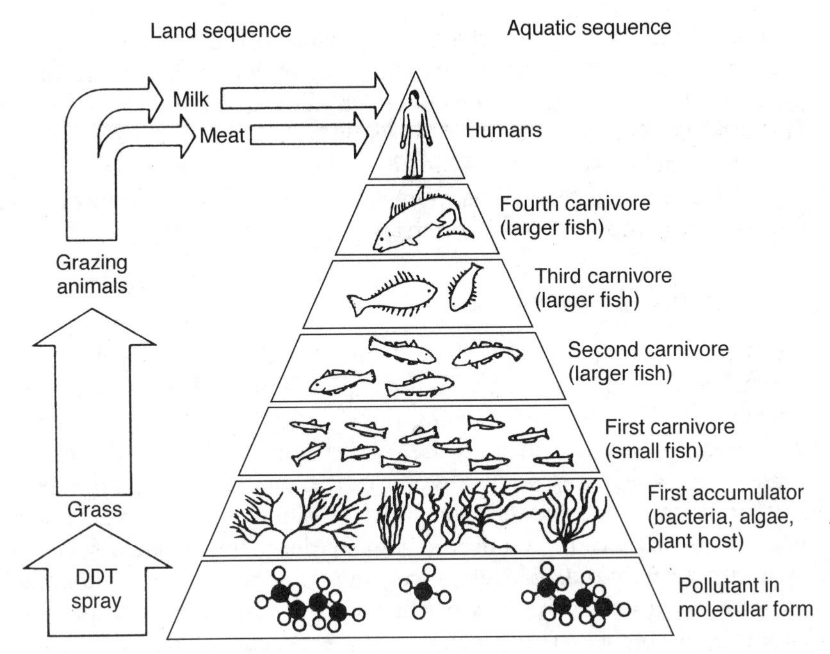

Figure 4.13 Pathways of a long-lived chemical such as DDT and its metabolites (e.g. DDD and DDE) through an aquatic food chain (i.e. of wild plants and animals) and through a land-based sequence of domesticated plants and animals. At each stage it is possible for an organism to accumulate more toxins than it excretes and so the concentration in the later stages of the web may be high enough to kill an animal or produce morbidities such as reproductive failure (from Goudie, 1981)

Gulf War of 1990–1991, about 49 km^2 of desert was covered in some 300 small lakes of oil. Even where these have been drained, the depressions are lined with a layer of tar which penetrates up to 1.5 m into the sand.

One net effect of land cover changes during industrial periods is a net addition of carbon to the atmosphere. In the UK, the 66 per cent of land that has remained in the same use over the last 50 years has accumulated 2.2 million tonnes of carbon per year; the 34 per cent which has changed use has caused the net emission of 0.9 million tonnes of carbon per year.

The effects of today's industrializations are felt well away from their major foci. Developing countries which supply materials to the industrial heartlands experience considerable impact from, for example, major mineral developments such as coal in Colombia or copper on Bougainville. Cash crop agriculture will often speed up soil loss and reduce biodiversity in such countries. Technology transfer is often part of such change. In early forms this may have been a simple matter, e.g.

the export of steel knives to hitherto stone-using cultures in remote areas of tropical forest or the Arctic; nowadays the bulldozer is a more likely implement, or the all-terrain vehicle which allows pastoralists to live in towns and take water out to their flocks on a daily basis. So the centre–periphery model established in the 19th century has still some validity, though it is becoming blurred as more states try to enter the ranks of the NICs, the newly industrialized countries.

The post-industrial world

What then, we may ask, of the advanced nations where the older combinations of manufacturing based on coal and oil are being replaced by a footloose segmental industry using only electricity and where services carried electronically are the 'sunrise' occupations taking over from the 'smokestacks' which are now the prerogative of the NICs and LDCs? Firstly, no country has made a complete transition to such an economy; rather, the new pattern has reinforced some of the characteristics of the old, such as city growth and redevelopment. Innovations such as home-based work with electronic linkages are still relatively rare and so exert little environmental impact. The greatest change probably comes from the immense demand for electricity for all purposes in these countries, from domestic as well as transport and industrial consumers.

In environmental terms, the results have not been spectacularly different. Nuclear wastes are the best example, along with the environmental manipulation needed to produce large volumes of cooling water at inland sites: in France, many rivers and canals have been diverted or constructed to supply water to EdF installations. In the medium-term future, an interesting set of events to watch will be the assessment of the environmental impact of the 'alternative' energy sources. These have usually been presented as requiring little environmental manipulation compared with mainstream commercial fuels and as producing a low environmental impact. This is not always the case. Wind farms, for example, have been criticized for their noise levels as well as for visual intrusion. Large-scale solar collection installations are also visually intrusive, though since they are mostly in deserts there is not a huge outcry. Geophysical sources such as hot water are often laden with sulphur, which has to be disposed of and which often smells of H_2S and mercaptans. Devices for deriving energy from tides and waves are often thought to be intrusive into the attractive scenery of the high-energy coasts where they are most useful, and large-scale tidal barrages have a high environmental impact upstream. Biomass energy capture probably fits best into existing visual and land-use patterns, though the prospect of farmland being converted to fast-growing trees such as willow in order to fuel wood-burning power stations is unlikely to please most agricultural organizations.

The outcome of 10 Ky of human occupation

In some atlases, as we remarked before, there are maps of the major natural biomes. The outcome of the processes discussed in this chapter, however, is that such biomes scarcely now exist. Very few parts of the world are free of the traces of humanity's occupancy of the Earth. The highest mountains have needed waste clean-up campaigns, as has Antarctica. Ice-sheets like those of Greenland have fallout materials such as aerosol lead (used as an anti-knock additive in auto fuel) stratified into the 20th century horizons. Even the famed biodiversity of the tropical forests is seen by some workers as having been in part created by land uses such as shifting cultivation. Perhaps only the deep oceans are true wilderness, though parts of the Arctic lands carry very little impress of their inhabitants outside actual settlements, like the higher reaches of many mountains. Nowhere is free from the fallout of aerosol material however, whether this be dominated by sulphur compounds, lead or radioactive isotopes. Hence the creation of a humanized world within and alongside the world of nature is more or less ubiquitous, though it is clearly wrong to suggest that it is complete. Meyer and Turner (1992) have suggested that between 1700 and 1980, cropland increased by about 400 per cent, with irrigated land burgeoning by 2400 per cent. On the other hand, closed forests have decreased by 15 per cent, as have other types of forest and woodland. Overall, the area of grassland and pasture has not changed much.

In any look back at the past, it is interesting to look for turning points. In this case, there are several important processes whose origins would constitute such points: the control of fire, the conscious domestication of plants, the shift from wood to coal in Europe and North America, and the achievement of controlled nuclear fission. But only in the last of these can any chronological precision be obtained and then only by the selection of one event among many that were critical to the outcome. Nevertheless, approximate eras can be designated for the middle pair and these give us a reasonable basis for dividing human history into periods of significance for human-related environmental change. Although the latest period is one of immense human impact (and the period after 1950 is especially so within it), nevertheless the millennia of agriculture in areas like East and South Asia and the Mediterranean effectively transformed whole regions into humanized landscapes and ecosystems. Terracing, for example, represents a very widespread imprint of human activity dating from that era. Agricultural-period achievements include, after all, the Great Wall of China, one of the few distinctively human features to be seen by the human eye from space.

Rates of change

We have been careful to emphasize that the natural world is subject to change under its own dynamics as well as being the recipient of human-directed impacts of the kind discussed above. The addition of human-led processes to those which are natural is of three kinds:

- those which accelerate the rates of natural processes and so usually add to the magnitude of them;
- those which diminish the rates of natural processes, either by trying deliberately to stop them or as an unplanned by-product;
- those which are entirely in addition to natural processes.

Accelerations and additions

One example of acceleration is the transport of species from one area of the world to others. This happens without human intervention, but we can speed up the process. Without such activity it is doubtful whether the proportion of the introduced flora of some islands would have reached its current percentage of 59 in New Zealand, 36 in the Falklands/Malvinas and 23 in the islands of Tierra del Fuego, for example. One of the best-documented of examples consists of the rates of soil erosion under natural cover and then under various forms of human-initiated disturbance, even when remedial measures are introduced (Fig. 4.14). In Colorado, USA, the rate of soil loss in the last 100 years has been 1.8 mm/year, which is about six times its rate in the previous 300 years. At the present time in Virginia, another study has shown that soil loss during construction was ×10 that from agricultural land, ×200 that from grassland and ×2000 that from forest within the same area (Goudie, 1993). In forests of the western Cordillera, construction of logging roads increased the magnitude of debris avalanches by factors of 25–344 compared with the uncut forest. Clear-cutting itself increased the avalanche magnitude by factors of only two to four. The expansion of the gravel industry in Britain allowed an explosion in the number of breeding pairs of little ringed plover by 6500 per cent during 1948–1960. This is a small addition compared with the increase of phosphorus in English rivers, which increased 40 000 times in the same period, and fallout lead in ice-cores from Greenland, which increased 20 000 per cent from 800 BC to AD 1950. Such examples can be multiplied almost indefinitely.

Diminutions

The same kinds of process bring about reduced rates in some cases. Heavy grazing, for example, diminishes the rate of rainwater infiltration into soils: from values of 30–45 mm/h to 17–30 mm/h in

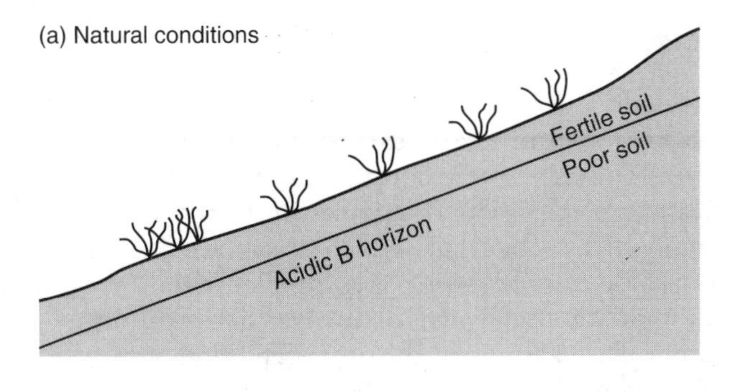

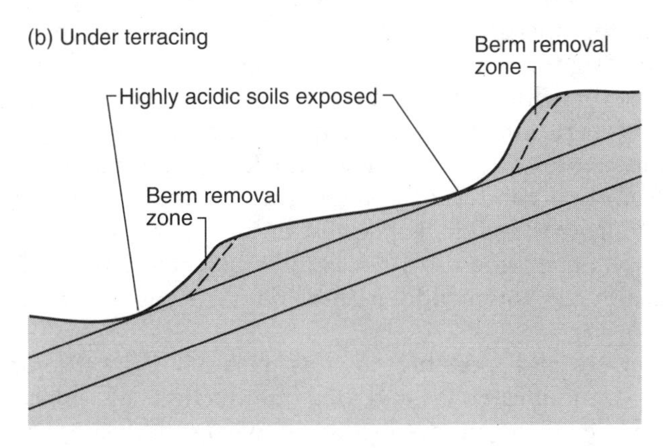

Figure 4.14 Soil loss may occur unpredictably, as when slopes have been terraced in Rwanda. Terracing is expected to stop soil movement downslope but in this instance it exposes the poor soils as well as allowing the loss of more fertile horizons from the leading edge of the terrace (from Johnson and Lewis, 1995)

the western USA, for instance. The capacity of forests and grasslands to regenerate is removed by transformation to cropland, which has happened equally to about 19 per cent of the grasslands and non-tropical forests since the industrial revolution. Regionally these data conceal increases such as the 214 per cent increase in the area of cropland in the USA between 1880 and 1980, and 273 per cent increase in Burma (Myanmar) in the same period. The increase in silt runoff from erosion may be counteracted if the rivers into which it flows are dammed: the South Saskatchewan River in Canada carries only 9 per cent of its pre-dam silt load and the Nile now bears only 8 per cent of the load of silt it carried before the building of the Aswan High Dam; similar values ranging from 8 to 50 per cent are common throughout the world.

Introductions

Human ingenuity, aided by science and technology, has meant that phenomena unknown in nature can be put together. Some of these are largely static. The city, for example, affects its regional environment considerably but is itself bound together in the most literal of senses. By contrast, many of the chemicals which are synthesized by the chemical and pharmacological industries are mobile within the environment, moving through soil and water pathways that were never foreseen by the innovators. The example of the organochlorine pesticides is probably the best documented, not simply because of the bioaccumulation but also due to the breakdown patterns which in some instances produced compounds even more toxic to life than the originals. In a less important example, the breeding rates of trout in British rivers are being affected by the poor performance of male fish in developing their testes; one explanation centres around the levels of female hormones in the water derived from oral contraceptives. (Sperm counts in male humans seem also be falling world-wide, especially in industrial nations, though the evidence was not so good in 1996 as to be beyond dispute.) The risks to human health are often hidden until damage has been done (Misch, 1994).

Even newer, and possibly just as pervasive, is the ubiquity of electronic communications and the production of information-rich societies in which the possession of knowledge is being touted as the route to power (Fig. 4.15), just as a high population and the steam engine were in a previous era. It is too soon to judge the correctness of all those who make these assertions but in terms of the global transfer of capital needed to launch at least some of the environmental transformations of which we speak, there seems to be some truth.

Chronology

These examples reinforce the earlier idea that the 18th century was a take-off point for human-induced environmental change. But as the data for energy use hinted (Table 4.1), recent years have seen much greater rates of change. These are summarized for a number of processes in Table 4.2, where the last column represents the total amount of change that had occurred by the mid-1980s. For each example, therefore, it equals 100 per cent. The preceding columns chronicle the progress towards that percentage, showing in some cases that by 1950, only half or less of the 1980s change had been accomplished. In the case of deforestation, earlier years had had a profound impact, but the human population was only 50 per cent of its 1985 size in 1950. For a 'new' substance like carbon tetrachloride, there was none in 1860, and by 1950 only 25 per cent of its recent use rate

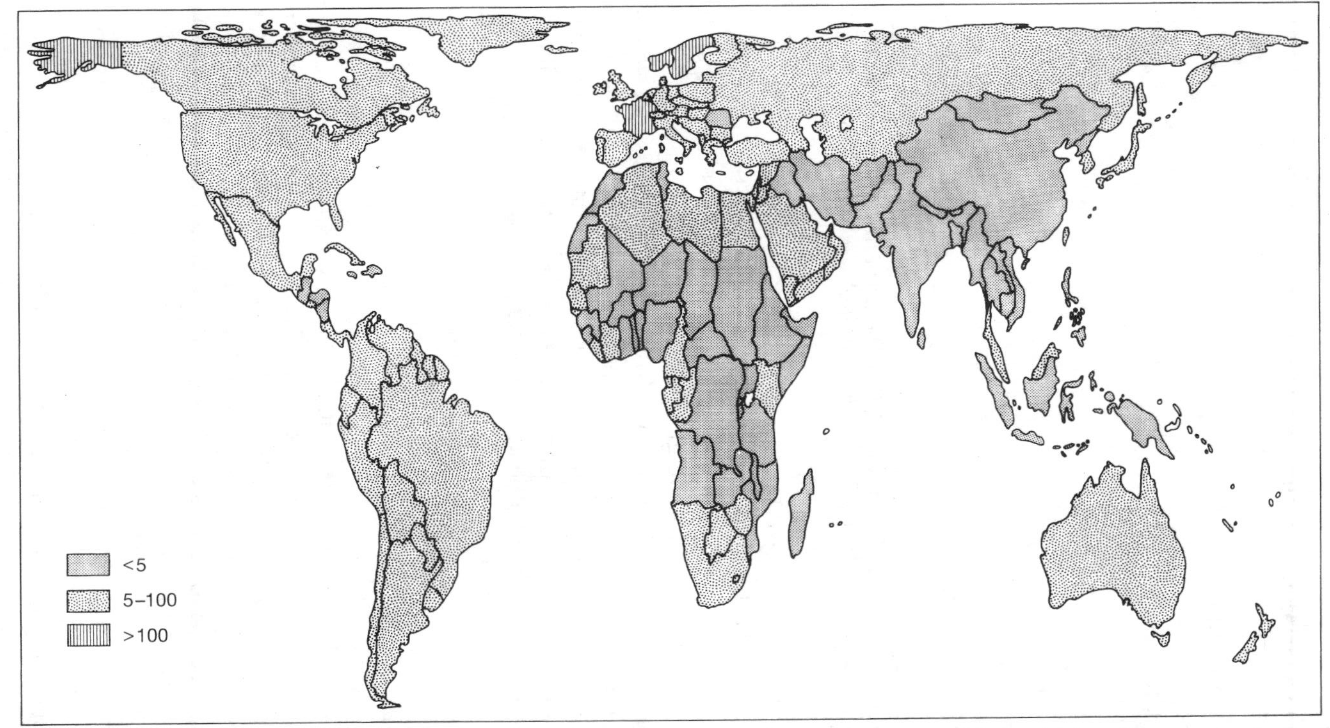

Figure 4.15 An attempt for the 1980s to measure information-richness on a world basis by the possession of data-transmission devices such as computers and fax machines. There are no real surprises. If access to a TV set were added, then there is some levelling-up, though in that case the data transmission is only one-way. For the majority of that period, 'Soviet Union' is the correct appellation (from Czinkota *et al.*, 1992)

Table 4.2 Cumulative impact of human-induced change

	per cent change		
Form of transformation	**1860**	**1950**	**1985**
Deforested area	50	90	
Terrestrial vertebrate species loss	20–50	75–100	
Water withdrawals	15	40	
Population size	30	50	
Carbon releases	30	65	
Sulphur releases	5	40	
Phosphorus releases	<1	20	
Nitrogen releases	<1	5	
Lead releases	5	50	
Carbon tetrachloride production	0	25	all 100 per cent

Source: modified from R. L. Kates *et al.* (1990) The great transformation. In B. L. Turner *et al.* (eds) *The Earth as Transformed by Human Action*, CUP, Table 1.3

had come about. So the imposition of the human upon the natural can be said to be largely very recent (i.e. the last 50 years are the most important) and, presumably, still growing.

These data have limitations. From them it is difficult to tell when, where, and how much new technology will produce what level of ecological change. We need to note that any changes take place within the framework of an already human-altered ecology in most instances. Further, we know relatively little about the resilience of either natural or human-transformed systems.

Further reading

The present author's works on human impact through time can be consulted: in short form as Simmons (1993) and in somewhat longer form as Simmons (1996). For a 'horizontal' and systematic approach to environmental impact, see Goudie (1993) and for an in-depth (but selectively broad) set of essays on the last 300 years, Turner *et al.* (1990). A summary account of land-use and land-cover change is very usefully seen in Meyer and Turner (1992). More specialized material on agriculture is found in Briggs and Courtney (1985) and Mannion (1995); companion volumes of similar depth but concerning other types of land use do not seem to exist. An excellent combination of culture and environment is found in David Harris's (1990) discussion of the origins of agriculture. Ideas about the changes that have occurred are frequently found in an implicit form that requires interpretation, but Worster (1988) contains an interesting although personal selection. The historian Simon Schama's (1995) book on landscape and history has much interesting material and good illustrations. (Though this seems a short selection, many of the works are quite long.)

Briggs D, Courtney F 1985 *Agriculture and Environment.* Longman, London

Goudie A S 1993 *The Human Impact on the Natural Environment*, 4th edition. Blackwell, Oxford

Harris D R 1990 *Settling Down and Breaking Ground: Rethinking the Neolithic Revolution.* Netherlands Museum of Anthropology and Prehistory, Amsterdam

Mannion A M 1995 *Agriculture and Environmental Change. Temporal and Spatial Change.* Wiley, Chichester

Meyer W B, Turner B L 1992 Human population growth and land-use/cover change. *Annual Review of Ecology and Systematics* **23**: 39–61

Schama S 1995 *Landscape and Memory.* Harper Collins, London

Simmons I G 1993 *Environmental History. A Concise Introduction.* Blackwell, Oxford

Simmons I G 1996 *Changing the Face of the Earth*, 2nd edition. Blackwell, Oxford

Worster D (ed.) 1988 *The Ends of the Earth. Perspectives on Modern Environmental History.* Cambridge University Press

Links with other Chapters

Since this point is something of a half-way mark in this book, it may be useful to review very briefly the linkages of the historical setting of most of the material so far. We shall look back at some of the connections with the matter of Chapters 1–3 and anticipate a few of the ties between the present chapter and the rest of the book other than the last chapter.

Links with earlier material

- The notion of environmental problems: in all eras some of the changes will have produced some apparent losses of resource or environmental quality, e.g. the silting up and salination of Mesopotamia in antiquity, as the surrounding hills were used by pastoralists; and the gross urban pollution by smoke-emitting industry. After hunting and gathering, no period has been free from the effects of externalities.
- Even natural phenomena have different rates of change. Climate, for example, may show very rapid change, as at the end of the Pleistocene; gradual secular change as in the cycle of an interglacial, or punctuated change as when affected by a massive volcanic eruption. In post-glacial times, there has never been an equilibrium climate to which all other natural phenomena could in some final sense adjust.
- Predictability is uncertain even where natural phenomena are concerned and even poorer where human societies are involved: at any stage of the events of Chapter 4, how much of the next stage could have been predicted, even with today's means of information acquisition and transfer?
- Human-induced environmental change can reduce the availability of resources, as when marine populations are permanently reduced or soils eroded at an accelerated rate. Some changes, however, can

produce resources: for example, the valued environment of the archetypal English countryside is a product of forest clearance followed by the enclosure of common fields. The highly attractive oak-grass meadows of Yosemite Valley in California are the result of burning by the aboriginal population in order to encourage oaks, valued for their acorn crop.
- The use of energy resources can be translated into amounts of environmental change. The data in Table 4.1 show the accelerating use of energy on a world scale through the 19th and 20th centuries, and this is proxy for (a) big increases in the human population and (b) considerable impacts upon the environment in both the industrial cores of the world and their peripheries.

Links with later material

If we look forward to the rest of this work, then a number of the issues raised in Chapter 4 will be subjected to some more discussion.

- Nearly every human society has given itself permission to change its natural surroundings. Rarely have societies wanted to produce total metamorphosis but few enough have worried about making no discernible impact.
- Can a developed science of human–environmental relations also construct a prescription for human behaviour in that field, i.e. an ethic? Is it feasible to translate the recipes for stability and resilience in largely natural ecosystems into ones for sustainability in human-directed systems?
- How then are resource use and environmental change to be managed for long-term viability: who is to control access to resources, for example, and who is to decide what intensity of environmental change is permissible? What kinds of human institutions will work? Is it in the end a matter of human population levels as much as all the other factors?

All such ideas and questions point inevitably to the complexity of the human–environmental nexus, where it might be possible to deal with one disharmony at a time but where in fact they are all simultaneous. This question is the subject of the last chapter.

CHAPTER 5

Cultural constructions

In this chapter the emphasis is upon the human mind. Especially it is upon the way in which the manipulation of symbols by that wonderful entity has addressed itself to the consideration of what we call 'environment'. The main categories were foreshadowed in Chapter 1 when we mentioned the particular contributions of the natural and social sciences. Here, we shall examine some of these in more depth, noting especially those that are changing rapidly to address new issues. Hence, a major new topic will be a consideration of the topic of environmentally oriented philosophy. This is followed by the way in which such thinking about our place in the world finds its way into ethics, which is concerned with notion of agreed standards of conduct. These particularly cultural constructions are composed of ideas; in the world of action, they come most obviously in the form of laws and are dealt with in Chapter 6.

The natural sciences and technology

The natural sciences have a reputation for producing knowledge of an altogether different quality from any other methodology; indeed it has been elevated by some of its practitioners to the heights of being the only knowledge worth having. 'Science' is now taken to mean 'Western science'. Nevertheless, until the Late Renaissance, that of China was as advanced as that of the West in astronomy, algebra and quantitative cartography, for example. It was allied to a technology that could produce the development of water-power, the ability to drill deep holes for brine and natural gas, the foot-stirrup and the wheelbarrow. Thereafter, however, the predictive capacity of the Western developments led to the explosion of 'reliable knowledge' which is now crucial to the survival of industrial societies. The reliability crucially depends upon the ability of science to produce accurate predictions.

Foundations of science and technology

To repeat an earlier assertion, we Westerners live in a world in which science and technology are so assimilated that we cannot, quite literally, bear the thought of what our lives would be like without them. In any such context we ought always to look critically at anything which deploys such power over us.

The basis of modern science

Part of the present power of pure science has come from its representation of **consensual** knowledge: statements are tested and agreed independently of any personal characteristics of the observer. In its most ambitious form this requires a totally unambiguous language, the provision of which is one aim of mathematics. In essence, the scientist as an observer and as a language user can capture external events in statements or messages that are true if they correspond to the facts and false if they do not. Any true statements or propositions thus have a 1:1 relation to the facts even if the facts are not directly observed. This latter condition may result from hidden events or properties, or even events that are distant in space or time. Such events are described in *theories* inferred from observation: it is assumed that hidden explanatory mechanisms can be discovered from what is open to observation.

Science is above all **empirical** in the sense that it starts by making observations which are based on the apprehension of the senses, an experience which may be much enlarged by technological means. Clearly, the validity of any empirically based truth must depend upon the quality of the observations and there is a lively debate about whether really objective observations can be made or whether they are all preceded by theory (Chalmers, 1982). In everyday terms, though, science proceeds as if it was arriving at the truth, making interim statements along the way, even though there is no necessary convergence of the conceptual framework of theories towards a universal truth. But any scheme of knowledge that makes strong claims for itself is bound to come under intense scrutiny: the same has happened to theology. In the case of science this is doubly so because to pursue it generally costs money for which there are competing uses.

Those who claim to have found flaws in science have generally been from two groups: (a) those who assert on grounds of logic and philosophy that not all the claims described above can be upheld; and (b) those who regard science as one cultural activity among others and that when it is placed in that framework, it exhibits certain qualities that societies ought to watch very closely since dangers might ensue from the continuation of such activities.

Technology

Technology is not easily differentiated from science. It has often come first in the sense that machines have been made to work without the inventor knowing much about the precise value of the gravitational constant or the laws of thermodynamics. In all cases, technology will work only if conditions A, B and C are satisfied. Therefore it is only attempted if conditions A, B and C are present. If they are not, then it may become desirable to change X, Y and Z into A, B and C. In this view of technology (with many implications for environmental change), technology is a neutral tool to be used for culturally determined human ends. These may be good ends, as in boring wells in dry lands, or bad ends, as in slaughtering migratory birds when they cross Italy.

This view of technology as a kind of impartial instrument is not universally held. Some think that there is very little control exercised: the test for society is, simply, if you *can* make it then you *will*, as with Concorde and nuclear weapons. These examples lead to the opinion that some technologies are inherently political: some lead to centralized authority, others to decentralized decision-making. In 1872, for instance, Engels repeated the example put forward by Plato that a ship at sea must have a single captain and an obedient crew. This leads to the belief that technology begins to shape human affairs, such as where the understanding of technology has merged with self-understanding, as in our use of terms such as 'interface', 'network', 'output' and 'feedback' and especially in the common comparison of the human mind with a computer (Smith and Marx, 1994). It steers us towards the possibility that technology can become a master rather than a servant.

Drawbacks or no, the natural sciences are the dominant way in which humankind frames the environment for itself. Indeed, the view can easily be found that the sciences are providing the 'facts' and that society's role is simply to adapt to them.

Ecology and climatology

If science has indeed regarded the evidence of the human senses as fundamental, then it is not surprising that the nature of living things and of the weather have been leading elements in scientific investigations of environment, though naturally they are not the only ones. They both recognize that explanations derived from the study of static patterns (e.g. plant distribution or continental rainfall maps) have severe limitations and that closer investigation of what happens between these time- or space-slices is essential, i.e. they are both concerned with the dynamics of the systems which they study.

Ecology as a science

A commonly accepted definition of ecology is the scientific study of the relations between plants and animals and their non-living environment. This definition is scale-free as far as space is concerned. Ecology can investigate the heat exchange properties of a small beetle living under the bark of a dead tree just as well as evaluating the role of living organisms in the total flow of carbon around the planet. There are invariably problems of complexity since a scientist's normal reaction is to look for explanations at the next lower level of organization (which may in this case be a smaller area (Fig. 5.1) or a smaller set of organisms), in a classic reductionist mode. Yet it is not at all certain that events at one scale translate simply into those at a higher or indeed lower level of organization and complexity. So knowing at what scale to observe an ecological system can present difficulties. Time is not neatly defined either, though it is customary to separate ecological time from evolutionary time, with the last 10 000 years being given over to the ecologists.

Given the complexities inherent in the rather ragged edges of time and space, it is scarcely odd that ecology should not have acquired a good reputation for making accurate predictions (Peters, 1991). J. Maynard Smith, an evolutionary biologist, has been quoted as saying that ecology is a branch of science in which it is usually better to rely on the judgement of an experienced practitioner than on the predictions of a theorist. Empirical work on successional change in forests in the USA suggests that there is no determinable direction of change and that it goes on without reaching a point of stability. The work indicated none of the expected features predicted by some ecosystem theories, such as an increase in biomass or diversification of species. A forest is a shifting mosaic of trees and other species without an emergent, holistic collectivity except in the physiognomic sense. The result is acceptance of the idea that disturbance and readjustment is continual and that outcomes of change are very poorly predictable. In this respect, these ecological views look forward to chaos theory.

The ecology of humankind

In this section we shall deal with ecological interpretations of the position of mankind in ecological systems, the notions of ecosystem stress and stability, and ecologists' approaches to ecological problems and solutions. In biological terms, *Homo sapiens* can be seen as an outside influence, rather like a sudden change of climate or a natural disaster. By contrast, humans can be characterized as a component of the system, indeed often as the dominant organism whose influence determines the nature of the

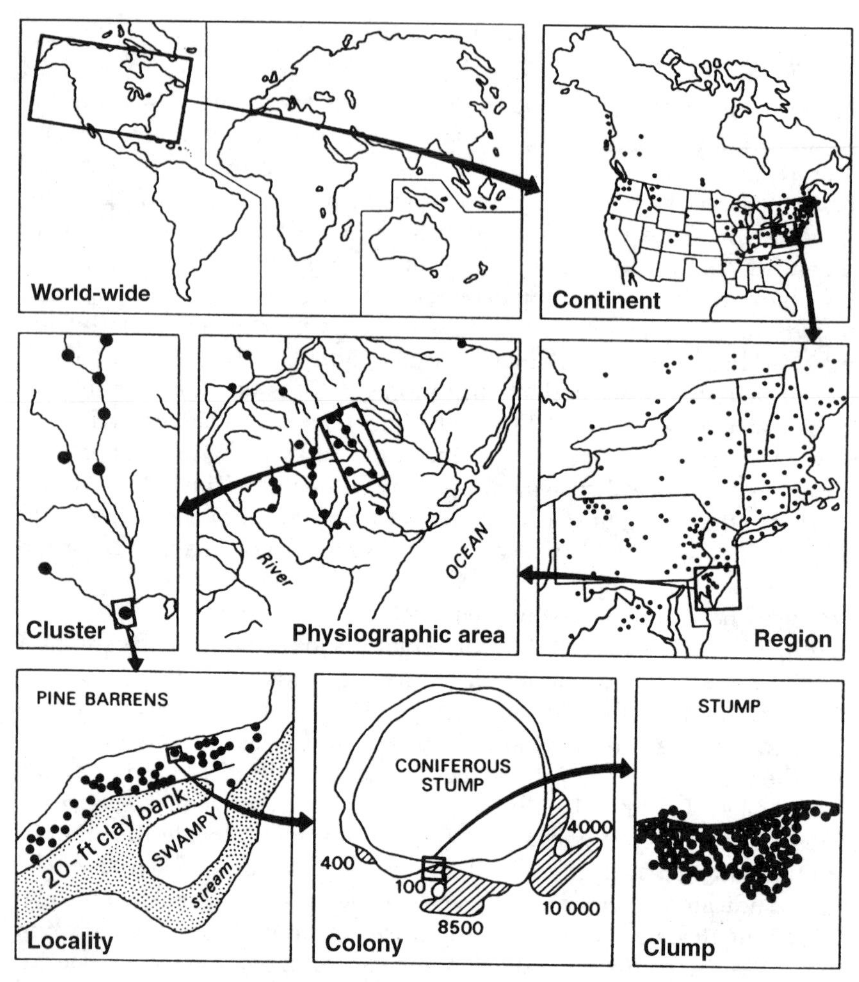

Figure 5.1 The scales of investigation at which ecology might be involved: some ecologists begin to get interested at the 'physiographic area' scale and pursue their work all the way down to that of the clump, though most would concentrate their explanatory efforts at the 'locality' and 'colony' levels. From 'cluster' upwards, the biogeographers would be the most interested group (from Simmons, 1982)

rest of the system, as do the trees in a forest. If a more socio-economic perspective is required, then features such as information transmission can be added to energy and matter flows as elements of the system's functions (Bennett, 1976). Finally, a whole human culture can be worked into the other linkages (e.g. as culturally determined choice-points) to try to encapsulate the variety of aims which humans may have in bringing about alterations in their surroundings.

In the 1960s and 1970s, the multivariate nature of processes such as the scarcity or imagined future scarcity of natural resources and the perceived degradation of environment caused by contamination with various toxins led to a great surge of interest in ecology. However, many professional ecologists felt that since ecology dealt to a great extent with uncertainty then they had to make the probabilistic nature of their predictions a feature of any involvement with public life. They pointed out that in any ecosystem there was natural variability, which could only lead to imprecise predictions. In spite of sometimes being wrong about outcomes, some ecologists became involved in policy-making. They saw that much information already known was not finding its way to legislators and that quite possibly the perceived integrity of scientific data could not be preserved during its passage from the field through the computer to the specialist report and thence to the committee hearings, the TV appearances, and the making of law and its day-to-day enforcement. Legislators and administrators feel that scientific data must be evaluated, authenticated, organized, analyzed and interpreted, which gives ample scope for the distortion rather than the fidelity of scientific information.

Climatology

In 1960, there were probably less than 50 climate modellers in the world. Now there are several thousand of them and they have a dominant voice in the constructions of environment being made by scientists who are called upon to comment upon the likelihood of human-induced global warming. Their remit is wide: the Global Atmospheric Research Programme deals not only with the atmosphere but also with the dynamics of the hydrosphere, the cryosphere, the land surface and biomass, past and present. The aim at present is to develop and refine models of the atmosphere of different types and to end up with powerful and predictive global climatic models (GCMs).

There are problems in dealing with the very large data sets that can be acquired (especially from satellites) and so there are problems of archiving information and gaining fast access to it. Even with all that is collected, GCMs are still plotted on a coarse spatial basis. And so far as prediction is concerned, the forecasts are never any better than the assumptions that are made about the relationships between past, present and future (Kemp, 1994). Even short-range weather forecasting can have quite severe limitations (Fig. 5.2). So both ecology and climatology deal, in their modern forms, with dynamic systems of great complexity. Climatology has been much more planet-wide in its initial scope,

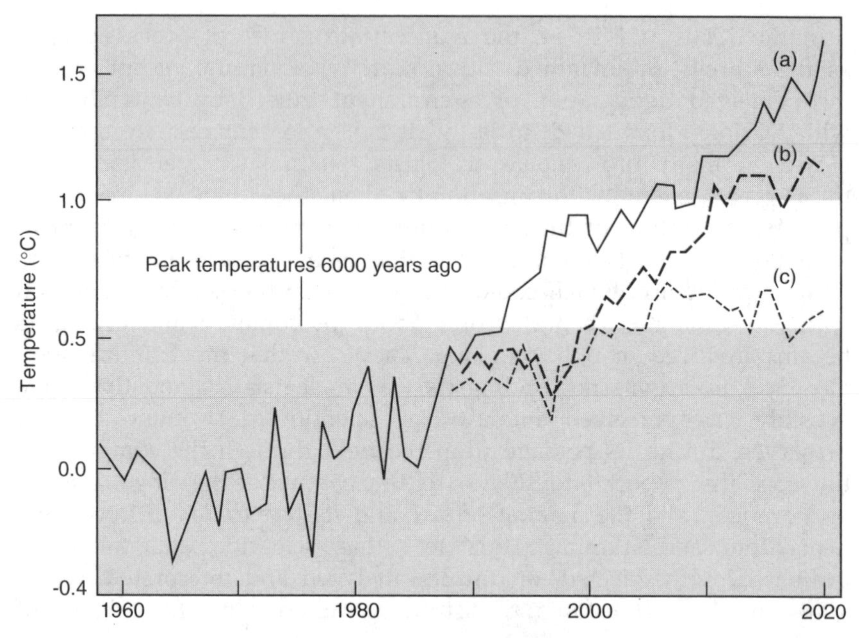

Figure 5.2 Predictions of the net global warming based on assumptions made in the 1980s. Two sets of assumptions are made: (i) that trace gases are the primary agents responsible for global warming, and (ii) that future actions will produce more or less proportional ('linear') results. (a) assumes growth of trace gases at 1.5 per cent per annum; (b) assumes growth rate slows so that the forcing effect remains constant; (c) assumes that drastic cuts in emissions cause the forcing effect to cease to grow by AD 2000.

although the regional effects of possible climatic change and sea-level rises are also studied. In common, also, are the suggestions for technical fixes: the suggestions of sowing the seas with iron filings to encourage phytoplankton to raise their NPP levels and thus sequester more carbon dioxide, or the planting of fast-growing and long-lived trees everywhere to achieve the same, are such measures.

Evolution and entropy

Two concepts give particular chronological depth to scientific constructions of the environment. The first is largely biological, namely that of evolutionary theory. The idea of organic evolution (associated with Charles Darwin (1809–1982) because of the depth of his evidence and cogent argument) asserts that species have come and gone during the millions of years of life on the Earth. It further suggests that life-forms have taken on increasing complexity during those epochs and that humans are a product of the same evolutionary processes which produced for example the other primates and even the AIDS virus. The variety that is produced by evolution is relentlessly culled by extinction and so the survival of a species or other group includes a large element of chance. Thus there can be no purpose in evolution.

The idea of entropy is not so easy to comprehend. It was firstly a measure of thermodynamics and related to the Second Law, in which it becomes a measure of the disorder among the atoms whose state constitutes the energetics of the system. An initially ordered state is bound to become random as time proceeds: high-quality energy able to do work must ineluctably become low-grade heat unable to do so. In a broader frame, the concept has been applied to the Earth as an open system which takes in ordered energy from the Sun but converts it to low-grade heat which is radiated back to space. In between, the energy has powered life and many other complex processes but in the long run, disorder must prevail: the universe will become randomized. We have to ask whether the concepts of evolution and entropy in any way form environmental constructions by themselves or whether they add to others; for example, do they amplify the ecological construction of environment or do they reach beyond it?

The absolute kernel of Darwin's contributions is the theory, not yet falsified by scientific evidence, that species die out and are replaced by others. But S. J. Gould (1989) puts us in our place (Fig. 5.3) by saying, 'Life is a copiously branching bush, continually pruned by the grim reaper of extinction, not a ladder of perpetual progress.'

The process of replacement also at one stage produced *Homo* just as it had other taxonomic groups. Out of an examination of the processes which result in these changes, a mighty set of constructions has grown, all the way from the purely biological to the intensely moral. With humans, of course, the cultural transmission of acquired information becomes ever more important and there is now some tension between (a) those who assert the primacy of the cultural, and (b) those who follow very closely in Darwin's footsteps by asserting the predominance of the very stuff of evolution in the shape of genetics. For some biologists, organic evolution itself will carry on in humans via genetic changes that 'improve the nature of man himself'; for others this is made self-evident in terms of extrapolation from the diminishing intervals between earlier stages of evolution. In the case of the field of socio-biology, the genes are said to hold culture on a leash, and so a genetically controlled evolutionary function is seen as a competitor with psychological processes such as motivation and indeed a loser to it: there is no free will here.

We can perhaps see evolution in a broader frame along with Eric Jantsch when he talks of evolution and the future for humankind in two

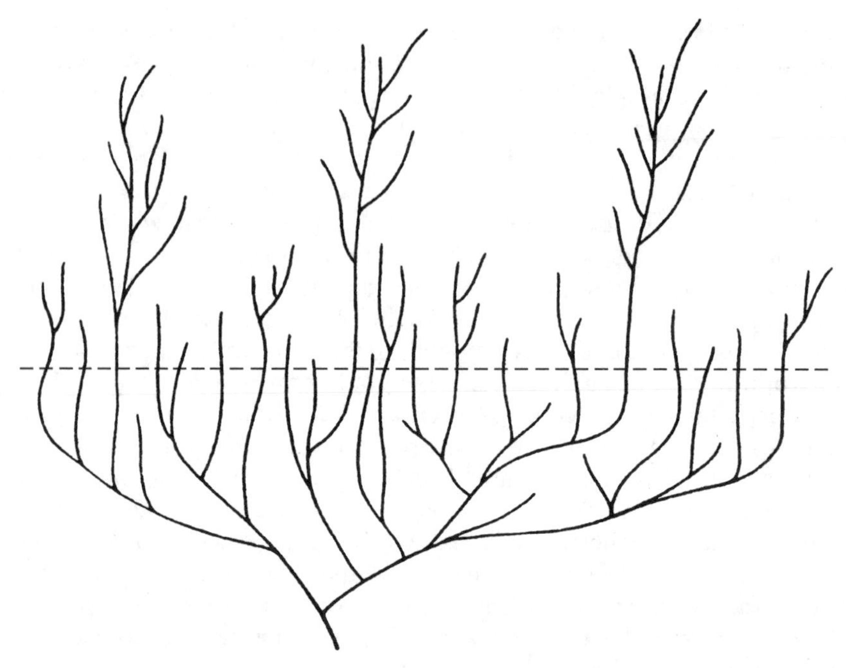

Figure 5.3 S. J. Gould's depiction of organic evolution as a branching bush. There often seems to be a surge of species formation early in the evolution of a particular group of plants or animals but this is pruned by extinction, leaving large morphological gaps among the survivors. This model is presented as a counterweight to those showing humans at the top of an escalator. It especially brings out the role of chance in the survival of any species (from Gould, 1989)

ways. The first is evolution by extension of the environment into space, the discovery of many New Worlds. The second is the opening up of new niches for humankind by the extension of our consciousness; this can naturally include information technology though not as part of a machine-dominated drive to conquest but as an extension of personal experience. Learning thus becomes a creative game played with reality, a co-operation between autonomous wholenesses, and is infected with the idea that evolution is never purely functional: there is always some extravagance. Do we want to live, asks Jantsch (1980), only in a predictable environment that can be rationally controlled?

Entropy and the validation of time

Entropy is of most interest in so far as it is a measure of energy in the universe and hence on Earth. The energy comes in two states, namely available or 'free' energy and unavailable or 'bound' energy. These are anthropocentric terms, meaning available or not available to humankind, and entropy is a measure of the conversion of one to the other. In coal there is free energy and low entropy which is converted by combustion to bound energy in the form of heat, smoke and ash, and high entropy is created. Also, the presence of free energy implies the presence of some kind of ordered structure whereas bound energy (and high entropy) indicates the dissipation of the energy into disorder. On Earth, there is free energy in the form of the rather dilute sunlight. It also comes in the more concentrated form of fossil fuels but this non-renewable stock in fact constitutes no more than a few day's supply of free energy as measured by the input of solar radiation.

This teaches us a few long-term lessons with implications for environmental matters. For all isolated systems the future is of increasing entropy and this confers a uniqueness on each instant of time since entropy is continually increasing and cannot be reversed. Closer to everyday concerns, economic processes must be seen as entropy-creating since they eventually degrade energy into low-grade heat (Fig. 5.4), and matter into junk and refuse. A growing economy is therefore increasing the rate of formation of entropy even though it creates complexity for a while at least. Hence, no technology can at present produce its own fuel but must rely on some form of stock; even solar collectors take a long time to garner enough energy to reproduce themselves (Georgescu-Roegen, 1971).

The Gaia hypothesis

The term 'Gaia' can be used as shorthand for the Gaia hypothesis, which is a holistic model of the outcome of global evolution and ecology considered together and associated with the name of J. E. Lovelock. It stands conventional ideas about the evolution of life on their head since its core is the statement that the Earth is a self-regulating entity with the capacity to keep the planet populated with living organisms which control their chemical and physical environment. However, Gaia should not be thought of as some form of sentient being; the term 'she' can be used as it would be employed of a ship, along with the connotations about such a vessel being rather more than the sum of its parts.

The basic concept

The Gaia hypothesis suggests that life itself controls the systems which seek to optimize the physical and chemical conditions for its existence, though not necessarily for human life. Life produces a non-

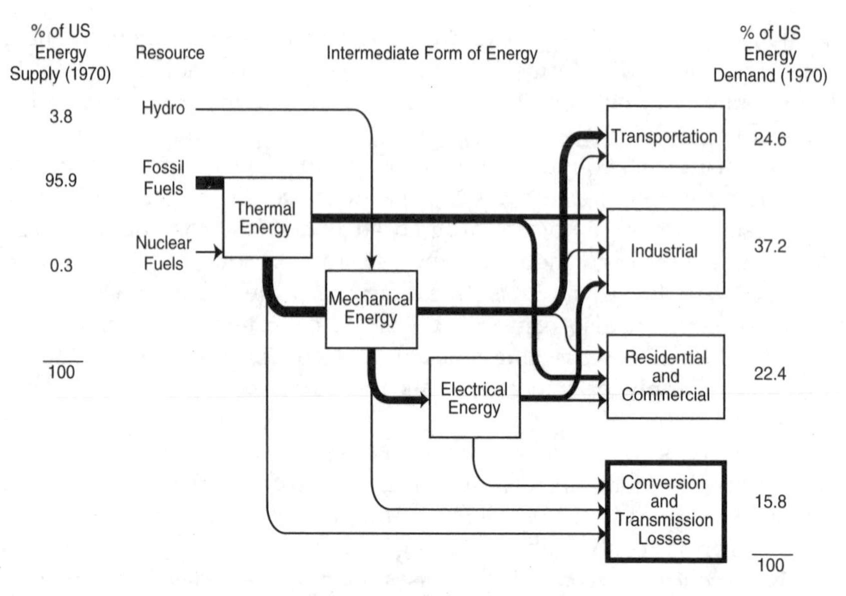

Figure 5.4 Energy conversion always entails loss of potential to do work. Photosynthesis, for example, is usually less then 1 per cent efficient. In human-directed systems, there may be several conversions between a resource and the end-use. This diagram is for the USA in 1970, when efficiency was not a primary concern and before nuclear power had attained its present significance. In the generation of electricity, for example, the conversion at the power plant is at best about 40 per cent and thereafter there may be transmission-line losses. So the key box in many ways is the bottom one, showing the proportion that is lost in conversion and transmission: nearly 16 per cent in this instance (from Summers, 1970)

equilibrium set of states of, for example, the composition of the atmosphere and the salinity of the seas, both of which are different from the predicted levels of a lifeless planet. It thus regulates the amount of carbon in the atmosphere by sequestering excess amounts in plants and at the bottom of the oceans and also regulates salinity by precipitating minerals in tropical shallow-water areas and to the deep ocean floors. The predicted connection between life and the atmosphere was thought to have been fulfilled by the discovery that phytoplankton emit dimethyl sulphide (DMS) in aerosol form which then forms the cloud condensation nuclei over the oceans. Thus precipitation amounts and locations would be governed by the activities of living organisms, but the story is probably more complicated than that (Fig. 5.5).

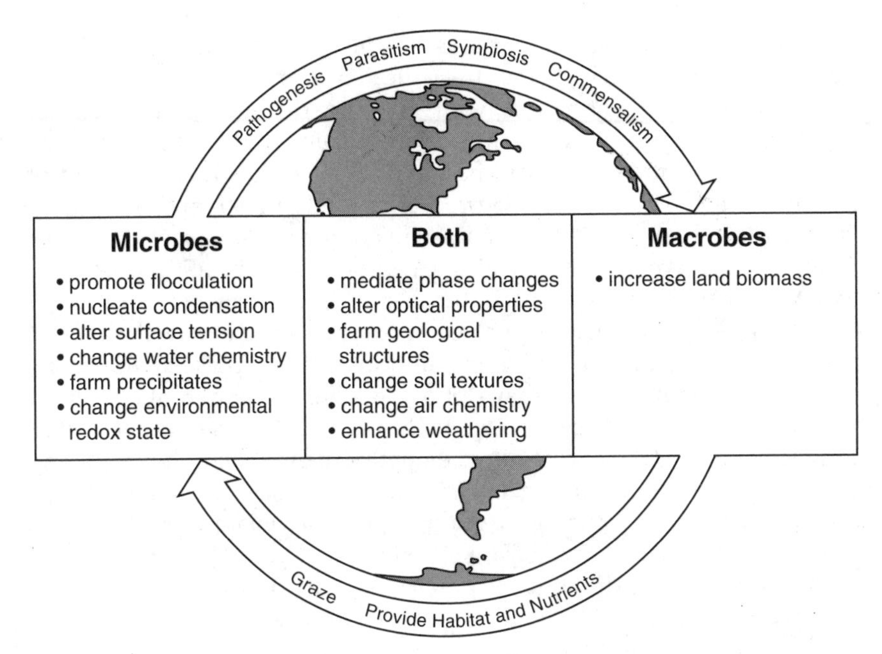

Figure 5.5 The linkages between life-forms is critical to the holistic nature of the Gaia hypothesis. It is possible that life-forms exert indirect biological influence that is out of proportion to their size, rather as hormones do in multi-cellular organisms. They may also operate on critical points such as the surface–atmosphere transfer conditions (from Boston *et al.*, 1991)

Gaia and environmental impact

During the 20 or so years of its existence, many commentators have looked for environmental lessons in the Gaia hypothesis. It seems as if the most sensitive parts of the feedback loops are the microflora of the oceans and the soils, which together with some forests are responsible for the turnover of about half the world's carbon. It is therefore human impact upon these systems which is likely to have the strongest effect upon the function of the Gaian system: the poisoning of the continental shelves of the tropics and the felling of tropical moist forests seem to be the obvious examples of such processes. It is argued that human impact on Gaian systems could, depending on its intensity and its location, throw the whole set of webs into a new phase of non-equilibrium, i.e. Gaia could 'flip' from one state to another rather than undergo any form of gradual transition. Any adaptation producing the perpetuation of life on Earth does not automatically include human life and indeed if humans violate the basic mechanisms then in the long run they are likely to be swept aside.

The status of the Gaia hypothesis is still open for discussion. If we think of a 'strong Gaia' theme, then the Earth is some kind of super-organism: an emergent whole with qualities that transcend those of individual components. A 'weak Gaia' postulate would restrict the idea to a series of complex feedback mechanisms on a global scale, a kind of geophysiology. The 'strong' view has the additional and intriguing characteristic that it is not strictly Darwinian: the behaviour of the whole is more purposive than Darwinian selection normally allows. That is, evolution is goal-directed: the optimal end-state might be more like a process, such as the maintenance of an overall structure in the face of change or an increase of order and information. At present the lowest common denominator is perhaps of a powerful metaphor, and as such by no means to be despised, for metaphor is a central descriptive device in most sciences. The idea of an emergent domain which links all living things and their surrounding environment, in a self-regulating system with an open but non-purposive nature, does not violate scientific norms of validity (Schneider and Boston, 1991). The attribution of 'purpose' takes us out of the realm of science as it is usually defined.

Complexity and chaos

The discussion of ecology has suggested that it has not managed to achieve anything like the success of physics in making temporal predictions. This is largely because it deals with indeterminate systems, i.e. those where a number of outcomes are possible no matter how similar a set of starting points. The intervention of chance produces unpredictability. The mathematics of chance are founded on probability theory, and strenuous efforts have been made in both science and business to apply this in their dealings with factors of the environment.

Chaos

In the world of science that was in place by the 18th century in the West, the prevailing model was that of clockwork, and so was basically immutable. After the discovery this century of the uncertainties of the quantum world, the universe takes on more of the aspect of a cosmic lottery. Stewart (1989) describes whole classes of stochastic behaviour in deterministic systems. In these, simple equations can generate motion so complex and so sensitive that it appears random. Simple systems, it appears, do not necessarily possess simple dynamic properties. The weather is one such example: if short-term predictions are strung together to form a long-term forecast then tiny errors build up, accelerating, until total nonsense is the outcome. This leads to the so-called 'butterfly effect', in which a butterfly beating its wings in Tokyo affects the weather in New York. In effect, the variety of possible states of the atmosphere is so great that a return to a state of 100 per cent of the initial conditions need never occur.

Chaos also occurs in ecological systems. The simplest models of population growth can engender periodicity and apparent chaos: some of this may be due to external influence but parts of it seem to be internally generated. Ecology thus may contain a different dimension from the equilibrium concepts that have underlain so much of its thinking. If chaos is involved, therefore, ecology is a study of disequilibrium. If there were a 'butterfly effect' in ecology, then how can any student of ecology possibly determine what is worth measuring? This throws into doubt any claims that ecology might have had as an over-arching narrative for the whole of human–environment relations. Like post-modernism in social and literary theory, such foundations are being questioned.

Constructions by ecology and the sciences

Here we shall review some of the generalizations that can be made about the ways in which ecological approaches (i.e. those emphasizing dynamic interactions between the living and the non-living) and then other sciences assemble constructions of the environment, and the human place within it.

Ecological approaches

Treated entirely within a scientific framework, a number of general statements can be made as a result of work in the last 20 or so years. The first group of statements is about humanity itself:

- We have the highest biomass of any one species: about 100×10^6 tonnes dry weight or 6×10^{14} kcal of embedded energy. Because of our longevity, the biomass turnover rate is low.
- Our population growth rate (just under 2 per cent per annum) is high by biological standards for an established population.
- The degree of structural organization is the highest in the animal kingdom: the exchange between groups of energy, matter, individuals, knowledge and traditions is very marked.
- Of the energy that we use, 90–95 per cent comes from the surplus of earlier ecosystems (i.e. the fossil fuels) and not from active ecosystems of the present day. In terms of the appropriation of the energy of nature, estimates suggest that 40 per cent of net primary productivity on the globe is used, diverted or reduced by human activity.

We can look at the major effects of humans upon ecosystems as producing the following:

- acute but transient perturbations from which recovery is more or less complete: a forest clearance for shifting agriculture then allowed to regenerate would be an example.
- chronic changes such as permanent land-use changes or the continuing extinction of species.
- different energy and nutrient relationships in ecosystems, e.g. by the use of fossil energy for 'subsidies', and by the opening up of nutrient flows, as by long-distance transport of food crops and by accelerated soil erosion.
- new genetic materials by plant and animal breeding, with even greater potential being conferred by the techniques of genetic engineering.

The scientific overview

Putting these together, we can make a number of general statements which result from the natural sciences about the relations of humans and nature, which in effect add up to a construction of the environment:

- Few human effects upon ecosystems are typically or exclusively human except by quantity and combination, i.e. the sheer amount of materials changed, their variety, and their combination in space and time. However, some of the syntheses of organic chemicals, the isolation of plutonium and the splicing of genes might make it defensible to erect a 'humans-only' category.
- Humans can therefore be considered as a normal though highly manipulative (partly through being comparatively unspecialized biologically) member of the components of an ecosystem.
- Virtually all the ecosystems of the planet are now open to widespread exchanges of energy and matter.
- The biosphere is now heterotrophic in the sense that many trophic levels receive influxes of energy capital. Some 8–10 per cent of the second and third trophic levels' energy is from fossil sources.
- There is increasing compartmentalization of the biosphere into ecosystems of different character and function, as into protected wilderness and park areas, agricultural areas, urban–industrial areas and analogous examples. The simpler systems, and the inert areas, are occupying more space at the expense of mature or 'climax' ecosystems.
- Many natural ecosystems have a high diversity of species and human activity matches this with a considerable versatility and complexity of organization which results, *inter alia*, in increasing quantities of energy going through just one species.
- The state of ecosystems throughout the world is now dependent on both the structural complexity and the integrity of human societies.

Added together, the conclusion is that there appear to be apparently irreconcilable differences between the long-term stability and maintenance of some ecosystems and the short-term requirements of some human economies.

> The last two of these statements are terse formulations of a wide variety of what are generally called *environmental problems* and as such have prompted many ecologists to put forward not merely pragmatic and *ad hoc* solutions, and but also to suggest that ecology as a subject leads inevitably to certain values which can form the basis of prescriptions for human behaviour. These will be discussed later in this chapter. They are examples of a linear approach to knowledge in which it appears as if scientific data are foundational and that human societies will then base their behaviour upon such findings. An intensely rational approach like that has been mooted in a number of societies since the 19th century but, in general, humans have introduced many other variables and vagaries. Some of the more methodical of these are the social sciences, most of which are also the product of the 19th century.

The social sciences

Given the authority of the natural sciences, it is no surprise that the interactions of humanity with the environment are regarded by some social scientists as suitable for the kind of distanced study that is carried out by biologists or physicists. The description and analysis of a society and its environment, for example, can be treated much as if it were that of a population of grassland prairie-dogs. So far as is possible, the language of description is mathematical and the search is for regularities of a law-like nature and for the building of theory. There has grown up a conventional set of procedures and categories of knowledge in most of the social sciences, usually offset by one or more 'radical' alternatives. Conventional views often find in favour of the present situation; the radicals are more likely to advocate revolution, either intellectually or politically or both.

Economics

The heart of the discipline of economics is the way in which it treats the behaviour of humans who find that their possession of means is insufficient to meet their desired ends. The emphasis on means is important, for economics is supposed to be indifferent about ends: the latter are the subject of the psychologist, the historian and the theologian, for example. The removal of intentions, however, does emphasize all the data-gathering and processing techniques of the sciences and their common language of mathematics. Economics is currently the major mediator between human societies and their use of their biophysical surroundings. The subfield of environmental economics in particular sees itself as a way of attributing measurable value (usually as money) to features of environmental importance. Examples might be the direct contribution of natural resources to economic growth, the role of the environment in the quality of life (as in the provision of beautiful places, wildlife or cultural heritage), and the negative connotations of a 'poor' environment, which may add to stress and bad health in the human population.

Conventional economics and the environment

By 'conventional', we mean here the Western capitalist economics that are found in the free-market economies of the world. The theoretical basis of this view of the world comes from a series of concepts arrayed round the setting of price as the central mediator between supply and demand. The core idea is of consumers and producers who both wish to maximize their satisfaction from a transaction. The first wants some form of contentment from the purchase; the second wants to make a profit (Fig. 5.6). It is assumed that both are in possession of perfect information about the state of the market and of sources of alternatives should they exist, and that the supplier does not have a monopoly of the product. There has been, therefore, a tendency to a theory which does not always fit one-to-one with the world of experience.

Environment has sometimes been seen as a set of inconvenient variables for some theoretical models in this type of **neo-classical economics**. There is first of all the question of externalities, which particularly affect our treatment of wastes. A price can be fixed for coal, for example, which reflects the costs of delivering it to the power station. Acid rain from its combustion then falls 1000 km away in another country and produces a cost there in terms of loss of fish and forests. In a neo-classical economy, the latter are of no interest to the supplier and consumer of the coal. In fact, these external costs are subject to government intervention, since there is political pressure from the sufferers of acid rain. The government then either subsidizes the system by paying for desulphurization of the stack gases out of taxes, or allows the generating plant to charge more for its product so as to be able to invest in pollution-control technology. The latter process is called internalizing the costs. Secondly, many aspects of the environment are not under private ownership and therefore are common to all who can achieve access to them. Under such circumstances it may be gainful for

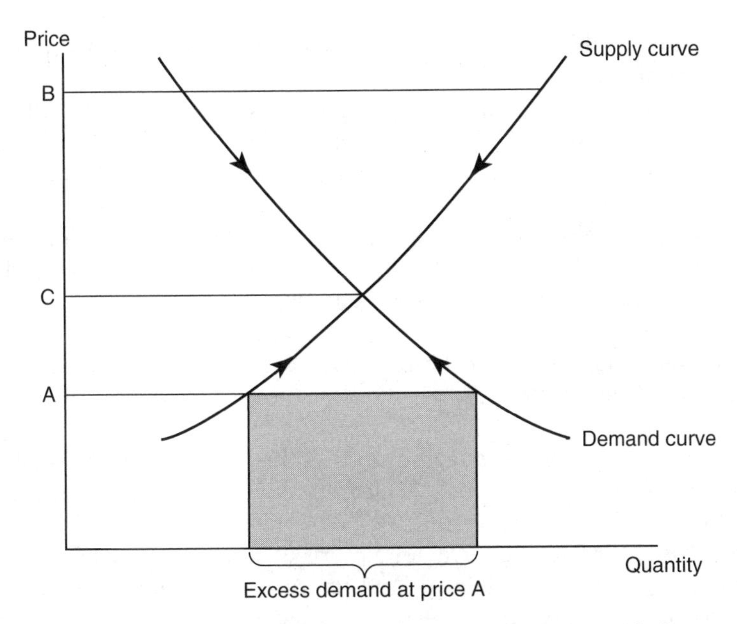

Figure 5.6 The classical supply and demand curves of free-market economics. The supplier increases the quantity available as the price rises; the consumer buys less as the price goes up. At price C there is an equilibrium at which both parties are content. The shaded area under the curves shows that at price A there would be an excess demand which is unsatisfied; at price B there would be an excess supply which nobody could afford

every consumer to make ever greater use of these resources even though this is a source of disbenefit to all the other users. Pollution of the atmosphere or the uncontrolled use of a fishery are examples that come to mind. Thus the whole thrust of the UN Law of the Sea negotiations during the 1970s was to move large areas of oceans and their beds under the aegis of national sovereignty so that they might have an 'owner'.

In the practical world, the techniques employed by economists to help with decision-making include cost–benefit analysis (CBA). This is primarily a technique designed to improve public decision-making by assessing the benefits and costs of a project to all groups involved, and can include intangible effects providing a price can be put on them (Pearce, 1983). A CBA has to consider which costs and benefits are to be included, how they are to be calculated, what the relevant interest rate for future discounting should be, and whether there are administrative and political constraints (Dasgupta, 1982). The types of economic values that can be used in CBA are as follows:

- *User values*, which derive from the current prices that are paid for access to them.
- *Option values*, which measure a willingness to pay to forestall some future probability, e.g. a payment to protect a landscape from development for a particular number of years or 'indefinitely'.
- *Intrinsic values*, which derive simply from the existence of, for example, a biological species. Such values derive from concepts of altruism ('they have as much right to be here as us') or stewardship ('it is our duty to ensure their perpetuation').

In thoughtful discussions of the true nature of CBA, these are placed in the context of such factors as

- *Irreversibility*. In the event of an undesirable outcome, is it possible to reinstate former conditions? A cleared slope can be reafforested but can a runway be un-concreted?
- *Uncertainty*. The future always holds risks of an unpredictable nature. Can any analysis build in a capacity to compensate for unforeseen effects of a given project?
- *Uniqueness*. Both organic evolution and cultural development have thrown up unique phenomena, such as a particular species or a monument like Angkor Wat. Should this confer on them a very special economic value?

The actual measurement of the value of environmental phenomena has been the subject of much exploration in the 1980s and 1990s. These contributions have been directed at collecting the various parameters needed for a proper valuation of environment which is not detached entirely from ethical concerns. They are concerned with the following:

- the value of the environment (natural, built and cultural) in promoting both materials, services and a contribution to the 'quality of life';
- futurity, both in the short to medium term of 5–10 years and in the longer-run future beyond that;
- equity, in placing emphasis upon the disadvantaged members of human societies ('intragenerational equity') and on not closing off opportunities for future generations ('intergenerational equity').

These items are now placed, as a package, where every economic evaluation must consider them. Future generations must be compensated for reductions in the endowments of resources brought about by the actions of present societies. In particular, the new work is adamant about the proper accounting of externalities. In the past it has often been possible to measure the benefits and 'efficiency' of a development project without regard to all the negative spin-offs. Conventional CBA has no doubt been manipulated to justify decisions made on other grounds (it is not difficult to specify the values of the inputs once it is decided what the result is to be) and it cannot measure a great number of the things upon which human happiness depends.

Only if the measures described above are adopted can we begin to talk of sustainability. Mostly out of concern for the Third World, but applicable also to the First, economists have attempted a deeper analysis of the notion of sustainability (Daly, 1991). It includes the following elements:

- resource harvest levels no higher than regeneration rates, and
- the input of wastes no faster than the receiving systems can assimilate them.

In more conventional economic language, these ideas can be expressed in the following way:

- sustainable economic growth means that real GNP per capita is increasing over time, and that the increase is not being undermined by negative biophysical impacts or social disruption, and
- sustainable development means that per capita well-being is increasing over time, subject to the same feedback constraints as in the case of economic growth.

Thus future generations should inherit a supply of wealth (artificial and natural) no less than the stock inherited by the previous generation, and their natural assets should be no less than the previous generation. One problem with sustainability in economic terms is that its essentially biological inspiration may not fit well with an economics whose stimulus was of a more mechanical and atomistic nature. Another is that it can be an elusive goal that provides a convenient excuse for an endless epistemological search while the forests burn and the estuaries choke. More deeply still, it is possible that the world, at both evolutionary and ecological time-scales, is far from being in equilibrium and so the biological underpinnings were themselves based on a mistaken interpretation of nature. Challenges face its practitioners in problems such as the conversion of East African wildlife parks to wheat lands because the lions do not 'earn' enough.

The argument that capitalist economics is basically a tool of the rich within a society has led to the continued attractions of socialist economics. As well, the movement to try to assert the primacy of the findings of ecological science over the dictates of economics in public policy has led to advocacy of alternative forms of economics, some of which are less determinedly Western and capitalist in their world-views.

But both operate at present within a global context in which free-market economics is dominant and indeed gaining ground.

Socialist economics

Socialist economics has its starting-point in a view of society and history as a whole and the fathers of European socialism saw their work as being the equivalent for human history and social development of Darwin's theory of evolution. These founders were Friedrich Engels (1820–1895) and Karl Marx (1818–1883), whose work is fundamental to the school of thought called Marxism. Its underpinning is an explanation of the operation of capitalist societies, with several interlocking models providing the initial keys. Of these the most important are

- the analysis of the circulation and accumulation of capital;
- the social organization derived from that type of economy and the way in which this leads to the exploitation of one social class by another and to the degrading of natural resources;
- the operation of the ideological apparatus, including for example science and neo-classical economics, in order to consolidate the position of the ruling class.

Critics of Marx and Engels have argued that Marxism is especially hostile to the environment, regarding it largely as something to be conquered and put to use as a resource. In particular, it assigns no intrinsic worth to nature: it is only the human transformation of it that counts in this model. In its extreme form, the new society is to benefit man alone, and there is no doubt that this is to be at the expense of external nature. Nature is to be mastered with gigantic technological aids. Thus anti-Marxists level that Marxism pits humans against nature, in precisely the way that modern capitalism harnesses technology simply to cater to immediate material demands. They also aver that Marxism denies any value to 'external' nature.

Overall, there is perhaps a trend from a more environmentally sensitive position in Engels and early Marx to a harder line in the later writings of Marx. Engels was particularly strong on the long-term ecological effects of human actions, and Marx warned that capitalist agriculture was bound to lead to loss of soil fertility. He also inveighed against the idea that land had become a mere commodity and thus passed from one usage to another without regard to its historical and natural characteristics (Schmidt, 1971). So if we accept that the earlier discussion still underpins the later views, then a case can be made for Marxism as being quite sophisticated in its view of nature. In particular, Marx did not want a return to some pastoral Arcadia with its connotations of zero economic growth, for he felt that nature was full of untapped potential which was held back from people by the chains of the class structure.

Integrating ecology and economics

Bringing together the methods and findings of ecology and economics is not easy. There is to begin with the deep suspicion with which environmentalists regard economists, often thinking of them as the hired hands who provide the justification for much environmental destruction. Getting beyond that, there are some difficult intellectual problems, fundamental to which is the development of common units of measurement. So-called 'marginal opportunity costs' have been one obvious route, but in general have come under the type of suspicion that has befallen CBA: you put in the numbers that will give you the right result. The most popular common denominator at present seems to be energy. Others have revolved around game theory, energy-plus-matter and even evolutionary theory.

Since energy flows through an ecosystem and also through an economic system, it is not surprising that there have been attempts to use it as a common measurement to link both systems, with the possibility of using the current prices of commercial energy to cross-value the energy present in nature. We can thus put a price on sunlight, for example, which is not simply plucked out of the air, so to speak. The key measure for economic systems is usually *embedded energy* or its alternative name of *embodied energy* (Fig. 5.7). One complexity is that ecological systems usually run off sunlight, and economic systems off fossil fuels. Yet the calories from these two sources are not directly comparable, because they are of different qualities, are suited to different needs and are capable of doing different amounts of work. Hence it is not necessarily a direct calculation to see how many calories of sunlight are embodied in the production of a book, for instance. But if we accept the assumption that all products of the human system, including labour, ideas and information, are evolved from natural systems (including fossil fuels), then we have a basis for comparability and cross-calculation – the work of M. J. Lavine (1984) shows that this is not yet a simple matter for the pocket calculator. However, it has sometimes proved an irresistible conclusion that if energy input is greater than energy output then the process must be undesirable. Looked at cosmically, such a view would preclude life altogether, since low entropy is consumed in building up the complexities that make life possible. Noting, however, studies which purport to show the not-too-distant exhaustion of fossil fuels or mineral ores, or a coming crunch between growing human populations and food supplies, some economists have tried to construct

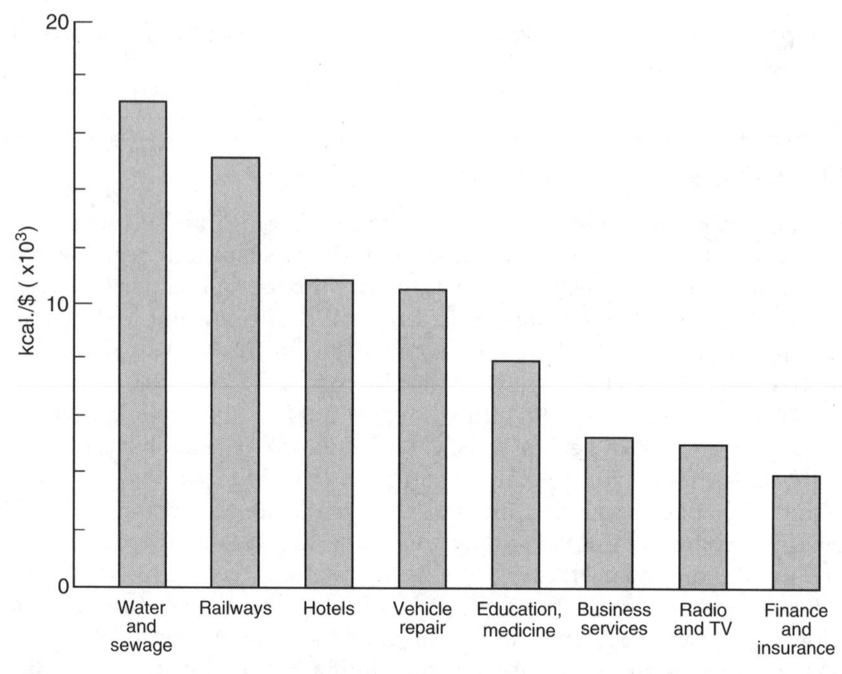

Figure 5.7 The energy embedded in some common services: bringing us clean water and disposing of sewage takes far more energy per unit value of product than running a hotel or being in education. The salaries of water company executives compared with professors probably reflect this trend also (from Hall *et al.*, 1986)

an economics of scarcity rather than of expected abundance. The distributional function of economics is thus intensified: the scarcity of resources is likely to be a matter of life or death (or at the very least of life-style in the industrialized nations), not simply of the price of petrol at the pumps. Economic adjustments mostly centre upon the role of governments in replacing the mechanism of market price with something more appropriate to an age of exponential rises in use rates of materials, in anticipation of the '29th day effect' (Fig, 5.10, p. 235), which, it is said, is not responsive enough to market mechanisms. This type of economics has been especially influenced by the work of K. E. Boulding, who used the metaphor of 'space-ship earth' to highlight the fact that we live in a limited sphere; he contrasted this view of our economic situation with 'cowboy economics', which emphasized the profligate burning of cheap candles at both ends. Boulding (1981) has also tried to integrate the basic thrust of evolutionary theory with the construction of human economies by narrating human history as the evolution of humanly fabricated artifacts.

One of the ensuing complexities is the realization that like any study of human–environment relations, the interface of ecology and economics has of necessity to recognize the existence and relevance of matters as diverse as the laws of physics (e.g. the second law of thermodynamics), law, individual psychology, sociology (e.g. the social aspects of carrying capacity), and spatial arrays of the kinds studied by human geography. So there is not yet a unified construction resulting from this particular interfacing. It has, however, succeeded in planting rather more firmly into economics the notions of limits in biophysical systems, or, put more vulgarly, there is still no such thing as a free lunch.

Other alternative economics

Marxism is not the only alternative system to the political economy of capitalism. Writers and thinkers have developed other sets of ideas, although it has to be said that they are nowhere in place at the national scale for inspection and evaluation. If we accept that the disentangling of ends and means is very difficult where the human use of commodities is concerned, then it is not surprising that non-Western value systems have been examined for their potential contributions to alternative economics. For instance, as much out of non-Western traditional values as out of today's context, E. F. Schumacher (1973) brought to our attention his concept of Buddhist economics. This combines the Buddhist ethic of detachment from the pleasures and pains of the world with the attitude of J. K. Galbraith towards the new corporate state. They meet in an exaltation of the frugal and thrifty, small-scale living and decision-making, and a renewed attention to the needs (as distinct from the demands) of individual people rather than corporate stockholders. This type of economics has appealed to groups of people seeking alternative life-styles, but not much to governments, who do not fancy isolation: in a world of jackals, do not lose your teeth, runs the proverb.

In a more directly Western context, the case for alternative economics rests on two major foundations:

- The inefficiency of the market mechanism for allocating resources through time. Equity and uncertainty considerations are at the heart of such problems and the answers are manifold in both the technical sense (e.g. different kinds of discount rates) and the political arena: what is a feasible time-horizon for a decision-making body? These views emphasize that 'efficiency' is only one thing to be optimized, along with, for example, population growth rates, democratic decision-making and the nature of income distribution.
- Marxists and other advocates of a world-view based on socio-cultural constructions of environment need on the one hand to recognize the

realist finding that there are indeed physical laws, and on the other that technological change of itself will not necessarily alter relationships such as poverty or the desire to change nature irreversibly.

Indeed, both lineaments may focus on the resurrection of the idea of an absolute natural scarcity, sometimes buried by theorists of classical and neo-classical persuasions. This time, the notion could encompass not merely stock resources that were no longer available for extraction, but widespread environmental degradation and ecosystem collapse, as with the worst forms of desertification, for example. Alternative economics then begins to advocate the efficiency and value of factor substitution (labour for energy, capital for energy and materials, increased efficiency of resource conversion) and improved organizing techniques which encourage the change to durables (Schnaiberg, 1980). The 'throwaway' economy would need to be replaced by long-lived products where design builds in not only long life *per se* but proper possibilities for repair and recycling. This is consistent with thermodynamic analyses, where the real savings are made in the longevity of the product.

Summary

What is the future for economics? Although it seems to preside over unemployment, inflation, capital scarcity for the poor and environmentally destructive notions of 'development', the immediate future seems very much like business as usual. Though some argue that the whole of Western society, economics and all, is in a phase change to another type of relationship with the non-human components of the planet, there seems little evidence of this in the pages of the *Wall Street Journal*. The suspicion has grown that much environmental destruction has been judged profitable by hired economists and that they must therefore be regarded as embodiments of some Manichean evil: to the contrary, every encouragement should be given to them when they show signs of getting greener, a trend which can sometimes be seen in some of the larger organizations such as the World Bank.

A wider view of economics (Turner *et al.*, 1994) would suggest that the tasks awaiting economists appear to be threefold:

- The interfacing of the ecological systems of the world with its economic systems. Neo-classical economics has always assumed that the economy is always below biophysical limits, which seems now to be turning out to be untrue unless a spectacular 'technical fix' for some problems of residuals like the 'greenhouse gases' is discovered. Feedback loops between ecological and economic systems must be a subject for special examination.

- The exploration of the bases of consumer demand. J. S. Mill's famous dictum, 'Men do not desire to be rich, but to be richer than other men' no doubt has its truth. But there are yet stronger pressures in the higher-intensity market-setting of industrial societies, where individuals can be led to misinterpret the nature of their needs and the ways in which they can be satisfied. There seems to be no sense of contentment or well-being in the higher reaches of abundance. Somehow the self has to be redefined, which is a task at which even economists balk.
- Some economists see that the metaphors of the approach need careful thought. Economics appears to have taken on the atomistic and mechanical assumptions of classical mechanics of the type developed by Isaac Newton. Yet economics interfaces with an evolving set of interconnections of a non-linear type, in the shape of the planet's biophysical systems. Indeed, some of them are best described by chaos theory rather than by mechanical orreries.

One assessment of econometrics suggests the need for economics not to forget its 18th century roots in the discipline of 'political economy'. Here the question of politics is not left out, so neither is that of the exercise of power and hence of values. Dignity, freedom and 'happiness' are not guaranteed by economic efficiency any more than by anything else, though we might gain some comfort from K. E. Boulding's view that in the end economics has done more good than harm.

Political science, sociology and anthropology

Like economics none of these studies of human societies is concerned only with environment. Yet the practitioners of all of them have in recent years looked at the questions of human–environment relations to see if they have something to offer to the debate (Eckersley, 1992). The types of analysis which they offer very often have much in common and here the disciplinary boundaries mean less than normal.

Political science

Politics is the deliberate and rational effort to direct the collective affairs of human societies. In that pursuit, then, politics is about power, and the ways in which that is conferred on individuals and groups and what they then do with it. In the case of the environment, political science also aims at an objective, non-judgemental study of what actually happens in terms of decision-making as to the allocation of resources. The agenda for environmental politics was in some ways set in the time of Plato and Aristotle, though admittedly in a largely pre-technological context. They and their contemporaries wrote, among other things, about population levels. This led to concern about how the numbers of people related to natural resources and also how population levels affected personal psychology and hence the imperatives of social organization, the *politeia*. Likewise 'nature' was of considerable interest, in both its 'natural' form and as 'human nature'. For both, the questions could be asked, 'what is permanent and what is changeable?' and similarly, 'what is the proper attitude to sudden eruptions of instability, whether tectonic or political?' And in particular, perhaps, 'are the desires of human nature insatiable?' How are they affected by the distribution of wealth or by human closeness to or remoteness from the natural world?

Conventional politics today

The normal political activity which concerns us here is that of the formulation of environmental policy. This can take place at many levels, from that of a local government deciding where to put its dump for householders to leave their rubbish, to the actions of UNEP in the face of desertification or the apparent problem of global warming. The management of resources and their residuals, as well as the designation of land and water for non-consumptive use, is always a political matter as well as being economic and scientific. In general terms, we can contrast societies with openness of information and decision-making with those which keep as many secrets as possible and involve only the bare minimum of power-holders in the exercise. The USA is a good example of the first category, the UK and France of the second. The LDCs are different again, for they mostly lack the institutionalized structures to enable them to tackle the complex problems that have a large environmental component.

But at present, we usually seem to have overwhelming central governments and TNCs against which the only significant countervailing force is the presence of interest groups, of which there are many in the First World with environmental concerns, like Friends of the Earth and Greenpeace. These groups may achieve some measure of success since participation in the processes of persuasion makes them highly professional and technically capable. Opponents argue that environmentalists' views could only be emplaced by highly authoritarian governments and so charges of 'eco-fascism' are levelled, with the implication that democratic change towards such different world-views is impossible.

Alternative politics

The land, industry, and rapid urbanization have for long been a fertile seed-bed for political alternatives, from pantheistic mysticism to varieties of socialism. The Western anti-growth movement of the 1960s and 1970s was primarily aimed at economics, but there was spin-off criticism of political science as being merely an agent of the *status quo*. As an alternative, community was to be the bedrock of decision-making, informed by ideas of design rather than planning, to allow more individual freedom, and by notions of stewardship rather than the primacy of consumption. The 'bioregion' is seen as a site for political action, cultural and spiritual expression, and personal change, all enhancing an ecologically based transformation of self and society. No doubt many individual humans enjoy being rooted in a 'local' community but for others movement, change and anonymity may be essential for creativity. And how could thousands of 'bioregions' mesh together to provide higher-level services like the more high-tech medical facilities?

Politics in a wider context

This interaction has always been a subject for comment by political theorists from Classical times onwards, with more than one Utopianist calling for an authoritarian body to oversee the introduction of new inventions into society, with a strong presumption that a conservative position was preferable.

There are direct instances of the effect of technology upon social order. The low bridges on the Long Island (NY) Parkway were meant to keep out buses and so keep out blacks. Hausmann's boulevards in Paris made barricade-building very difficult. The invention in California of the tomato harvester produced a decline in the number of growers from 4000 to 600 in the period 1960–1973 and by the late 1970s job losses in that industry totalled about 32 000. So technologies can be not merely symbols of a social order which contains power structures but more literal embodiments of that order. Technologies are thus ways of building order: choices made by people with power are strongly reinforced by investment patterns, materials and equipment and social habit. To be opposed to these patterns is generally seen as being not merely anti-technological but anti-progress as well.

The main danger, perhaps, is that a model of factory or firm is taken as the epitome for society as a whole. That is, if a firm makes most profit by being structured in a hierarchical and highly authoritarian way, why should not the same apply to society and, by extension, to our use of resources and hence our manipulation of nature? This is the theme of much of the work of Murray Bookchin (1982), who postulates a hidden prehistory of the development of a landscape of domination by and within human hierarchies. Not only is this expressed in our visual world of experience but it seems to be embedded in the depths of our psychic apparatus. Part of this has involved our disembodiment from an 'external' nature which some now see as mineralized and inorganic. The alternative (which is usually incorporated with Green politics) is more mutualism and self-organization, with subjectivity of experience having as much validity as 'objectivity', together with spontaneity and non-hierarchical relationships. The language, at any rate, is reminiscent of the theory of evolution and of the entropic theories of organic organization.

Sociology

Sociology undertakes to observe and describe social phenomena and to formulate theory in a scientific way. As it has developed, it has addressed itself to questions such as, how does a society hang together, i.e. what are the forces of its 'social physics'? The analogy with the physical sciences and in particular a Newtonian view of the relationships between entities is apparent here. Sociology has also been concerned with social evolution; in the main this has been underpinned by a progressive view of history, though a few pessimists have surfaced in every generation (Yearsley, 1991).

Mainstream sociological enquiry

Sociologists are the authors of numerous reports on people's attitudes to various environmental issues, from the purely local (do you want that meadow covered with houses?), to the national (are young people sufficiently interested in the environment?) and the global (are you willing to delay industrialization in order to protect the ozone layer?), and many more. The work of S. Cotgrove (1982) goes beyond this in identifying two social groups with particular environmental attitudes. These he calls the Cornucopians and the Catastrophists. The first put their faith in technology and economic development and assert that increased quantities of resources can easily be available for all, provided that investment in technology is high and that social structures encourage enterprise. These are the dominant group in the world at the moment. The second group think that there are physical limits to resources and that the planet's life-support systems can be badly degraded by environmental contamination; reform needs attention to wastes and to a lower level of material consumption in the industrialized nations. The viewpoint of the sociologist is that all these views of the future are rooted in systems of meaning which are themselves social constructs and lack any basis of objective certainty. They are to be identified with faiths and doctrines just as much as

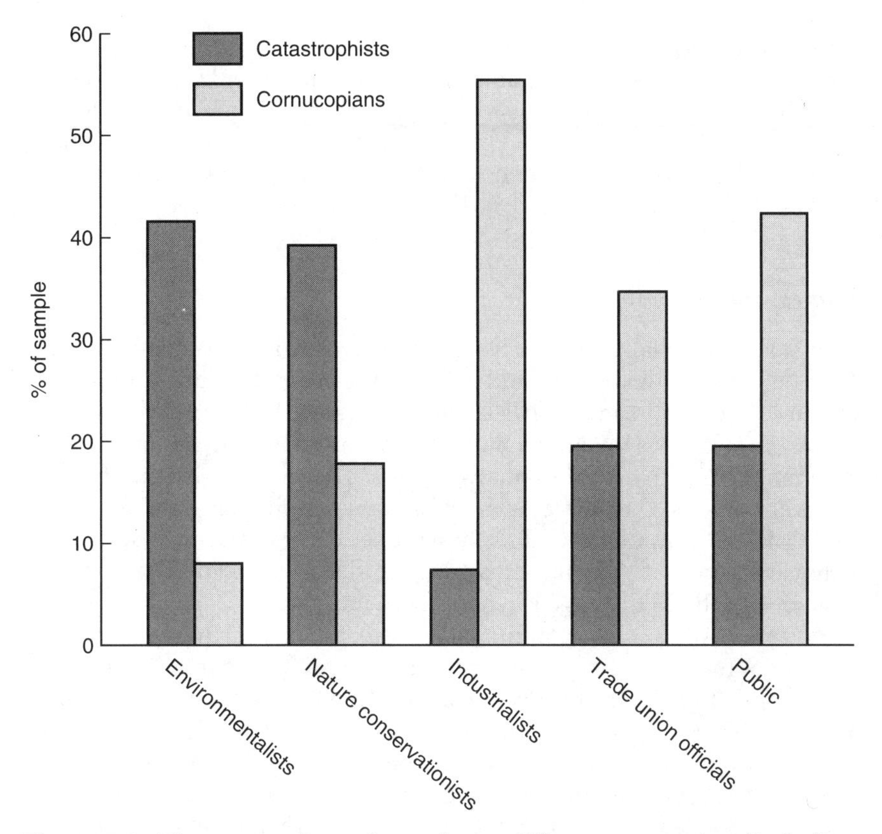

Figure 5.8 The proportion of people in different categories (including the 'public') who espoused Cornucopian or Catastrophist views in work published in 1982. (There was a third, intermediate, category so the sums of the two columns do not total 100 per cent). The UK trade union officials reflect the public rather well; the gap between the perceptions of the public and those of the environmentalists is very large (drawn from tabulated data in Cotgrove, 1982)

any religion. (Cotgrove adds that since the Cornucopians are dominant, as seen in Fig. 5.8, their views need the most stringent examination.)

In the critique of the Cornucopian position, the lead has been taken by the 'limits to [physical] growth' school of the Club of Rome and its associates. A notable study, however, parallels this with the ideas of social limits to growth. In them, it was argued that economic liberalism promises prosperity to all, provided they obey the rules of the game, but in fact it unleashes demands and processes that no Western society (or, we suppose, the LDCs) can contain. Because so many goods lose in comparative quality if they are open to all (those of the environment, e.g. open space or a second home, are good examples), keeping ahead of the pack is still important for many: only that 'undiscovered' island in Greece will do for holidays. Thus some people are always striving to improve their position relative to others and the resources experience increasing pressure.

Radical social theory

This side of Utopia, visions of societies which are in a greater position of harmony within themselves and with nature have their attractions. As a counterweight to the industrialized state of the late 19th and 20th centuries, a process of change towards a more decentralized and de-urbanized set of communities is advocated by many writers. Theodore Roszak has gone on to encompass the Gaia hypothesis within his social thinking and to suggest that Gaia can communicate directly with her citizens and thus adapt their behaviour so as to be life-enhancing in the long term. The mode of reception of this knowledge is what we call intuitive.

Whereas some social commentators see gradual change as the only way in which a true 'greening' of society will occur, others follow Marx and Engels in foreseeing revolution. This is the position of Herbert Marcuse, who sees the necessity for a technology of liberation which will succeed the present technology of repression, and that the one must succeed the other very rapidly, i.e. at a revolutionary rather than an evolutionary pace. Bookchin, on the other hand, identifies the problem in a much more specific way. The core is seen as hierarchy in human societies: thus the pyramid of dominance enables us to negate so much potential for liberty and freedom (Bookchin, 1982). In such a structure, the ethical meaning of the community has been replaced with what is termed the 'fetishization of needs', i.e. a society devoted to satisfying, at whatever cost, the demands for consumer goods. In such a situation, even the state of nature becomes a commodity, as witness the bookings needed to get into wilderness areas in North America. As far as the environment is concerned, Bookchin's central message is that we must stop dominating each other if we are to stop wreaking havoc on nature.

Green Politics

Green political parties are devoted to moving towards the Utopias of the environmentalist cause and are Catastrophists in Cotgrove's terminology. In particular they advocate a low-consuming economy, the use of renewable energies, decentralization, the revaluation of the roles of, for example, work and women, and especially the peace campaigns which aim at the removal of nuclear weapons. They are, however, different from other political parties in the sense that they wish to put themselves out of business, having converted all the other parties to their world-view. In general, though, they are not anarchists: they are about participation in a structured society, albeit one devoted to different ends from those of the present in the industrialized nations (Spretnak and Capra, 1986). Many see it as inevitable that conditions will get worse before the mass of people are converted to their way of thinking.

Strains within the Green movement are often evident, however. These are usually between the purists who want to move directly to the new consciousness and values of the 'deep ecology' (see p. 229) type, and those of a more 'shallow ecology' type who will join coalitions with those who accept the prevailing patterns of production and consumption. The latter are engaged in a form of social engineering which aims to bring about better short-term conditions for humans rather than a change in our whole world-view. They do, nevertheless, wish to lessen our impact upon non-human systems and to reduce contamination levels.

Anthropology

Anthropologists have taken very much to heart Alexander Pope's dictum that the proper study of mankind is man. Founded in the 19th century wake of Darwinism and perhaps also in the desire to show that the theory of evolution can be applied to human culture as well as to species (and indeed to clarify the relationships of 'civilization' and 'savagery'), anthropology has always focused in the end on the human community itself. That is not to say that environmental factors have been ignored, but for a large number of anthropologists they were for many decades subordinate to factors such as kinship, ritual and material culture.

Cultural ecology

At the heart of cultural ecology was the biological notion of the carrying capacity of an environment for people. Culture might then be seen as a way of adapting to an environment's limitations, and the unit to be treated was a population (as in a natural ecosystem) rather than the more traditional social order of social anthropologists. This latter approach was called 'neofunctionalism' and, like its biological equivalent, focused on regularities in ecosystem-level processes. In particular, the ways in which human populations functioned within ecosystems were discussed, usually as an examination of the

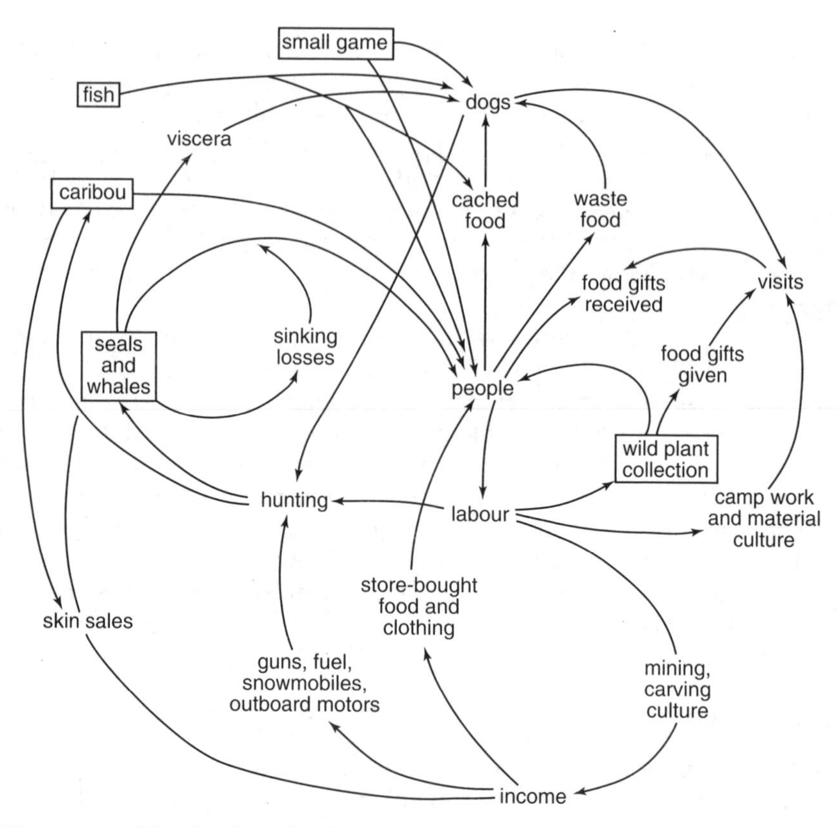

Figure 5.9 The kind of diagram that is the basis of cultural ecology, in which the natural world (boxed) is linked by a set of material flows with the cultural world of a human society. This is based upon quantitative work in the 1960s by W. B. Kemp and this drawing is an amplified version of a non-quantitative sketch in a 1995 book (from Johnson and Lewis, 1995)

mechanisms which linked social structures and material culture to the environment (Fig. 5.9).

The emphasis on process led to such studies as the relation of demographic variables to agricultural productivity as in the work of Esther Boserup (1994), the formation of adaptive strategies in extracting environmental resources (concepts like the niche were transferred from biology), and on the lessons of history as elaborated by Marxists. But the major change from the past, occurring mostly in the 1960s, was a focus on ways in which behavioural and external constraints influenced each other: environment was no longer simply a background that provided enough necessities for people to go on living.

Applications

These newer views, coming as they did in the 1960s, interacted with the then very high level of public concern over 'the environment'. Anthropologists pointed out that humans were not behaviourally homogeneous in the way of many animal species, and that an understanding of the ecological required a knowledge of the social. For example, where environmental decline and degradation was perceived as a problem, then anthropologists might be able to help with the social matrix of its causes. Notably, they might be able to identify the economically rewarding and psychologically satisfying behaviours that led to environmental decline and contamination, and suggest less harmful substitutes for them. Essentially, they brought forward the idea that the human use of nature is inextricably bound up with the human use of humans; the remedies for the destructive use of the environment must be found within the social system itself. Obviously, this leads towards a high valuation of indigenous knowledge in non-industrial societies.

Attempts at integration

From time to time there emerges the problem of integration in looking at the human–environment relationship. In complex societies, many ways exist. There are the 'hard' findings of the natural sciences on the one hand, and the more personal response found in say poetry on the other. Each of these, and those in between such as the social sciences, has its own language and these tongues may be mutually unintelligible. Hence, any attempts to bridge the gaps between existing disciplines or established modes of discourse are of substantial interest.

Ecological economics

For years it has been customary to point out that economics and ecology had both divergent aims and different languages (Perrings, 1987). In particular, economics was about the satisfaction of human demands on a short time-scale whereas ecology was about long-term stability. Some shifts in both areas of study have taken place in recent years; for instance, economics has shifted its thinking on the use of natural resources. The conventional view was that any income derived from the use of natural resources could be invested in other assets that would yield a higher economic return. The (probably) non-renewable natural resource would be converted thus into non-exhaustible or 'reproducible' capital and so produce a steady stream of economic benefits, which had

Table 5.1 Comparison of conventional and ecological economics

	'Conventional' economics	**Ecological economics**
Basic world-view	Mechanistic, atomistic	Dynamic, evolutionary
Time frame	Short, <50 years	Multi-scale, days to eons
Space frame	Local to international	Local to global
Primary macro goal	Growth of national economy	System sustainability: ecological and economic
Primary micro goal	Max. profits for firms and utility for individuals	System goals require adjustment
Assumptions about technology	Very optimistic	Prudently sceptical
Academic stance	Disciplinary	Transdisciplinary

Source: adapted from R. Costanza *et al.* (1991) Goals, agenda and policy recommendations for ecological economics. In R. Costanza (ed.) *Ecological Economics: the Science and Management of Sustainability*. New York and Chichester: Columbia University Press, Table 1.1

the additional virtue of standing free from the vagaries and discomforts of the natural world.

Economists are now learning that some natural functions are irretrievably lost with the conversion of a resource, and equally that the full range of economic benefits (present and future) conferred by for example biodiversity, the global climate or the resilience of fragile ecosystems, is rarely explored and measured. In these intellectual shifts, the ways of measuring environmental benefits (a specialized affair not probed here) are being closely examined in order to bring a closer convergence between ecological and economic thinking on topics such as instability, resilience and the insights from thinking ethically. The contrast between older economic aims and those represented by the newer ecological economics are set out in Table 5.1; these are much nearer to the concepts that result from natural sciences such as ecology. How these newer findings will inform ways of environmental management that are backed by economics, such as the charging of green taxes and the trading of environmental permits, will in due course be seen.

Geography

If any academic discipline was made for the interactive study of humanity and environment, it ought to be geography, with its history of interest in that interface. Several geographers have made substantial empirical contributions to the historical evolution of a humanized world but the number who have contributed to the debate over the nature of the relationship is small. The most notable have been R. W. Kates, who drew early attention to the role of human population growth in creating environmental pathologies, and T. O'Riordan, who has explored the development of human environmental perceptions in the context of the extent to which different ideologies (in the sense of the sociologist Cotgrove, discussed above) are adopted in different groups of society. Geographers seeking to infuse social theory with real-world knowledge of place have been few, with exceptions such as M. Redclift's (1987) questioning of the likely contradictions in the concept of sustainable development in poorer countries. The once-famous ability of the geographer to reach out, octopus-like, and bring together disparate information from many sources and modes of investigation seems to have diminished. No doubt this is in part due to increased differentiation of languages, so that no one person can understand what is being said by the groundwater hydrologist *and* the structuralist-trained anthropologist, and nobody is likely to pay for a team of research assistants for a geographer, as distinct from a particle physicist or a biomedical pharmacologist. Perhaps grand syntheses are out of fashion anyway.

> Not all the social sciences operate in a way which imitates the natural sciences. Some of them have schools of thought that are primarily concerned with human experience in all its diversity as experienced by the individual from within as it were. The term for this 'inside-out' view is phenomenology and the attempt then to understand not simply the dynamics of measurable phenomena but their meaning is part of the study of **hermeneutics**. The latter takes for granted that all information is culturally interpreted, even before it is gathered so to speak. Thus hermeneutics has to grapple with the double difficulty of how we understand the meaning of meaning itself. Transforming phenomenological data into policy carries the difficulty of translating the understanding of one individual into that of a larger group and of finding ways of truly understanding the 'others' in our lifeworld, whether these be human or non-human. Nevertheless these ways of thinking can be holistic rather than reductive since they seek to encompass the whole richness of humans and their surroundings rather than to isolate one or two threads from which to hang a theory, some laws and a few models.

Philosophy

Philosophy is now understood as the study of the nature of things and ideas in the most general of ways; thus any idea (including those we find repellent) is a fit object of discussion. Ethics, on the other hand, is an investigation of how we ought to live, both as individuals and in societies at various scales. Philosophy may therefore discuss what a principle actually is; ethics tries to apply one to daily conduct. But ethical principles can, of course, be ignored.

Pre- and non-literate philosophies

In the sense that a philosophy can be a codification of attitudes towards or identification with the natural surroundings of an individual or society, then it can be transmitted orally. Thus societies which never gained literacy, or present-day attitudes which eschew the shackles of writing, are by no means debarred from having an environmental philosophy. To the extent that the thinking may include a supreme power as one of its elements, then philosophy and religion may not always be distinct.

Oral traditions

There are numerous examples of pre-literate societies that have developed practices which involve environmental tenderness. Pacific island communities, for example, often designated a kind of conservation warden whose job it was to ensure that over-fishing within lagoon areas did not take place. Several native North American groups knew that animal-bearing areas had to be rested from time to time to allow the populations of the hunted species to recover. In the sense of a developed philosophy rather than a set of pragmatic practices, the world-views of Native Americans have perhaps been most recorded. They were a diverse set of peoples but some generalizations are apparently possible. They centre around a rejection of the notion of hierarchy in the cosmos in which humans hold a superior place. J. Baird Callicott summarizes it thus:

> . . . most American Indians lived in a world which was peopled not only by human persons but by persons and personalities associated with all natural phenomena. In one's practical dealings in such a world it is necessary to one's well-being and that of one's family and tribe to maintain good social relations . . . with the nonhuman persons abounding in the immediate environment.
>
> In sum, I have claimed that the typical traditional American Indian attitude was to regard all features of the environment as enspirited. These entities possessed a consciousness, reason and volition no less and intense and complete than a human being's. . . . All creatures, be they elemental, green, finned, winged or legged, are children of one father and one mother.

Communication of a direct sort was also possible: animals about to be killed were addressed, to apologize to them; dreams were an important form of consciousness in which the non-human personalities might speak. Additionally, the shamans were a channel of communication between all the members of the cosmic dance. This was, no doubt, an ideal situation which not everybody would always live up to: in certain circumstances overkill would happen, and the intrusion of Europeans made it harder to keep to the traditional ways since the immigrant's world-views were so different.

Modern mysticism

In an age dominated by ideas of reason and the practice of science, we do not expect to find lineal descendants of world-views in which humans express a metaphysical continuity between themselves and the non-human world. But it happens: in Frijthof Capra's influential book *The Tao of Physics* (1976), he describes a mystical experience on a Californian beach, where he felt and heard that he was at one with the waves and with the energies of a cosmic dance. As T. S. Eliot puts it in *The Dry Salvages*,

> . . . the moment in and out of time,
> The distraction fit, lost in shaft of sunlight,
> The wild thyme unseen, or the winter lightning
> Or the waterfall, or music heard so deeply
> That it is not heard at all, but you are the music
> While the music lasts.

In a wider context, writers have wondered whether intuitive knowledge (in which we claim we just 'know' something without the intervention of normal learning processes) could represent communication between the environment and humans. There are people who think that Gaia is neither a metaphor nor a model but an actual sentient living entity. They go further to aver that she can communicate with her denizens and that intuitive knowledge and the emotions are the vehicles. This type of un-verbalized relationship is highlighted by some feminist groups, who claim that only women can be carriers of the truth about the Earth since it is women who identify in an isomeric fashion with the rhythms of the planets and with the fruition inherent in life; so women are by their very nature the conduit for the only true knowing about how we may successfully inhabit the Earth. Similar claims are made by groups usually labelled as 'New Age Pagans'. To call these views 'mystic' is not to apply a denigratory label, however, but simply to contrast them with the post-Enlightenment mode of thought based on human-nature duality and the primacy of the written word. Such a contrast applies to religious thinkers in the West who base themselves not on the Bible (with its minatory concepts such as the Fall and Stewardship) so much as on a celebratory attitude to the plenitude of the Earth and its diversities in which the dance is as appropriate an act of worship as any.

The Western philosophical tradition

To summarize over two thousand years of human thinking in a few paragraphs is an awesome task. One initial generalization to be stressed is that it is a written tradition and so the exact meaning of words has always had a crucial place; furthermore these meanings have not always been the same as those of everyday language but more like technical terms. A second generalization is that almost every age since that of Classical Greece has produced significant thinkers about what is generally called 'the relation of man and nature', with the possible exception of the century between 1850 and 1950, which can be seen as significant in terms of the apparent control over the natural world which was promised at that time.

The ancient world

Though the tradition is dominated by the output of Classical Greece, other civilizations have exerted some influence down the ages. The religion of ancient Persia, called Zoroastrianism, apparently started the Western tradition of dualism by espousing a fierce dogma of the battle between good and evil and the overriding necessity for good to triumph. Some echoes of this can be heard in Plato: for example, in the idea that humans are dualistic themselves, having both body and soul. Thus they are set apart from the rest of nature both in their very essence and in being the only creations open to the possession of morality. Another main stream of ideas associated with Classical Greece is that of atomism. The concept that the cosmos was composed of separate atoms (which lasted until well into the 20th century) encouraged a view of the world which was reductive, particulate, inert, material and conducive to understanding via mathematics. It seemed possible thus to change one thing without changing any others. Overall, the world-view thus constructed went well with the agenda for science and technology when that emerged in the Renaissance, which was after all a re-birth of Classical ideas (Glacken, 1967).

The Judaeo-Christian traditions

Out of an interaction between the beliefs of the ancient Near East, Classical Greece and Rome, and the emerging religion of Christianity, came a set of ideas that eventually dominated the thinking of the West until the Enlightenment of the 18th century. Aside from the ecclesiastical dogmas, this creed implanted a number of elements of a world-view, many of which confirmed earlier (mainly Greek) ideas, but others of which were novel. In the latter category, for example, was the central notion that God transcends nature, which becomes a profane artifact of a divine creator. The human species, on the other hand, comprises the only species which is made in the image of God and is therefore set apart from the rest of nature. Humans are therefore given dominion over the Earth, to multiply their numbers and subdue the rest of nature, taking their place in a hierarchy which reads God–man–nature, with 'man' usually meaning just that. So non-human entities of all kinds lack the intrinsic value of God-in-man, and nature exists largely as a support system for humans. One enduring result of these views has been the denial of moral value to anything but humans, leading to the forbidding of the foundation of a branch of a Society for the Prevention of Cruelty to Animals in the Vatican in the 19th century, on the grounds that animals had no right to moral treatment.

All these profoundly influential views were irradiated by the pessimistic view of humanity as a fallen entity. The expulsion from the Garden of Eden meant that no action of humans was free from the taint of inborn sin and that any undesirable consequences of human action were merely to be expected. Nature, too, was vitiated by Adam's sin: hence the presence of natural evils like famines, diseases and death in the world. Commanding figures like St Augustine, Tertullian and St Cyprian implanted the view that nature was senile and the world destined for an early end. Not until our own day has this attitude been strongly assailed from within the intellectual community of Christians, with the development of 'creation spirituality', associated especially with the name of Matthew Fox (1981). This set of arguments, by contrast, is inclined to regard the Fall as a growing-up process akin to the acquiring of responsibility, and the Earth as being not senile but full of wonder and re-creational potential: once more the dance rather than the scourge.

Renaissance, enlightenment, science

The 16th and 17th centuries are associated with the evolution of the kind of science we recognized in Chapter 2. The patient accumulation of evidence from the senses and from their extensions via instruments, and the formulation of generalizations, which relied on these data and not on for instance a pre-ordained theological formula, was the hallmark of men such as Copernicus (1473–1543), Galileo (1564–1642), Descartes (1596–1650) and Newton (1642–1727). The foundation they built was essential for the movement known as the Enlightenment, in which reason was held to be a guiding principle not simply in the investigation of the world but in running it as well. The world-image which they constructed has been credited (or indeed debited by some recent writers) with being mechanistic and totally materialist, with the

Greek characteristic of atomism finding an outlet in separations and dissociations of spirit and matter, value and fact, subject and object, and most famously of all, mind and body, following René Descartes' '*Cogito, ergo sum*' ('I think, therefore I am'). In this construction of the world, more is better, especially in a utilitarian philosophy, which seeks the greatest good for the greatest number of people. In many ways, all this is integrated into the **positivism** of Auguste Comte (1798–1857), which postulates that all value judgements are subjective and unreliable. In many ways, it suggests, they are not 'proper' knowledge at all and it is certainly not possible to infer 'ought' from 'is'. One consequence has been the separation of ethical considerations into another set of separate compartments: worthy but isolated criteria essentially unrelated to political or economic matters.

Science in the 19th and 20th centuries

The last 150 years have seen remarkable changes in the world. During much of that period, major philosophers were not much concerned with what they would have called 'man and nature'. In part this was because technological advance appeared to render nature no longer problematic in the industrializing nations and would not do so in the poorer parts of the world if they adopted European ways.

Less directly, ideas about humanity and nature were revolutionized by the writings of Charles Darwin (1809–1882). Building upon the work of geologists, who were coming to see that the Earth came into existence well before the 4004 BC calculated from the Bible, Darwin produced evidence for the evolution and extinction of species. Most important of all for our theme, he became convinced that *Homo* had been subject to this process and that we were cousins to the great apes and thence to the rest of the animal kingdom. The underlying themes were those of constant change and of mutation of living matter. Darwin, of course, did not know the mechanisms by which organic evolution proceeded: that came later with the discoveries of geneticists. Later interpreters have also emphasized that evolution may not proceed smoothly: that there may be periods of intensive species creation and extinction due to natural causes. Chance played a great role in what survived.

The role of chance (in its dressed-up form as probability theory) plays an important role in the most important development of 20th century science, that of quantum mechanics. At the subatomic level it appears firstly that our knowledge of any particle is affected by the way in which we observe it: if we determine its position, for instance, we cannot find out its velocity. Further, the behaviour of the subatomic particles is not determinative but subject rather to probabilities: at the heart of matter there is uncertainty. This has given impetus to constructors of models of complex systems (such as ecosystems, the weather and economics) to believe that chaos theory, for example, is well-founded. But such knowledge does compromise one of the major thrusts of the natural sciences, that of being predictive. The more complex the system, the lower the chances are of being able to predict its future behaviour, in which case, the future for humanity and the environment, as seen by the natural sciences, is even more open than we have ever dreamt.

The eco-philosophers

In response to the growth of concern over the future of the natural environment (in reality, of course, just as much a concern over the future of the populations of the developed countries), the period since 1960 has seen a growth in the number of philosophers who are trying to develop wisdom about nature. Since their work is relatively new and not winnowed by time (though well raked over by their fellow practitioners), it is difficult to give the kind of distillation now possible for earlier centuries.

One group of thinkers has been much influenced by the findings of ecological science. The mediator has often been the North American ecologist Aldo Leopold, who talked of land use in 1949 in terms of 'a thing is right when it tends to preserve the integrity, stability and beauty of the biotic community. It is wrong when it tends otherwise.' In other words, value is attached to the whole system and not simply to individual parts. The ecosystem in its entirety places, produces and promotes (in an evolutionary sense) things, including human beings. This is taken much further in the Deep Ecology movement associated with the Norwegian philosopher Arne Næss (b. 1912), which espouses among other things a biospheric egalitarianism where everything in the biosphere has a right to flourish, and a comprehensive holism in which organisms are temporary knots in an all-enfolding web of being. Hence natural diversity has its own value: there is no sense in equating value simply with human ideas. A more radical-sounding extension of this position is the extension of the human self to embrace all those phenomena with which it interacts, which in the case of denizens of the industrial world is more or less everything on the planet. Moral agency is thus extended to mountains and rivers because the entire system is necessary for each and every thing to realize itself. Nature is then valuable because it is the setting for all things to evolve into their own true natures, a concept worked upon by Baruch Spinoza (1632–1677) and called by him the *conatus*.

All these extensions of thinking are, obviously enough, in the Western post-Enlightenment tradition of being founded on the idea of reason and expressible in language. Also in the Western tradition, however, is a continuing strand of mysticism, as discussed above. Here, the human being identifies non-verbally (we might use the words 'intuitively' or 'emotionally' to describe the state) with the rest of the cosmos or at any rate a part of it. This is too much for the inheritors of rationality and in the best-known of these counter-arguments the Australian philosopher John Passmore (1980) describes as 'rubbish' the notion that mysticism can successfully address our problems where technology cannot. Instead, Western rationality is fully capable to adapting itself: its democratic institutions, the liberal attitudes and the spirit of enterprise, he argues, can all be put to the service of a viable future for both humans and nature.

The debate over the 'best' kind of ideational underpinnings for human–nature relations is currently lively and multi-stranded. In this, it is not even confined to its regional past, for a resurgent interest in the philosophy and religion of Eastern cultures has been evident since the 1960s.

Non-Western philosophy

In spite of arguments from philosophers such as Passmore, the attraction of non-Western ideas about the human place in the cosmos has continued. In the present context, the interest began to build up strongly in the 1960s as an element of the counter-culture but it has now become a focus for scholarly evaluation as well. The initial focus was on Zen Buddhism but the ambit of enquiry is now wider; this has brought the inevitable realization that 'Eastern philosophy' is no single entity but contains a considerable diversity of ideas. Further, its relations to the West are not free of problems (Callicott and Ames, 1989).

The first of these is the use of categories. The idea of an Eastern *philosophy* as such may be misplaced. Philosophy seems to emerge from the idea that it is rational to seek for certain knowledge. Many Eastern traditions, on the other hand, look for an aesthetic more than a scientific sense of unity, with a significant resistance to abstraction. Thus we have images, not theories, experience not argument, metaphor not logic. Where Eastern thought is seen through the lenses of Western philosophy, then considerable distortion has occurred. Hence the second problem: that such ideas are not suitable for export to the West. It can be argued that the differences between the Western atomistic–mechanical traditions of thought and the holistic–organic Eastern legacy are so great that cross-fertilization from what is now the politically and economically weaker strand is unlikely. But we should also note that in the past, there have been imports to the West that have been quickly assimilated. The example of the Chinese garden may well have paved the way for a heightened appreciation of nature which emerged in Europe in the 17th and 18th centuries.

The 'tariff-barrier' model may also be inappropriate when global cultures are emerging. It may well be the case in the next few decades that we move from the model of a Western culture absorbing (or rejecting) Eastern ideas to a global synthesis in which both are synergized in something new. Here, environmental philosophy becomes a broad openness to the world about our shared ecological and cultural relationship with it. With this in mind, a few examples of the way in which Eastern traditions of thought (which encompass Western categories of 'philosophy' and 'religion' rather seamlessly) may be inspected for their potential as intellectual resources. Among other things, we will be looking for a cosmology which denies that any one perspective is 'better' than any other, whether it be divine, human, material or imaginary; a method of discussing human actions which are not founded upon the principles associated with reason and reasoning; and a language which would not denote the existence of any object free from some experience which gave it meaning as well as a classificatory category. A visual image often helps. Hua-yen Buddhism gives us the image of the net of jewels created by the great god Indra, described on p. 276. In it, there is no theory of a beginning time, nor of a creator, nor of a purpose. A human, for instance, is in some sense boundless: we are not 'in it' but we (and everything else) 'are it', and everything counts (Cook, 1977). The *Avastamsaka Sūtra* ('Flower Ornament Scripture'), in a Chinese text from the 7th century AD, also puts it more simply:

> The forest exists dependent on the earth,
> The earth remains solid based on water,
> Water depends on wind, wind on space
> While space does not depend on anything
>
> (Cleary, 1993, pp. 982–983)

Another example comes from the Japanese concept of 'nature'. In the west, the Latin *natura* has been employed in various languages. In most uses there is a common feature of meaning that which is outside: that tree over there; that person's nature which has developed without our involvement. In Japanese this European *natura* is translated by *shizen*. In one set of meanings this indeed means a way of being which exists without human intervention. But there is in its history a greater subtlety. In this reading, the term is used as an adverb, with a meaning akin to 'as something is of itself, as it becomes thus of itself'. When humans, animals or inanimate things manifest their own particular manner of being themselves, then they show their own *shizen* so are whole and are free. In one East Asian tradition, therefore, it is possible to refer to the 'buddhahood of plants and trees'.

It is naïve to suppose that the immense changes being wrought in Asia by the impress of technology can be redirected by studies in comparative philosophy. But at the very least, such traditions of thought can help the critical evaluation of the West by providing an alternative set of viewpoints; in addition they provide a rich resource of myth, symbol and metaphor to be used on their own or possibly to be taken up into a synthesis which is appropriate for a world that in some ways at least will soon possess a global culture.

Philosophy is for most of us a daunting word. Partly this is because most of us spend most of our thinking time engaged with material objects rather than abstractions. Also, academic philosophy often has its own language, which seems as impenetrable to the non-specialist as that of the more complex reaches of mathematics. But many of our actions are underlain by philosophical concepts: we assume the cosmos possesses order, that it is not an entirely random set of processes, for example. Or, following centuries of scientific thought, we agree that fact and value can always be separate: 'that is a value judgement' becomes as condemnatory as 'it's not economic'. One philosopher who has consistently tried to make such ideas comprehensible to a wide readership is Dr Mary Midgeley, some of whose works are discussed in the Further Reading section of this chapter.

Environmental ethics

In this context especially, ethics can be understood as the pursuit of goodness in a way which transcends individual interests and human imperfections (Pitt, 1988). Within the idea of ethics (Table 5.2) are a number of components, collectively called *norms*:

- **rights**: those conditions and actions which their holders ought to have access to if they so choose, such as being free of externally inflicted pain;
- **justice**: founded upon the idea of fairness, in terms of equality of treatment, either in opportunities or in outcomes;
- **ethical systems**: depend upon, for example,
  - (i) obligatory rules such as 'do not steal';
  - (ii) maximizing human welfare or Utilitarianism;
  - (iii) situational aims which promote the flourishing of each individual in each specific setting;
- **values**: derive from basics through to cultural and aesthetic preferences.

Table 5.2 Components of ethics

Norms	Equality	Opportunity
		Distribution
	Rights	Individual
		Communal
	Ethical systems	Rules
		Utilitarian
		Situational
		Personal
	Values	Socio-political
		Economic
		Cognitive
		Cultural/aesthetic

Source: adapted from C. A. Hooker (1992) Responsibility, ethics and nature. In D. E. Cooper and J. A. Palmer (eds) *The Environment in Question*. London: Routledge, 154

There are numerous complexities which beset such categories. The most obvious among them is the conflict between goods. The right to free assembly might trample a fragile ecosystem, and maximizing welfare of humans might very well leave wild animals in a very poor situation. So ethics is unlikely to provide a single correct answer to all the questions raised of the kind, 'what ought we to do?'

Scholars engaged in the study of environmental ethics have been much concerned with what is and what is not worthy of ethical consideration. On one side is the position that only humans and things that make us human fall within this ambit. On the other side is the argument that responsibility should include whole systems, whether local ecosystems or the whole cosmos with which we interact. The historical position concentrated upon the individual human but has been widened by environmentalists who would also include animals, for instance. This can be seen as ethical progress on the part of humans but is essentially an extension of human concern rather than an acknowledgement of the animals' intrinsic given-ness in the world.

In many ways, therefore, environmental ethics will reflect the real world in the sense that there will be complications; just as environmental management can only make complexity manageable in particular places at particular times, so the ethics will always involve some simplification and hence there will be conflict and tension. Granting rights to animals, for example, will in several contexts conflict with granting rights to whole ecosystems; with human rights, as of free assembly; or with maximizing the welfare of another species or of humans. Promoting artificial (yet desirable) environments like gardens may contend with the protection of wilder places; maximizing human pleasure in the present may be at odds with maximizing it for future generations; and above all there are the discords between human wealth and the wealth of the natural world.

One of the ways in which these disharmonies can be lessened is through design. By this we may mean the translation of environmental ethics into fields of morals (for example, arraying society's rewards and punishments so that they encourage individuals to act in a way favourable to the environment). As well, design in the technological field has a part to play. But above all, there is the need to design good institutions, whether of government, education, law or economics (to name only a few) which boost the attitudes towards the environment which we see as essential. Since governments at all scales play such an important part in most people's lives in advanced nations and increasingly in others, a good portion of the next chapter will be devoted to them. Nevertheless, it is a feature of today's global community that there are many processes of a political and economic nature that lie beyond the control of national governments. The turnover of a multinational company, for example, many well exceed that of many developing countries. Many companies and governments, too, are concerned that they continue to 'grow', where growth is measured in a yearly increase in throughput of energy and materials. Such is the nature of exponential growth, though, that the '29th day effect' is always a possibility. As Fig. 5.10 shows, if the lily pad doubles in size each day so that in a month it covers the entire pond, then on the 29th day it has covered only half the water area.

> It may have seemed obvious that 'the environment' or 'nature' is external to us: it is 'out there'. But the thrust of this chapter has been to complicate that model. The example is quoted of a smoker looking for an ashtray in the house of a non-smoker. She eventually encounters a seashell, which serves the purpose. But she did not 'find' an ashtray: she constituted one; to say 'I found one' is an interpretation. But the 'ashtray' exists in her experience as well as her mind: it is 'out there' as well. So the external world is indeed there in lived experience, which is much more than a purely subjective report of an objective reality. So, in many ways, nature/the environment is as much a social creation as it is an evolutionary product. It is a fascinating discussion (not alas to be pursued here) as to how nature would be perceived by a human individual with no cultural constructions in her mind.

Figure 5.10 Snapshots of a water lily which doubles in size each day so as to have covered the pond in 30 days. On the 29th day, the lily has covered only half the pond

Further reading

Over such a wide range of material, it is difficult to single out a few works which take the reader further without embarking on a whole year's reading project. Given the centrality of economics in the human–nature relationship, then probably Schnaiberg (1980) gives the clearest account of how capitalism has fashioned resource use and environmental impact. Perrings (1987) is harder to read, but gives a good idea of new explorations which are taking place. Most of the other books deal with a particular discipline and expand on the summary accounts in the present volume. Two summaries of much of the thinking in the ecology–economics–society interface are Hayward (1994) and Trzyna (1995). The non-linearity of human–environment systems is explored in Jantsch (1980) and Stewart (1989). Worster (1993, Chapter 13) explores the ecology–environment–predictability complex from the angle of chaos theory.

Hayward T 1994 *Ecological Thought: An Introduction*. Polity Press, Cambridge

Jantsch E 1980 *The Self-Organising Universe*. Pergamon Books, Oxford

Perrings C 1987 *Economy and Environment. A Theoretical Essay on the Interdependence of Economic and Environmental Systems*. Cambridge University Press, Cambridge

Schnaiberg A 1980 *The Environment from Surplus to Scarcity*. Oxford University Press, New York and Oxford

Stewart I 1989 *Does God Play Dice?* Basil Blackwell, London

Trzyna T (ed.) 1995 *A Sustainable World*. Earthscan, London

Worster D 1993 *The Wealth of Nature*. OUP, New York and Oxford

CHAPTER 6

The 'real' world and its alternatives

In ways not realized by many societies, philosophy and ethics spill out into public life, but they are also very much associated with what is left of the world's ivory towers. In this chapter, we consider how those philosophies and ethical ideas are put into action in the 'real' world of everyday affairs. So there will be an emphasis on the formulation of environmental policy and its translation into law. This is then followed by the very important matter of the institutions erected by various societies to put law (and other types of agreement) into practice. But there is nearly always a subversive element in human communities that seeks to put forward visions of alternative ways of relating to each other and the world. These individuals and groups may well promulgate the ground rules for a country or a community free from any blemishes, where all relationships are free of disharmony: a Utopia. Others confine themselves to a critical commentary, or to prescience about oncoming change through the media or the fine arts; we shall look selectively at all of these.

Environmental law

A competent body, often a government at one or another scale (local, regional, national), formulates environmental policies. The government acting on behalf of its people and on behalf of their environment will have certain ends in view (Sagoff, 1988). It may wish to move towards a so-called sustainable economy, or to reduce the emission of a pollutant or to protect the last remaining specimens of a rare animal, like Stoddart's peccary. All such a variety of ambitions are legitimately within the concerns of government. In a democracy these general aims are translated first into policies. At that stage there may well be discussion and consultation with numerous parties, with the publication of documents and perhaps public hearings or enquiries. A special commission of the great and the good may hear expert testimony. Eventually, though, a law will be enacted that will have the force of compulsion upon the humans of the territory involved, though less so on recalcitrant animals like the peccary.

The nature of law

Many laws relate to the environment in an indirect way. Those concerned with taxation, for example, may encourage or discourage the rates of use of various raw materials and hence the rate of change of the environments from which they come. Environmental law in a specific sense, however, is mostly about the protection of the environment in some fashion. Though we tend to think of the law as representing a set of prohibitions, it can also set out a framework for behaviour which may encourage in a positive way; it may thus move from a rather traditional reactive stance (of trying to remove an established source of pollution for instance) to one in which the prevention of harm to people, property and the environment is the main aim. This is the principle of prudent behaviour.

Environmental law has a long history. Pre-literate societies often enforced certain types of environmental behaviour through a variety of mechanisms, so that for example game species were not over-hunted, or irrigation water was fairly shared. Monarchs in medieval cities attempted to regulate noise (Julius Caesar tried to regulate chariot noise in Rome) or smoke (coal use in the London breweries was intermittently forbidden from the 14th through to the 17th centuries by the kings and queens of the time) and many nobles ensured that peasants did not poach the species of animal that were to be hunted as an aristocratic pastime. The 19th century saw the establishment of the first modern laws. These were often to protect certain species of wildlife by then becoming sparse, and in favour of which there was some public agitation, or to control some of the more noisome side-effects of the industrial revolution. Thus in the UK, the first modern pollution control came with the Alkali Act of 1863, designed to restrain the worst effects on public health of the manufacture of large quantities of caustic soda. On a world scale, this law was quite advanced as was indeed the comprehensiveness of physical planning envisaged in the UK's Town and Country Planning Act of 1947.

Much of this type of law-making has two characteristics in common the world over. Firstly, in keeping with the Western framework of ideas in which it is usually conceived, the environment itself has no rights. If a person's property is damaged by environmental contamination then there will be recourse to compensation at the expense of the polluter. But this is payable to the owner of the property and nothing is 'available' to any component of the environment that is affected. So if a portion of the environment is not 'owned' in some way, no compensation for misuse can be paid. Common-property resources, then, are apt to get damaged to a disproportionate degree, though it is not at all impossible to find ways of managing them for a common good. Secondly, whether and how the law is enforced is almost as important as what the law states. Thus the practice in a given place is critical. It is no use enacting a comprehensive body of law forbidding (let us say) the discharge of sulphurous gases to the atmosphere unless there is a body of trained personnel to record the levels of sulphur away from the emission point and then to have the legal expertise to take on the polluters in court.

In much environmental law, the state is the key body, for few environmental processes are so restricted in spatial scale that lower-level bodies can be solely entrusted with them. In the last 30 or so years, the international level has assumed an ever-growing importance, as the trans-boundary scale of many environmental processes has been realized. Hence, many bodies of domestic law are now nested within international legal frameworks. These may have general principles which they try to apply to all the states over which they have influence: the EU has adopted a 'polluter pays' principle in which a producer has to pay all the external environmental costs, including those incurred in avoiding pollution. Though it sounds comprehensive, it turns out be merely a guide in many countries in which it is supposed to apply. In the UK, for example, it is modified by the principle of BATNEEC, the Best Available Technology Not Entailing Excessive Cost, which can easily be seen to be an elastic addition.

Administrative frameworks

A good deal of resource use and environmental management in the world is done by individuals or small groups of people, often acting in accordance with customs that are orally transmitted. But beyond them, there is a great body of written policy and law which affects the environment and its resources. Here, action is taken in a fashion that reflects a holistic outcome of all the many factors of ecology, economics, philosophy, politics, fear and aspirations (and very likely others, like ignorance) laid out in the form of written statements about policies and laws. Institutions then put them into effect.

World-views: a reminder

The existence of certain principles of outlook and behaviour are critical in determining patterns of resource use. Dominant among these is the set of ideas that comprise the Western world-view. This contains the core idea that the Earth is a set of materials for human use, that material conditions are expected to get ever better for people, and that technology is the key to providing that affluence. In general, more of most materials (and certainly more choices), constitutes progress. If we want a simple label, then perhaps **technocentric** would be a reasonable word. It is worth noting that this world-view is common to both free-market (capitalist) economic structures and to socialist economies. Further, with the spread of Western imperialism in the 19th century, reinforced by the economic and cultural dominance of the West in the 20th century and now consolidated by the primacy of the industrial nations' media moguls in the field of global communications such as satellite TV, these Western views hold sway over much of the Earth. They influence (and seem likely to do so even more in the next 20–40 years at least) the outlooks and attitudes of people even in remote places and living in relatively simple economies. While it may have been possible in the past to maintain economies rooted in other points of view like the sacredness of nature or the fulfilment of only a limited range of material demands, it seems unlikely that these will persist at state level, even though dissenting individuals and small communities survive.

National governments

In the mesh of organizations that the modern world has erected to deal with its environmental processes and problems, it is not easy to find a logical starting-point. But since the nation state is at present such a fundamental building-block of most legal structures, we will begin there and thereafter look at both smaller and larger spatial scales.

The importance of the role of the state is plural. It rests, nevertheless, on a legitimacy in which it works for the interests of all, and is not seen as 'owned' by one interest group in the population. If a state government is accepted as legitimate, then it is a major power within its territory and is a crucial authority for making law. It normally has a monopoly on the use of force to implement the law. The sovereignty of the state extends to all individuals and over associations of individuals within it, and applies equally to all of them. Its servants are recruited and trained according to their skills and not their parentage. Lastly, the state has the power to tax its inhabitants and hence to apply some of that revenue to environmental protection and enhancement or alternatively to destruction and degradation.

The basic process is the adoption of a set of policies which are then enacted into law. The law is then administered through a series of regulations which derive their character and validity from the law, since they can be amplifications of the law but must not be in conflict with it (Polden, 1994). One great difference between different kinds of government is the degree to which they attempt to influence a society's use of resources and the resulting environmental change. Those with a commitment to collective modes of action, such as socialist governments, generally wish to regulate most phases of the systems, by having laws about emissions for example, or by having major resource-using processes such as energy and agriculture under national government control. By contrast, free-market capitalist governments tend to have an equally fervent belief in as small a role as possible for central government and so environmental management in all its phases is left to the mandates of the market. Ideally, a properly costed system would also deal with the external expenditures of any resource-using process. In reality, it is usually necessary to legislate for gaseous emission levels and for the control of toxic wastes, since market mechanisms cannot easily be made to apply. In the case of radioactive wastes, no government can abdicate its overall responsibility, since the danger to life of some of them is likely to outlast the lifetime of any private company, whereas we assume (possibly wrongly) that the nation state is everlasting.

State policies can then be divided roughly into two groups:

- Those which are *for* things, e.g. an energy policy that encourages the use of one source rather than another, or which supplies it at particular levels by means of rationing, as in Romania in the 1980s, where power and heating were supplied for only a few hours per day. Food policies might do the same kind of thing and could be organized to encourage the eating of less meat in the West or by contrast can be oriented by means of subsidies to preserving, for example, sheep rearing in marginal environments. Similarly governments can try to influence family size by pursuing pro- or anti-natalist policies.
- Those which are *against* particular kinds of processes or events. Environmental pollution is the obvious example, and most governments have some laws and regulations (not always obeyed) in this field. The location of industrial development is another matter which may not always be left to the free market: there may be incentives to locate industry (and with it the complex links of resource supply and environmental metamorphosis which are accompaniments) in areas of high unemployment or as a way of claiming territory.

Few nation states, however, are isolated islands and so they have to manage their territorial resources and environments in accord (or not) with neighbours. Many environmental problems in fact cross national boundaries and so regulatory levels above those of the state are often needed.

Supranational organizations

Nation states come together for many purposes but in the present context there are two main sets of purposes. The first of these is the conclusion of one-off agreements for a specific purpose, e.g. managing an internationally distributed resource. The second is membership of a more general-purpose supranational body like the European Union which has to deal with environment just as it must act on foreign policy, trade or transport (Holl, 1995; Kramer, 1995). In addition, a block of nations may be the subject of special concern by a subdivision of a higher unit, as with the Regional Economic Commissions (e.g. for Europe and Africa) of the United Nations Organisation (Caldwell, 1990).

Focused agreements

Focused agreements have a long history of concern with managing common resources, i.e. those without a single owner but to which many groups have access. The oceans are a particularly good example. Even in the 19th century, Pacific countries concluded agreements to prevent the extinction of fur seals, and there has been a subsequent record of catch limits on certain kinds of fish; these bilateral or multilateral accords were especially necessary before the enactment of 200-mile Exclusive Economic Zones (EEZs), when the oceans might be exploited by anybody with suitable technology. When a common resource like the open ocean has existed, it has been vital for the main users to regulate themselves so that newcomers could be policed or excluded altogether. Something similar has been true of anti-pollution agreements, as with the attempts of Canada and the USA to clean up the Great Lakes; neither one could achieve this alone and a co-ordinated approach has proved to be essential. Analogous processes underlie discussions about the control of acid deposition in the northeast of these two nations but agreement is proving hard to achieve.

Wider frameworks

Multipurpose organizations can have a great effect on environmental management, both by means of direct policies and indirectly as the results of other actions. The EU, for example, may have direct effects on drinking water standards through community-wide pollution control laws but may have even greater indirect effects via agricultural policies which encourage intensive agriculture and hence the application of large quantities of nitrogenous fertilizer to the land. Environmental change is likely to be speeded up as the result of regional policies which bring about investment in areas otherwise relatively poor: the industrialization of parts of northern Portugal, with effects on land use, water quality and shoreline management, may be quoted here. Such organizations behave like national governments but on a larger scale: they can of course deal with transboundary problems better than single-purpose concordats between several states since frameworks already exist and do not have to be specially set up. On the other hand, they may be slower to respond to new or newly discovered environmental problems than an individual state, not least because they have to be careful not to contravene national sovereignty, except where rights have explicitly been given up (Sands, 1993; Holl, 1994). However, supranational organizations may extend 'best practice' regulations to laggardly members: the introduction of the formal Environmental Impact Assessment and Statements (EIA and EIS; Fig. 6.1) came to the UK in the 1980s only by virtue of membership in the EU.

As with nations, policies and laws may be both for things and against them. Pollution is the most obvious target for the direct exercise of the 'against' variety: standards can be set and then enforced by means of community courts. Air and water quality are good examples of early targets in such a context. In terms of promoting the supply of certain resources, supranational organizations are usually more cautious but the EU, for example, has quotas for agricultural production which aim to eliminate over-production of certain foodstuffs and an agreement to share energy supplies if there is a sudden shortage resulting from, for example, political decisions outside the region. The outstanding example in all these fields is the EU but other blocks of nations are coming to perceive advantages in both resource and environmental matters.

Global organizations

Many companies are in fact global organizations but here we shall deal with public structures, of which the main example is the United Nations Organisation (UN). Also important are international finance bodies such as the International Bank for Reconstruction and Development (IBRD), usually called the World Bank. We can classify the activities of these world-wide phenomena into roughly two groups: (a) those that get involved on the ground showing how things can be done, and (b) those that store data and transfer information or money and thus act as catalysts but do not themselves get wet and dirty.

The first group is typified by some of the agencies of the UN which are concerned with development. They maintain databases and produce innovative research and publications but they are also concerned with sending out teams of experts who advise on the ground as well as in government offices. An outstanding example is the UN Food and Agriculture Organisation (FAO), which focuses its concern on the improvement of nutritional standards. It maintains a central HQ and

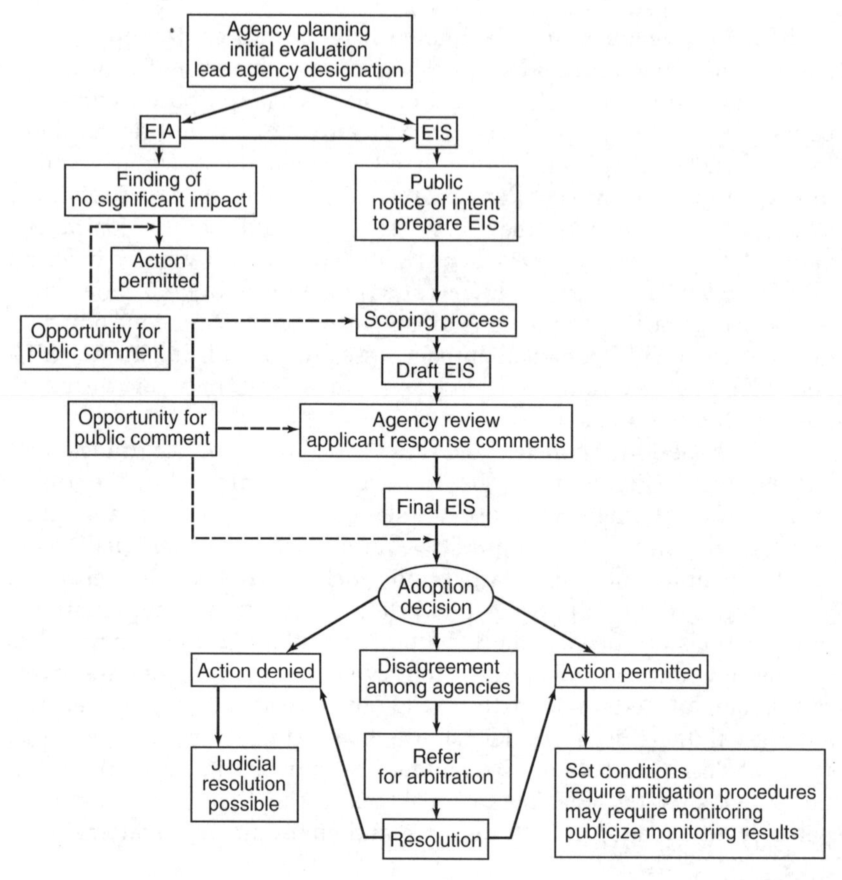

Figure 6.1 A flow chart for the carrying out of an Environmental Impact Assessment (EIA) leading to an Environmental Impact Statement (EIS). This is, of course, an ideal and the procedures at each step may not be perfect. To make a proper assessment might require years of ecological monitoring, which in most development proposals is unlikely to come about (from Roberts and Roberts, 1984)

regional offices, gathers information and conducts research, and puts together teams of appropriate experts to improve, for example, soil erosion problems, protein deficiencies, or livestock genetics. The UN World Health Organisation (WHO) is analogous in its field and is closely allied in structure and methods to the UN Fund for Population Activities (UNFPA), which is concerned not only with spacing births but with maternal health and sex education. The tasks of all these groups is very difficult, not least because they can be seen to represent 'top-down' development in which Western technology (be it plough or pill) is applied to the LDCs as the answer to the problems as perceived by the agencies themselves.

Behind these on-the-ground agencies lie those who set a broader context by trying to get global pictures and wide overviews before advising UN agencies or national governments. Set up in 1972, the UN Environmental Programme (UNEP) is of this type. Its main function is to co-ordinate the efforts of all UN agencies with an environmental concern, to extend the range of data available on environmental matters and to highlight areas of problem or potential problem. It operates through established channels and does not have local teams. It does, however, act as the lead agency for environmental treaty-making. A series of treaties focused on, for example, wetlands, or the polar bear, has been supplemented in recent years by agreements to tackle global problems such as carbon and CFC emissions (Fig. 6.2). The negotiations are, as always, thorny but it is probably a sign of some changed attitudes that such all-embracing subjects can even be contemplated.

More directly involved at one level is the IBRD, which provides capital for development in the LDCs. Many large projects, especially those concerned with water control and electricity generation, have been funded by the World Bank. It has thereby been heavily criticized for its

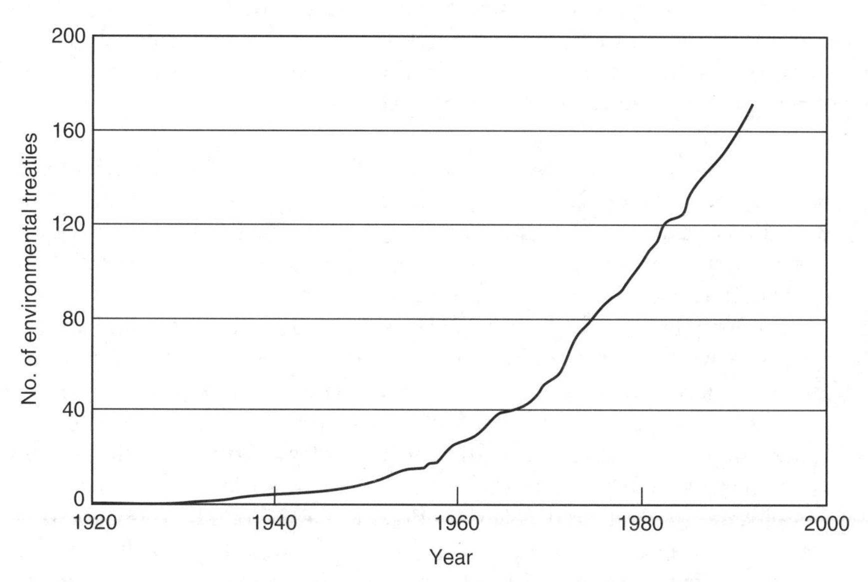

Figure 6.2 The gathering pace of accumulation of environmental treaties. Some of the most recent (such as the climatic and biodiversity conventions signed at Rio in 1992) are potentially global in their effect, rather than regional or aimed at particular habitats or species (from Brown *et al.*, 1995)

ignorance of the environmental and social consequences of some of its projects. One response in recent years has been to abandon the financing of large dams, for example, and to look with greater sympathy upon needs generated by, and expressed at, grass-roots level. But overall, the IBRD has been an instrument of technology transfer from North to South. Overlapping the concerns of both of these has been the International Union for the Conservation of Nature and Natural Resources (IUCN) which initially was largely devoted to the international aspects of species conservation but which has become the protagonist for sustainable development based on renewable resources, without distinction between North and South.

Local government

In many cases, the local government units of a national system see the sharp end of resource and environment issues (Blowers, 1993). They are often in the front line, so to speak, of any conflict between the citizens who are to experience (for good or ill) any changes that occur. Yet they may not have initiated these policies and indeed may be actively opposed to them (as when, for example, a nationalized energy industry proposes to open a new strip mine for coal in an agricultural area) and so are the unwilling deliverers of policies enacted without their consent. Local government in DCs is a complex entity, varying enormously in its structure and responsibilities from nation to nation, but it often carries out two particular tasks germane to our theme:

- *Development control.* This is usually part of the physical planning process now common in HIEs. Here the local government unit, acting within the context of national or supranational legislation, decides on matters such as land-use zoning: which areas are to be devoted to productive uses like industry, housing and commercial development, and which are to be in a protective use such as watershed areas, parks or nature reserves. This may not all be as negative as it sounds, since a large measure of economic planning may occur whose function it is to develop particular resources. The very concept of planning, indeed, is an invitation to a positive attitude to the use of resources. In HIEs these will most likely be social (e.g. to provide employment), but may also involve environmental considerations such as siting small factories in attractive surroundings. It may be necessary for a number of local government units to combine together if a feature such as a river transects their boundaries (Fig. 6.3).
- *Waste management.* In general, national governments do not wish to get involved in the day-to-day aspects of these processes except where the most dangerous materials are concerned, such as radioactivity and unused munitions. At the same time, the private sector has often been

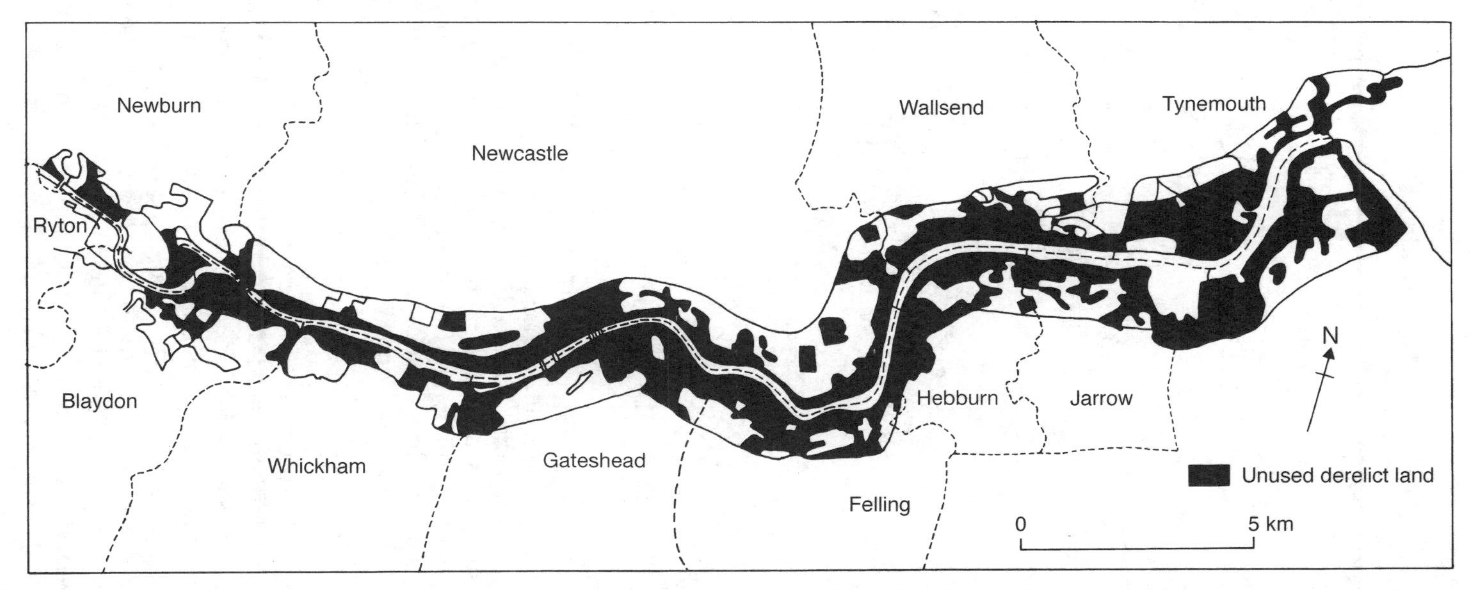

Figure 6.3 A land-use zoning plan for the corridor of the lower Tyne formulated in the 1960s, when it was not foreseen that the idea of raising the industrial use of the river frontage from 75 to 85 per cent would seem unreal in the face of de-industrialization of the region. Thus bodies set up to formulate such plans have to be careful to make them as flexible as possible in the face of unforeseen changes and are therefore cautious about committing capital which might produce irreversible effects. Today's plans are more likely to focus on leisure and shopping (from Anon., 1969)

Figure 6.4 Figure 1.6 put into its regional context. Note the large number of small disposal sites 'for operator's private use', which suggests a diversity of materials, all of which may need careful handling, and the need for large-scale public waste disposal sites: County Council 'intended' locations (from Durham County Council, 1984)

judged incapable of sustaining a commitment to the proper environmental considerations needed for the safe treatment and disposal of 'normal' wastes, which include some very toxic by-products of industrial processes. Local authorities, then, have to deal with municipal wastes, sometimes sewage, some industrial wastes and possibly specialized wastes such as those from hospitals. Protection of local waters, the finding of local burial sites and local air quality are thus likely to be the responsibilities of the town, county or regional council (Fig. 6.4). As such, they are constantly being badgered to improve their standards by citizens and citizen groups who want higher environmental quality. On the other hand, major taxpayers such as industry and commerce, or sources of subsidy like a national government, want less money spent on these matters so as to keep down corporate or public expenditure. Smaller local governments, too, are not likely to attract either the elected representatives or the professional staff to deal well with such complex interactions of public perception, scientific information and financial constraints.

The private sector

So far we have looked at the role of various organizations of a public character. Much of the actual processing of resources is, however, carried out by firms in the private sector along with their analogues in centrally planned economies like those of Cuba and North Korea. So companies are the resource finders, developers and suppliers. Often they operate on a very large scale across several nations, if not continents, and so the label transnational or multinational company (TNC) is applied (Fig. 6.5). Examples can be found in the areas of metals, timber, rubber, food and energy as well as many others, and international waste disposal is now entering the list of flows thus controlled.

The great virtues of TNCs lie in their capacity to absorb rapid change. In the energy field, for example, an oil company with developments in many different places can ensure a constant flow to its customers. If one source dries up temporarily because of a revolution or because a refinery is out of action, then there will be surplus capacity elsewhere which can be diverted to keep up supplies. If the coffee harvest in Costa Rica is blighted by frost then Global Foods Corporation can supply its Austrian outlets with coffee from Kenya without any hiccup in the flow other than a slight increase, no doubt, in the price to the consumer.

Because of their size, the TNCs can also cause difficulties for the nations in which they operate. For instance, a mine in a small LIE may provide a great proportion of the local employment and gross domestic product (GDP) but for a TNC it is a small operation, to be closed if the price for the metal falls slightly, or the costs at the mine get higher due to environmental protection measures or higher wages. Equally, TNCs may be able to manipulate demand, as was shown classically by J. K. Galbraith. Advertising especially can be used to create and maintain a demand for a product (like powdered milk in LIEs or frequent changes of car in DCs), which must increase environmental impact somewhere along a chain of resource–environment interaction.

The same can be true of many large firms even if they are not multinational. They too can manipulate demand and they too are

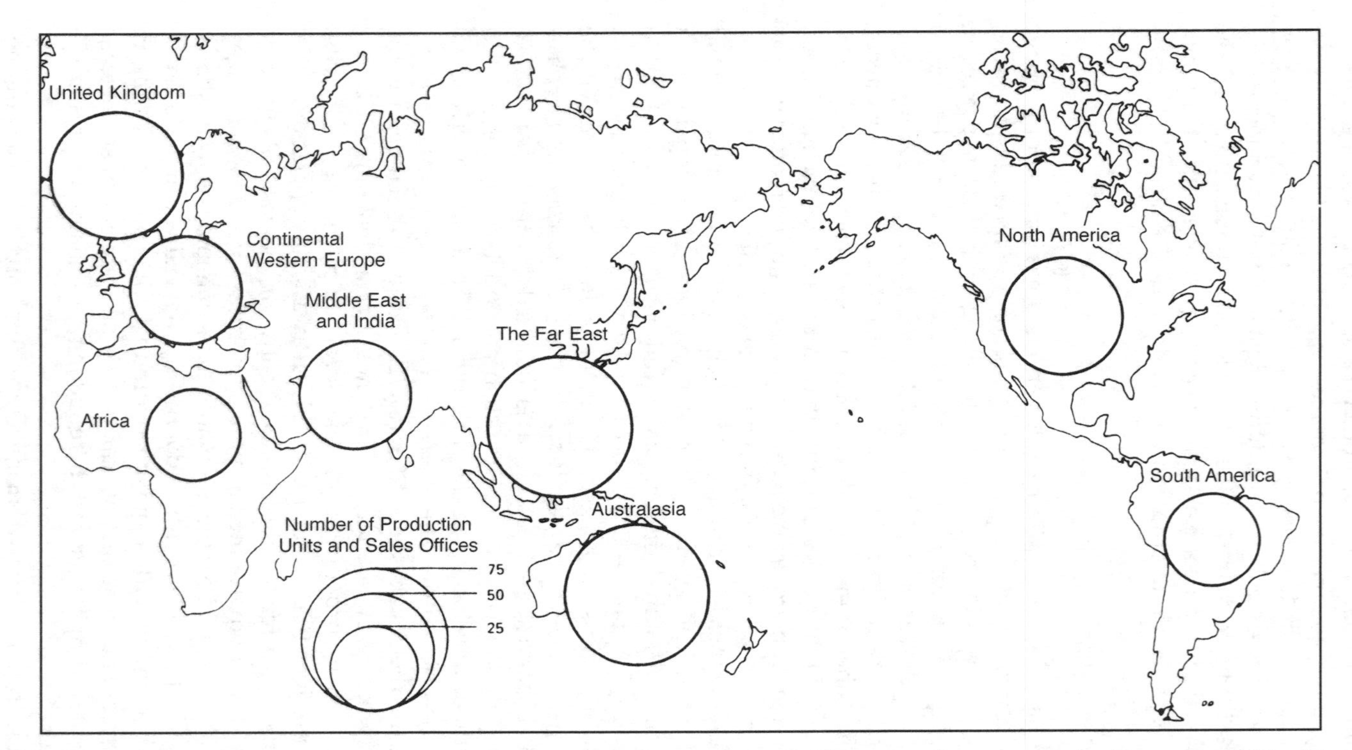

Figure 6.5 The world-wide division of production within a major multinational company in the early 1980s, based on the number of units of production and sales offices. The company was ICI, with headquarters in the UK. In general, though, there is a very even spread of numbers, though not necessarily of volume or profit (from Clarke, 1985)

responsible on the ground for resource supply and environmental change. They can also be the technological innovators, more willing to take risks than big and possibly sclerotic businesses. They may pioneer new processes which lead to lower environmental impact, or to greater energy efficiency for example. It is a mistake to see all private corporate involvement with resources and environment as being detrimental to everything except the company profits, just as only the ideologically blind would deny that all levels of government have a key role as arbiters of the common good.

NGOs (non-governmental organizations)

Ranged around the twin columns of the modern world economy, those of state and business, are the organizations founded by private citizens to pursue other ends. They pride themselves on freedom of thought and action and see themselves as collective consciences for those organizations (of any type) who seem to lack such a facility, and generally as goads of all those who are deaf to all voices except those of profit or ideology (Spretnak and Capra, 1986). They are by no means confined to the environmental field (development has a particularly dense representation) but it is an area which particularly spawns them. Like other organizations, they can be small and oriented towards a single issue or be virtually multinational companies. We can distinguish two main types:

- Those that are involved upon the ground with managing particular resources or ecosystems. In the UK, examples are County Wildlife Trusts which have nature reserves, and national organizations such as the RSPB which also manage land and water resources for the perpetuation of wildlife. These bodies are usually involved in pleasure and protection rather than production, though angling clubs may take home some of the catch. All are willing to engage in campaigning when their interests are threatened by actual or proposed environmental change.
- Those that do not own or lease property themselves but are campaigning bodies which harry governments, industries and commerce for better environmental practices. The Worldwide Fund for Wildlife, Greenpeace and Friends of the Earth are examples. Such groups have usually built up great stores of expertise so that they can mount technically sophisticated challenges to the accepted view on any particular topic; for example, on the issue of civil nuclear power they can produce their own economists to dispute the costings of the industry itself; on whaling they can argue that 'scientific' whaling is unnecessary for population studies by producing biologists who affirm the existence of non-destructive

methods of age–sex determination of minke whales. Some, like Greenpeace, apply the non-violent direct action techniques employed by civil disobedience movements though the years. The example of Mahatma Ghandi's independence movement in India is still a powerful one.

Those who envy these groups their independence, but who feel rooted in a more establishment-minded tradition of change, tend to set up permanent or *ad hoc* bodies of the eminent to produce reports which catalyse transformations of attitudes and practice. The International Council of Scientific Unions (ICSU), for example, has set up SCOPE (Scientific Committee on Problems of the Environment), which has produced a series of volumes of expert studies: those on global biogeochemistry and 'nuclear winter' have been especially influential. The number and scope of bodies engaged in research on global change, past, present and predicted, is very high, with 10 core projects in 1994; the impetus here is the possibility of global climatic change caused by human impact on the planetary ecology. Following the Brandt Commission's look at the economic context of North–South relationships, the Brundtland Commission tried to do the same for the environmental aspects.

The individual

It is a commonplace of today's world that the individual feels powerless. Complex governmental structures remove people from the sites of key decisions and even where they have representatives, such individuals may themselves owe their position to a party whose demands are more powerful than those of the electors. Equally, the idea of the sovereignty of the consumer is arrant nonsense, when many of their demands in the HIEs are orchestrated by large companies. The recovery of individuality is a complex matter and not one to be solved by formulae in textbooks. But the ways in which we define ourselves are somewhere at the centre of the issue of the relations of humanity and nature. Individuals are not totally powerless, of course. They can try to invest in environmentally responsible companies, for example, or they can adopt attitudes towards material consumption which reduce their environmental impacts. In aggregate, such actions may encourage governments to act, often first at the local level: the collection and recycling of wastes like paper and glass are instances. From time to time, individuals or small groups of people 'drop out' and try to live in an environmentally harmonious life-style, though this is less common now than in the 1960s and 1970s. The less radically minded can agree to live more simply.

Justice

The whole point of the law is the delivery of justice. Justice is founded on the idea of treating people fairly in a calculated way. Guidance from morality and ethics produces a system which provides for fairness, consistency and impartiality (Wenz, 1988; Hooker, 1992). There are various ways in which these concepts are applied: in this context the field of distributional justice (as distinct from retributive justice, for example) is especially relevant. This is basically about who gets what, and when and where) and deals with such matters as human rights and basic needs. It also entails a membership in society that enables everybody affected by change to have a meaningful voice in decision-making, noting the effect of distance in the intensity of moral concern and the notion of justice towards those who are regarded as 'other' or somehow not belonging. These are also the obligations we have to future generations bearing in mind the difficulties of prediction, and attitudes towards differences in society. Should, for instance, any change be made only if it benefits the worst off?

These principles are largely formulated as a system of social justice and hence are directed primarily if not exclusively at humans. Some of them have clear resonance for environmental components, however, and all have some kind of linkage to environmental systems. Should animals, rocks and trees be entitled to the same standards of just treatment as humans: do they have basic needs and the equivalent of human rights, for example? Are not plants and animals frequently the 'worst off' in processes of decision-making?

The environmental component in a judicial system is in some ways analogous to an absolute idea of justice rather than a relative one. Some authorities have always held that justice is founded on absolute principles that apply to every society, whereas others have argued that justice can only be relative to a particular group and that they have the right to formulate their judicial principles for themselves. Environmental components are possibly analogous to a relativist system while there is no great pressure upon them from human use, but when the limits are approached, then perhaps they become more absolutist in character. Yet human ingenuity in stretching the limits appears to be more or less unbounded (with some notable exceptions such as changing the composition of the atmosphere) so uncertainty prevails. One key element may be that if environment and society are inextricably linked, then any absolutes in the nature of environmental systems must find their way into human structures; justice may then not be about the preservation of the current social order in the way it so often appears to be.

This is scarcely a set of abstract concepts. The development of resource use and its associated environmental impact produced a

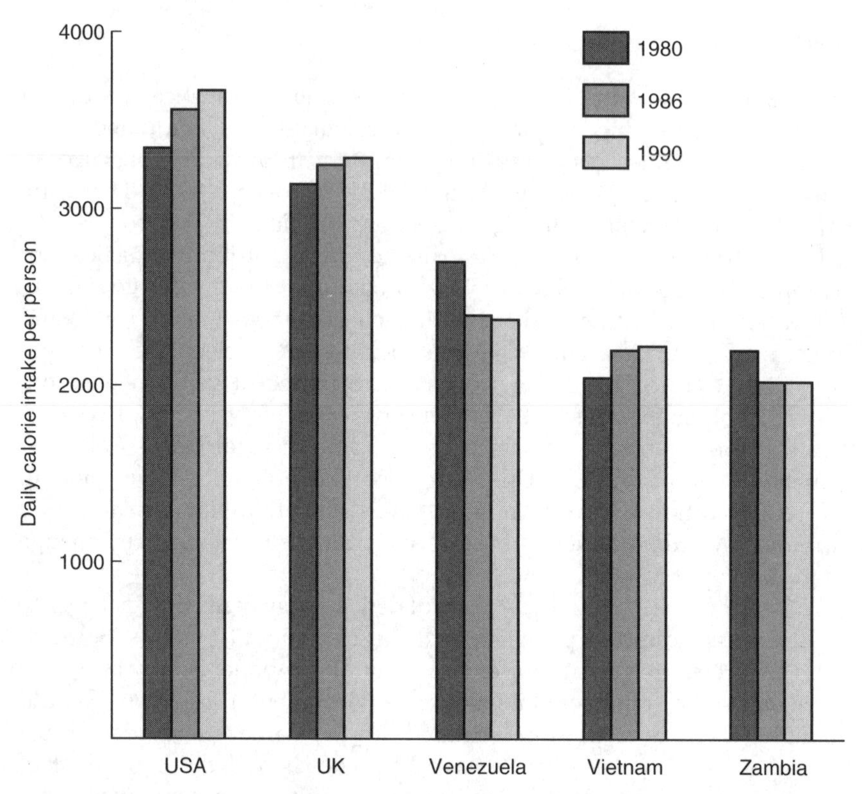

Figure 6.6 A simple measure of difference between rich and poor: the trends in calorie intake for a number of countries. The rich are getting richer diets, the relatively poor are barely holding their own. Vietnam probably illustrates the fact that war is bad for food supplies, as well as most other things (drawn from UN data)

fivefold increase in global economic output between 1950 and 1990. Yet such was the distribution of that growth that in 1960 the world's richest 20 per cent of nations absorbed 70 per cent of global income, whereas in 1989 the same percentage absorbed 83 per cent. In 1960 the poorest 20 per cent had access to 23 per cent, whereas in 1989 their share dropped to 1.4 per cent. Clearly there is a widening gap between systems that favour growth and consumption, and those which are devoted to equity and the alleviation of poverty. This chasm seems to produce very high rates of consumption at the top of the income ladder (Fig. 6.6) and all kinds of environmental degradation at the bottom. In Africa, some 22 per cent of all vegetated land has become 'degraded' since 1945, well above the world average of 17 per cent. North and Central America have only 8 per cent.

We are dealing with a complicated picture, with many overlapping layers as well as non-communicating sectors. World-wide, the number of organizations concerned with resource use and environmental change must be very large and themselves consume a great deal of energy and materials, especially paper, which we may hope will be recycled. What is extremely important, however, is the provision of widely available information (Spretnak and Capra, 1986). As ever, the possession of good information is a key to power and all the participants in these manoeuvres know it. So there is still much secrecy in the whole field of environmental impact and alteration, especially on the part of businesses and governments. If nothing else, that confirms the importance of the whole matter.

Alternatives of the imagination

The human mind is unique (so far as we know) in being able to construct a world entirely out of symbols. These are frequently verbal and the result is a written version of what we think the world is like or indeed would like it to be: law is one example. But humans represent the world in other ways as well; in the various creative arts such as music, sculpture, painting and drawing and in today's media such as photography, film, TV and video. The relevance for the present work is simple: these constructions of the mind may sometimes be the spur to action. Nowhere is this more likely than where an ideal world is deemed to be possible by an influential group or a charismatic individual, who then tries to move others to join them in creating such an ideal world. So we start here with the notion of Utopia.

Utopias

One of the many legacies of Classical Greece has been the Epicurean school of thought. It firmly launched the idea that the Earth was senescent and would in due course cease to be habitable to humans. The real time of plenty and harmony was past, in a Golden Age. Ever since then, there have been people dissatisfied with the present and the immediate future, and who want to recreate the wonderful times of the past.

To them can be added those who think that only a particular type of future is worth trying to build: one in which chance, contingency and disorder are eliminated in favour of a wonderful social structure in which there are no more problems. Both groups therefore are engaged in the search for Utopian ideals; in such societies the ideals are mostly social and political but sometimes they engage environmental matters as well. The classical example of the reclamation of the pristine virtues of a previous age is shown by Sir Thomas More, whose *Utopia* of 1516 was one of the first to harness the new ideas of scientific study to the desired ends. In this case the aim was to regain the Garden of Eden, no less, for the expulsion of Adam and Eve had clearly been caused by ignorance. More's work has some resonance today, for one of its lineaments was the control of technology: all new inventions were scrutinized by the guardians of Utopia and few were licensed. This has echoes in the mechanisms for the control of, for example, research into genetic engineering and fertility control and even more in the pleas for the deceleration of technological and hence social change in industrialized societies.

The basis of much Utopian thought and action is a speculation about the consequences of a fundamental transformation of society–environment relations rather than a simple extrapolation from the present. Thus it embraces imagination and focuses on the openness of the future rather than on the constraints of the present. But Utopias are never pure fantasy: they can challenge the findings of the sciences but not banish them. Few, in fact, aspire to the totally radical state envisaged by the writer of the Book of Isaiah where, in the last days, men were to 'beat their swords into ploughshares, and their spears into pruninghooks: ... The wolf also shall lie down with the lamb and the leopard shall lie down with the kid ... the desert shall rejoice, and blossom as the rose....' A mélange of the more familiar is more common: the Classical Greek author Euhemerus (*c.* 300 BC) invented an island Utopia where the blissful inhabitants lived in a climate that was eternally a soft, warm summer, where fruits fell into people's mouths so that there was no need for work. There is an echo of this in the holiday brochures. A technological approach to the subject is sometimes seen in today's environmental writing when an author tries to breathe life into a piece of prediction or advocacy by setting out a make-believe journey through his or her Utopian landscape. A 1993 example talks of 'A story of an imaginary town where the political will was found to apply sustainability principles to the design and operation of buildings, utilities, parks and transport' (Blowers, 1993).

The device of fiction is often used by thinkers to gain a wider audience for their ideas of an ideal society. B. F. Skinner, for example, laid out his reward-seeking, harmony-achieving community in *Walden II* (1948). With an environmental tilt, the novelist Aldous Huxley produced his idea of heaven in 1962 as a counterweight to his more famous vision of hell in *Brave New World* (1932). Heaven resides on the *Island* (as do many Utopias where they can be to some extent isolated from the corruption of the outside world) of Pala, which has no oil, no appetite for consumer goods, co-operative production of agricultural produce, population control and eugenics. There was also a high value put on conventional ignorance since, as one character puts it, 'habitual symbol-manipulation is an obstacle in the way of concrete-experiencing', the latter of which is very important in the predominantly Buddhist culture of the Palians. Alas, it all falls prey to their more modernizing neighbours, who take over while parroting the slogan of 'truth, values, genuine spirituality, oil.' Like several of Huxley's novels, the ideas are often rather better developed than plot and character, but as an example of environmental awareness entering a hitherto mostly social and political realm, it is a good read.

There is a moral in the take-over of Pala in the name of oil, for the assumption of Utopian ideals is that there is no need for novelty or change and that humans have a fixed and unaltering nature and that these can be aimed at universal, common and immutable goals; once these goals are realized, human nature is fulfilled. In the end, of course, the zealots who are the essential protagonists try to reduce the variety, innovation and inclinations of human life to a monochrome and static despotism. But to achieve this state, those who are clearly wrong can be destroyed, as in George Orwell's famous *1984*, published in the bleak Britain of 1949. It looks very much as if muddle and unstable equilibria are better for humans, which is also to some extent the lesson of ecological theories based on chaos and complexity.

The creative arts

The arts are usually defined as imaginative, creative and non-scientific activities of humans and divided into the practical and the creative arts. The former include fields such as architecture and design; the latter include a wide variety of visual, oral and literary constructions. The boundary between practical and creative need not, however, be fixed: the architecture of a monumental building will probably aspire to some degree of creativity and imagination as well as suitability for its immediate purpose. That the arts interface with nature is obvious, as is the antiquity of that combination, dating at least as far back as the Upper Palaeolithic cave paintings of southern France and northern Spain.

But in the setting of a 'cultural ecology' what might the creative arts tell us? Search a dictionary of quotations under the heading 'art' and one theme emerges strongly: that the arts enlarge our experience. This means, we might think, that they force us to consider ourselves in relation to nature in ways which otherwise might have escaped us and which science does not pretend to encompass. This may not always be a comforting experience: the painter Georges Braque (1882–1963) delivered the apophthegm that art was meant to disturb whereas science reassured. (This would be less accepted nowadays.) But if we combine disturbance with enlargement and assume that nature is often the subject of art, then it seems likely that here is a powerful source of environmental 'knowing' that must form part of any inclusive cultural ecology. Indeed, the poet Wyndham Lewis (1884–1957) said in 1927 that 'If you want to know what is occurring *inside*, underneath, at the centre, at any given moment, art is a truer guide than "politics", more often than not.' So how might the creative arts contribute to the inside of our knowing, the *outside* of which is the positivist view of the sciences and social sciences? It is, though, important to realize that in telling a story, any artist is selective in making his or her representation of the world they perceive. Hence, what they leave out may often be as much a story as what they put in. Likewise, the devices used to tell the story may reveal or conceal a great deal: the use of metaphor is probably the most important in dealing with the environment. It is often likened to something familiar (a mechanism, a family) when it may really not work like those things at all.

Prose fiction

The transition from scholarly analytical prose to creative prose is easy to make, for fiction is the 'most like' day-to-day writing. But here we are after what the poet Gerard Manley Hopkins (1844–1889) called the 'inscape', which in this case is the nub of the relationships between humans and their environments. Many imaginative writers attempt this task, often by using landscape, weather or biota as parts of their narrative. However, there is a difference, difficult to explain but obvious when it occurs, between using those elements simply as scenery against which the humans move (as if on a stage) and a condition in which their essential selves are bound together. The English novelist (and poet) Thomas Hardy (1840–1928) is generally thought to have exemplified the latter achievement.

Hardy appears to have needed a topography in which his characters were to move. His imagination did not work unless there was some solid ground under his feet, even though the land had to be distanced (perhaps even mythologized) by the renaming of places. The connection with the earth is emphasized by Hardy's habit of writing, while in the countryside, on large dead leaves, wood chips and pieces of stone or slate. His art, however, was in making the inner meaning of landscape visible to the reader. In some of the novels, this centres on rural change and even decay. His late (1895) novel *Jude the Obscure* takes the decay of the countryside as a constant theme and the village of Marygreen is cut adrift from its roots, as is the chief character: the village has no history beyond that of the recent months. More often, Hardy reads a landscape through its physical evidence and the connection of objects: the neck of a person may also be the neck of a jug and thus tied to a landscape reading in a physical sense. This type of scene formation is similar to that often used in the cinema and it is not surprising perhaps that several successful movies of the novels have been made.

Although many of the novels (e.g. *The Trumpet Major*) use the description of change as a reminder of time, the full panoply of his conviction that there was something fundamentally wrong in the scheme of things is deployed in *Tess of the d'Urbervilles* (1891). This allegory of humanity in a cosmos with no use for its desires or feelings might be said to echo the 19th century distancing of humanity and nature as human power increased. In part, this is brought to earth by looking critically at the perceived corruption of traditional agriculture, most famously in the description of the threshing machine, with the engineman who had nothing in common with either the region or its inhabitants:

> He was in the agricultural world, but not of it. He served fire and smoke: these denizens of the fields served vegetable, weather, frost, and sun.... The long strap which ran from the driving-wheel of his engine to the red thresher under the rick was the sole tie-line between agriculture and him.
>
> [Thomas Hardy: *Tess of the d'Urbervilles*]

Yet it is the engine which is 'the *primum mobile* of this little world.' More than that, the thresher which it powers is the 'red tyrant that the women had come to serve' and it is during the threshing that Alec d'Urberville reappears at the beginning of the sequence which leads inexorably to his death and Tess's execution, a fate which Hardy had earlier in the book prefigured in the dissonance between the social law and 'no law known to the environment in which she fancied herself such an anomaly.' Indeed, through most of Hardy's novels characters struggle against the dead-weight of the human landscapes in which they are trapped, and to which the 'natural' landscape is largely indifferent.

Poetry

All the issues of representations and of the discovery of 'inscape' present in creative prose are found in the intensified and re-angled form of creative language we call poetry. 'Nature poets' abound, as do those of place, but the harder task is finding how humans and nature 'are' when together. Here we shall look briefly at a poet not normally considered in this connection but whose output prefigures quite a lot of today's Green thinking: W. H. Auden (1907–1973).

We start in a landscape underpinned with calcareous rocks. Thomas Hardy portrayed the chalk downlands of Wessex sometimes as lush and warm, and at other times as meagre and unforgiving; Auden's childhood included the harder Carboniferous limestone of the northern Pennines:

. . . those peat-stained deserted burns
that feed the WEAR and TYNE and TEES,

[W. H. Auden: *In praise of limestone*]

which became for him the memory of innocence and then by transference to all that is good,

... when I try to imagine a faultless love
Or the life to come, what I hear is the murmur
Of underground streams, what I see, is a limestone landscape

[W. H. Auden: *In praise of limestone*]

If, though, the limestone country (with all its scars from centuries of lead mining) can represent a kind of heaven, other landscapes are more redolent of a humankind that is in dire need of moral guidance and reform, summarised perhaps by the title of one poem of 1933, *Paysage moralisé*. That poem itself ends in a rejection of a Golden Age. The gods visit us from islands but the desire to leave our wretched valleys for their abodes creates such troubles ('so many, fearful, took with them their sorrow') that joining them in a Utopia is no answer. Instead, we need the presence of water (not always the case in limestone country) to irrigate all our *ecumene*,

And we rebuild our cities, not dream of islands.

[W. H. Auden: *In praise of limestone*]

That landscape might thus express the moral condition of a society (taken up much later with characteristic spleen by Phillip Larkin in *Going, Going*) is demonstrated at a smaller scale in the seven-poem sequence *Bucolics*, and in particular the second poem *Woods* (written in 1953 or 1954), where

The trees encountered on a country stroll
Reveal a lot about a country's soul

[W. H. Auden: *Woods* (In *Bucolics*)]

With the perhaps inevitable finding of

A small grove massacred to the last ash,
An oak with heart-rot, give away the show:
This great society is going smash;

[W. H. Auden: *Woods* (In *Bucolics*)]

and the take-up of this country walk into the macrocosmic

> A culture is no better than its woods.
>
> [W. H. Auden: *Woods* (In *Bucolics*)]

(quoted in the Vignettes near the beginning of this book), which almost inevitably includes (in *Streams*) some praise of water in limestone terrain: in this case Swaledale in North Yorkshire.

A case can be made for the heart of 'inscape' being found in one of Auden's reactions to air travel: the 1954 poem *Ode to Gaea*, in the last word of which we recognize the eponym of the Gaia hypothesis. This long poem prefigures the kind of image we now associate principally with satellite-based remote-sensing pictures,

> it is the old
> grand style of gesture we watch as, heavy with cold,
> the top-waters of all her
> northern seas take their vernal plunge
>
> and suddenly her desolations, salt as blood,
> prolix yet terse, are glamorously carpeted
> with great swatches of plankton,
> delicious spreads of nourishment,
>
> [W. H. Auden: *Ode to Gaea*]

which is followed by a dissertation on the works of humans, of whom the wise 'wilt in the glare of the Shadow', leading to an assessment of our activities being,

> to Her, the real one, can our good landscapes be but lies
> Those woods where tigers chum with deer and no root dies,
>
> [W. H. Auden: *Ode to Gaea*]

a final state of affairs which might only be achievable in the end days when lions lie down with lambs and swords are beaten into ploughshares. The identification with the Gaia hypothesis of the J. E. Lovelock kind comes in the earlier line, 'and Earth, till the end will be Herself. She has never been moved/Except by Amphion...' This is a direct foreshadowing of the current Gaian notion that if humanity transgresses certain limits, then the tendency of Gaia to maximize for all forms of life may render human habitation impossible. It seems unreasonable to ask more of a poet writing in the 1950s than to express the core of a scientific idea that emerged in the 1970s.

Painting

'No problem', we might say, 'except there's so much to choose from', and indeed the shelves are full of critical studies showing how many artists (and especially those who concentrated on landscapes) have provided a text which cannot only be interpreted but also used as a basis for socio-political construction or indeed feminist deconstruction. Let us take the slightly harder example of a 20th century painter who started out in a figurative mode (which included landscapes from his native Netherlands) but ended up with a highly reduced abstraction: Piet Mondriaan (1872–1944).

Mondrian (who dropped the second 'a' when he moved to Paris) was much influenced by the Cubism of Picasso and Braque, but this often meant a more linear and angular approach to landscapes and trees, for example. This can be seen anticipated in some of his paintings of tall buildings like mills and towers from the first decade of the 20th century. He was, he said, not very keen on depicting solely the hidden relationships that were evoked by nature, even though the permanent structures of nature might reveal profound and solemn values: the painting 'Woods near Oele' of 1908 appears to embody in its paint and brushwork the movement and energy of nature on a grand scale. But Mondrian averred that appearance was not the totality of reality since humans also consisted of *inwardness*, i.e. soul and mind, as well as outwardness. All had to converge upon the general rather than the specific, and naturalism was too much tied to the specific (Blotkamp, 1994). This was what justified the form of abstraction (which he called Neo-Plasticism) that resulted in paintings such as the 'Composition with Two Lines' of 1931.

In fact, the later (from the 1940–1944 period, when he lived in New York City) work such as 'New York City III' could be regarded as a simplified map, in oils and tape, of the rectilinear grid of Manhattan. In this it takes to an extreme the earlier influence of the Cubists as seen in the 1912 pictures of trees that seem to be projected onto a plane of glass, rather in the manner of the roundels and rainbows hanging in some windows today. All perspective and projection have already been lost. The later paintings seem to move towards an extreme of Newtonian physics where there are only two lines of force striving in different directions (at 90°) with only one place of meeting and that eccentrically placed. In its convergence upon the general and degree of abstraction, another reading might place such a painting alongside the text of a scientific theory where both are exercises in applying the maximum loss of detail and the maximum gain of generality. Yet another view might deny that the picture could be seen at all as a totality, for it is always perceived in some context, if only (but in fact never only) in the gallery of the Stedelijk Museum in Amsterdam where it is exhibited. In that setting the intertextual connotations come into play and two lines might evoke in the viewer, for example, the whole of the totalizing enterprise of modernity.

Gardens and sculpture

The garden is present in most forms of human culture; usually adjacent to the home and usually enclosed. Since it may fulfil either or both of two purposes (beauty and amenity on the one hand; food and medicine on the other), it is more likely to approach high art when it belongs to the tradition that derives from sacred sites. In the West, the enclosed garden was early identified with the Persian word *paradeiza*, and indeed the Paradise of Eden was, notably, a garden. All examples share one characteristic: the garden is an enclosed space where nature can be intensively moulded in a cultural image. The size depends upon the labour and more recently on the technology available, but the desire is there: to express a culture in a combination of living and non-living materials.

The best examples of the garden as high art are found in Japan, where inherited Chinese traditions have been transformed into distinctive icons, at least since the Heian period (AD 794–1185). Although elements of the garden symbolize natural features such as mountains and water, there is generally a symbiosis between natural form and the right angle. Much of this is ultimately derived from Chinese geomancy, where a square Earth is surrounded by the circle of heaven. The earliest gardening manual dates from the latter half of the 11th century and already expresses a 'Japanese soul' as well as Chinese models for layout and components. It is therefore perhaps to be expected that the Japanese garden thereafter shows its qualities in a sensitivity towards nature (Fig. 6.7) which is coupled with the kinds of rigidities that have outwardly typified Japanese society ever since.

In the West, one of the best examples of the garden as a cultural emblem is the development of the English landscape garden in the 18th century. In the 17th century, the English renaissance was much occupied with neo-Classical models, rigid and formal in their lines. In the 18th century, the influence of Romantic thinking induced a shift to emblematic gardens which told stories. Poems, satires, genealogies and manifestos might all be present. One of the most famous is Stourhead in Wiltshire, which in part tells a story from Virgil where Aeneas is led into the underworld. Gardening in the 18th century was a form of high art that was ranked with poetry and painting and one interaction took place in the theory of the picturesque: landscapes became prized for their resemblance to paintings. If they were suitable for painting or reminiscent of certain styles of painting, then that added to their value. The logical heir to that relationship today seems to be 'environmental art', which is usually a form of sculpture in the open air, avoiding all the

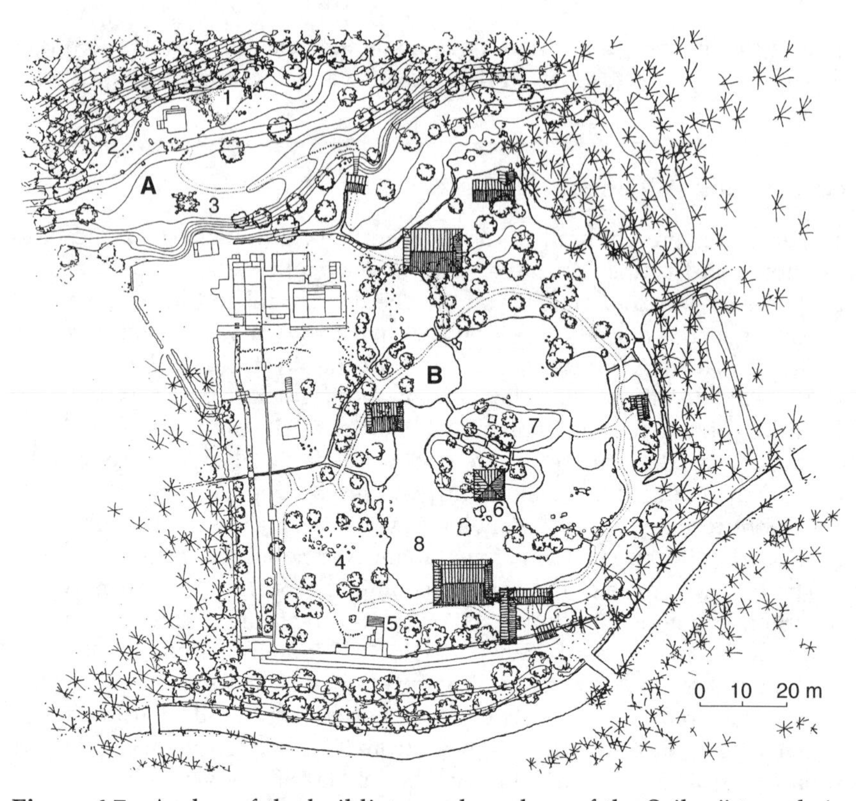

Figure 6.7 A plan of the buildings and gardens of the Saiho-ji temple in Kyōtō, Japan, started in 1334. The contrast between A and B is that between a moss- and a rock-based garden (1 = rock waterfall; 2 = stone for *zazen* meditation; 3 = turtle island of rocks) and a pond garden (4 = a special stone; 5 = tea house; 6 and 7 = islands of the evening and morning sun respectively; 8 = golden pond). Note that all the buildings, past and present, are rectangular, are on the same alignments and do not seem to follow the 'natural' curves of the ground (from Nitschke, 1991)

spatial and cultural constraints of the gallery. There are various forms: some are performances on a massive scale, even if only temporary, like Christo's $24\frac{1}{2}$ mile 'Running Fence' in California. This was taken down after a few days. Ephemeral in another sense is the work of Richard Long, starting in the outdoors in 1963. His work entails responses to a site which are meant to complement the landscape but which will soon be erased. The only record of them is then photographic. At its most thorough, this type of art may talk of 'co-authorship' with natural energies and structures, as with Alan Sonfist's work in the 1970s, including the growth of micro-organisms on canvases (Ross, 1993).

Television

TV is now a powerful medium of validation of social behaviour, as once were preaching and print. It does not concentrate on being an art form, however, since it has many functions in conveying news, facts, opinion and entertainment rather than the initial appeal to the senses that is common to the fine arts. At an overt level, engagement with the environment comes via the medium of wildlife and environment programmes. Less obvious is the endorsement of the Western world-view through programmes which emphasize the successes of science and technology, especially in medicine.

The closest to an art form that TV gets seems to be in its advertisements. Like 18th century oil paintings that 'sell' a view of ownership and property (whether of land or other people), the advertisements are concentrated doses of a message. Though the product is overt, the context (or subtext) often has a world-view message. This reinforces the desirability of the items to be sold, and the environment can be pressed into service as can other icons of today's world, e.g. the white-coated scientist or the be-jeaned youth. Some advertisements sell environments directly, in the manner of a holiday brochure: often indeed the same product is involved. There is no great subtlety needed in reading them: skies are always blue, everybody is happy, everything is safe. More interesting are those advertisements in which the environment plays a supporting role, though given the need to fill the screen, it may occupy the viewers' retinas for a greater time and space than the product being sold. Somewhat artificially, a number of categories of environmental involvement may be recognized.

In the first of these, the environment is presented as a danger to be overcome. Cars pass undamaged through curtains of fire that are sweeping the roadside landscapes, for example, or they may successfully negotiate a mountain pass with poor surfaces and precipitous drops. In extreme cases, cars are driven without problems on ski slopes or glaciers; all of this is intensified where four-wheel-drive vehicles are being sold, even if mounting the kerb of a Chelsea sidewalk is likely to be their hairiest off-road experience. There is a complete alliance with the Western world-view that celebrates in a guilt-free manner the dominance of technology over nature. A second, commoner, category puts the product in environments that are good to be among, such as rural landscapes of high value or picturesque towns full of old buildings. Cars again feature here, moving smoothly along winding but empty roads to dawn destinations in romantic port villages; to get at the top brand of food, dogs of all breeds bar the ugliest stream across open lands and through sparkling rivers without even pausing to add nitrogen to the ecosystems; soft drinks are consumed in orchards dripping with fruit or on gleaming white yachts that cleave the seas without making anybody seasick. In the UK, the visitor centre of the nuclear reprocessing plant at Sellafield appears at the end of a long pastoral idyll in the mountains and at the coast of Cumbria; one version includes the increasingly rare barn owl as a symbol of identification of environmental values. All these are examples of the transference of virtue or goodness by association. The unspoilt rural landscape is a *paysage moralisé*: it is charged with virtue and some (no doubt, the companies hope, a great deal) of this rubs off on the advertised goods. Lastly, there is the inevitable Golden Age. We are asked to pretend that the merchandise is not the outcome of 20th century technological processes but part of the past in which everything was natural and, very likely, hand-made. To reinforce this legend, the environment appears as part of an idyllic past: again, rural landscapes of high perceived value and small towns are brought into service. Given recent trends to 'healthier' food, bread has been a particular item for this treatment: we can now get bread just like it was when baked in the village ovens from locally harvested corn. Inevitably, the automobile gets this treatment: old-established marques are placed in 'quality' settings so that graciousness of place and past time can be rubbed into the bodywork along with the wax polish. Both crust and cars can be reinforced with the kind of music that paints a wash of nostalgia over the whole: in the UK, the more pastoral bits of Elgar, Ralph Vaughan Williams and Dvořák have been extracted for this purpose.

The importance of these communication forms, as with many others, is impossible to assess. Among other things, though, let us remember that advertisements are designed to sell goods and that every skill will be deployed to subvert any viewer who is considering reducing their consumption. We have to acknowledge that TV is a more pervasive, and presumably persuasive, medium than anything else around at the moment or indeed through the whole of human history. Via satellites, it is coming to have a global reach yet the miniaturization of technology (seen first in CCTV) may give it a strong local role as well. It is obviously well suited to be the outlet of computer stored and generated information, thus linking it to another *sine qua non* of validity; look how many TV ads of all kinds feature a person who is in the presence of a computer screen even if not actually using it.

The movies

In many ways, the cinema is the precursor of TV and video. Obvious differences are the length of the item, which is usually at least 90 minutes in a feature film, and the telling of a complete story in a self-contained unit; serials are now rare. There are now few documentaries of the kind that used to precede the main film (along with news and advertisements) since their role has gone more or less completely to TV.

So the focus is an act of creative narrative and there are examples of stories which are overtly ecological in character. If made in the USA, then Native Americans (Bierhorst, 1994) are a frequent choice: *Dances with Wolves* is an example.

The agenda may be more hidden, however, with the moral having to be inferred rather than spelled out directly. Many movies in which some monster of human creation runs out of control are taken to be paradigms of the enslavement to technology. Many start in the Jewish legend of the Golem, a clay creature who is brought to life. The robots who take over were the theme of Karel Čapek's (1920) stage play *Rossum's Universal Robots*, which in turn was the basis of the 1960s films *Westworld* and *Fantasy World*. The idea of the runaway monster created by humans but no longer responsive to their command is transferred to an actual machine in hundreds of formula horror thrillers but the star example for us is the nuclear power plant in *The China Syndrome* which threatens meltdown, China being antipodean to the eastern USA.

Intellectually more challenging (and possibly more effective since the 'message' is less conspicuous) are films where the surroundings endorse or convey or conflict with the central theme, as in TV advertisements or in Hardy's novels. The films of the Japanese director Akira Kurosawa (b. 1910) are excellent illustrations and *Dreams*, first shown at the Cannes festival in the 1990s, is most like an environmental parable. It consists of a two-hour set of eight episodes which are different in subject matter, each with an 'I' character who is central to the setting and action of each passage. The sections start with childhood and end with old age and death, though the 'I' character only ages to his mid-20s. Several of the scenes abound in environmental references; 'The peach orchard' sees the young boy being led into the hacked-about remains of an orchard and is shown it (by a group of traditional Japanese dolls) when it was in its full glory because he had cried when it was cut down. In 'Mount Fuji in red', the scene is of panic-stricken flight from a volcano which is melting down after being triggered by an exploding nuclear reactor. Yet in 'Village of the watermills' the film ends in a peaceful rural village with no electricity and only water power. The central event is the joyful funeral of an old woman in which the coffin is preceded by dancing children. Other episodes are less obvious in their references. There are the near-deaths of some mountain climbers, some weeping demons of Buddhist persuasion in a volcanic crater and the unquiet ghosts of a wartime platoon in a Wilfrid Owen-like tunnel. We also meet Vincent van Gogh after walking through some of his later pictures.

How do we read Kurosawa's understandings? Is he, as some critics have suggested, reflecting a core of alienation of humans from nature as well as from each other? Are the cutting of orchards and the presence of nuclear power going to lead to an inevitable perdition? Are we tampering with something forbidden, as in the first scene, when a little boy sees the foxes' wedding procession and they invite him to kill himself? There are some signs of hope: the boy's mother tells him to look for the foxes at the end of the rainbow and apologize; the nuclear power executive accepts responsibility in traditional Japanese fashion; and the peach orchard has one tree that has survived mutilation enough to bloom. Whenever things are getting better, Kurosawa uses Western music on the sound track – solemn quasi-Elgar, quasi-Walton type music. Is this, we might wonder, implying that new ideas and transformed relationships will come out of mainstream Western cultures and not from Asia, or from 'alternative' subcultures? Yet in the end we are back into the world of traditional values in a pre-industrial economy where there is a reconciliation between the good life and the good death. And this clearly will not do as a practical message, for there can be no going back, even if a lower per capita energy consumption is highly desirable, and soon. But then again, the title is *Dreams*.

Music

Look up music in any dictionary of quotations and the common item will be from the English essayist and literary critic Walter Pater (1839–1894), who said that all art constantly aspires to the condition of music. The explication of this usually revolves around the strength of music's emotional appeal ('Is it not strange', says Benedick in *Much Ado about Nothing*, 'that sheep's guts should hale souls out of men's bodies?'), and its abstraction from time and place. We can reach out to a shelf and instantly hear the whole world's music, from yesterday's composition back to the earliest of which we have some knowledge. Like most creative arts, music is heterogeneous and here the most consciously constructed form is discussed (i.e. art music) rather than the more spontaneous and sometimes ephemeral forms of folk music, jazz, rock and pop. This is probably unfair to folk music, for instance, which can preserve its essential features through many metamorphoses. Also, it can get taken up into art music as being emblematic of place and landscape. But how many people remember the dominance of skiffle in the 1950s? Yet since, like poetry or painting, art music is a conscious construction, we need to ask whether it too can express the inscape of human–environment relations.

At one level, like the cinema, there is the evocative but basically descriptive level. The music of place is generally of that category: the 19th and early 20th century spawned a number of *rhapsodies* which were tied to place or nation: 'España' and 'Fenland' typify the genre. The work of Jean Sibelius (1865–1957) is generally held to be so Finnish (though without using folk music) that it could simply not have come from anywhere else. Whether that would be so if we were not given the cues from the Finnish national epic poem (*The Kalevala*) in some of the

Figure 6.8 An extract from R. Vaughan Williams's Fifth Symphony: just after the opening bars. The lower part is played on the horns, the upper on violins (from Hill, 1949)

tone poems, is open to discussion. Olivier Messaien (1908–1992) incorporated bird song into highly abstract religious music so as to give the listener some sense of how to react, perhaps. For English (*sic*) listeners, the sound of horns and slightly wistful strings places us in rural and largely unspoilt sourroundings, like the beginning of Ralph Vaughan Williams's Fifth Symphony of 1938–1943 (Fig. 6.8). But the environmental anxieties of the time seem to be present in the sense of hesitation and uncertainty: it is not all watercress and cowpats.

A less obvious level can be seen in the theme of water and gardens in the work of the Japanese composer Tori Takemitsu (b. 1930). Many of his works have titles which elicit natural processes, such as *Rain Coming, Rain Spell,* and *riverrun* (*sic*) which use both Western and Japanese modes of musical discourse. Overall, it seems as if the music wishes to express not the romantic idea of the composer's soul but the more distinctive collective feeling characteristic of Japanese culture. Yet the language is often Western. A more complex exploration of the interaction between humans and landscape belongs to the US composer and insurance executive Charles Ives (1874–1954). Some of his work is overt in title and references: *Three Places in New England* (1908–1914) uses American hymn and marching tunes, for example. But Ives tried to get beyond that kind of structure to a dialectic in which the mind of the listener has to grapple with more than one idea at a time. Ives likens this to looking at a landscape,

> ... trying out a parallel way of listening to music, suggested by looking at a view (1) with the eyes towards the sky of the tops of the trees, taking in the earth or foreground subjectively – that is, not focusing the eye on it – (2) then looking at the earth and land, and seeing the sky and the top of the foreground subjectively. In other words, giving a musical piece in two parts, but played at the same time.
>
> (Quoted in Serio, 1978)

The result is Ives's unfinished *Universe* symphony (1911–1920), which needs to be heard to judge whether this translation of a basically visual way of environmental cognition to a sound-sequence can be successfully made. If it can, then composers can try not merely to depict the basically static elements of a scene but to think about the processes as well, even without using words. Alas, the symphony requires six to ten different orchestras on different mountain-tops and the score exists only in fragments.

There is an even wider implicit agenda, little explored by commentators. This links to the philosophical notion that humanity today is enclosed by science and technology. In that sense, there is no longer human control over those cultural activities: we live in a world in which they are as much a part of the surroundings as the air or the soil. Thus music can be an expression of a technological world without our realizing it. Examples which can be quoted suggest that traditional jazz was much influenced by the railway, with the basic four-square rhythm echoing that of the wheels over the rail-ends, a sound no longer heard in days of welded track. Another tack is to see a response to the invention of sound recording in the terse utterances of the second Viennese school: Anton Webern (1883–1946) is the obvious choice. Writing music in, for example, the 1900–1910 period, one horizon was the four minutes or so that could be got onto a recording cylinder. Whether the LP of the 1950s produced longer works in art music, as distinct from collections, is another matter. The extreme of this argument seems to be in the amplified world of rock and pop: this is not powered by windmills.

The arts seem more than one world away from that of the natural sciences, and at least one from that of the social sciences, law and ethics. Yet there are some parallels which we might hold in mind against any time when we try to make sense of the whole issue of humanity and environment rather than its parts. The first is that the creative arts all hold in common a relationship between order and variety. Too much order in the arts (say in the rhythm of music) leads to boredom; too much variety leads to a sense of disorder or even chaos. The same is true of painting and equally legitimate of an ecosystem. A single-species ecosystem possesses order; a randomly populated piece of early succession may have a lot of species but has not yet acquired much of the order characteristic of later stages. So in terms of their structures, there may be more in common between some of the phenomena of the real world and the creations of the human mind than we normally realize. This is partly explained if we recall the importance of metaphor in description and explanation. In explaining processes in scientific terms, recourse is frequently had to metaphor, just as a novelist will use it in talking of scenery or people. Consider only the example of 'the conquest of nature', so popular in the 19th century and until very recently in the Communist slogans of China. To what extent are military metaphors (a) accurate and (b) appropriate in this field?

A world I never made

The poet A. E. Housman (1859–1936), wrote of somebody being 'a stranger and afraid' in such a world and to come to terms with that emotion is one of the functions of religion, though some religions now place some emphasis on Housman's more celebratory words that,

> . . . malt does more than Milton can
> To justify God's ways to man.

In general, though, religions look for ultimate explanations of what humans do not understand and so are not afraid to look for explanations in the field of the supernatural rather than the natural or the purely human (Gottlieb, 1995). In the field of environment, religion has impinged at a number of places:

- the idea that the Earth was created by a god and that humans therefore have a special responsibility to look after the planet even though they are just one piece of that creation;
- the belief that humans are at the apex of a Creation and therefore have the right to order the Earth as they wish;
- the notion that imperfection is bound up with being human and that the resulting desire for goods and for power is bound to result in human and environmental disaster;
- the feeling that nevertheless the Earth and the Universe are places of wonder and that their qualities should he celebrated, and thanks given to their creator.

The diversity of belief in the various world religions is quite wide and so in the present context only a brief discussion of the explicit attention given to environmental matters can be given for some of the major faiths.

Christianity

The beliefs held within this religion are not uniform, but the following are central to most orthodoxies:

- A hierarchical construction of the Universe in which God is supreme, presiding over a descending list of the type angels–men–women and children–animals–plants–non-living matter. Value is assigned according to position on this scale and so endorses patriarchy in social structures.
- There is a fundamentally retrospective view of perfection in the account of the Garden of Eden. This flawlessness was destroyed by

> sin and replaced with an innate tendency to destruction. In extreme versions, only the baptized might eventually be reconciled and all else was not valued.

These orthodoxies have been challenged in recent years: Adam and Eve, for example, are held to have disobeyed as part of their coming into a full humanity. Anything less would have been a sub-human act. Thus there is scope for a theology of nature rather than of human nature in which a great deal more value is given to the non-human world, and the Creation can be celebrated in its entirety. Along with such an observance goes a diminished belief in hierarchy in human institutions as well. The new ethics which result from such beliefs include an extension of conscience to consider the natural world, the recognition of limits to the human exercise of power and control, acknowledgement of our status as tenants rather than owners of the Earth and indeed the rejection of what has been called the 'ultimate heresy' of thinking that we are somehow separate from nature.

Along with new theologies comes the demand for new liturgies and perhaps for new saints. Francis of Assisi (1181–1226) has been adopted as an inspiring icon: not just because of the legends surrounding his relationships with birds and wolves but on account of his adoption of what would now be called a low-impact life-style. Also from the 12th century, the German mystic Hildegard of Bingen has gained recent fame for her music and her 'green' poetry:

> the spirit of God
> is a life that bestows life
> root of the world-tree
> and wind in its boughs
>
> She is glistening life
> alluring all praise
> all-awakening
> all-resurrecting.

Judaism

As the immediate ancestor of Christianity, Judaism has some claim to be treated before it but, at present, its tenets are probably far less influential in practical terms or in formulating more abstract world-views: there are fewer Jews (*c.* eight million) than inhabitants of Mexico City.

Like Christianity, Judaism is much influenced by a written tradition: in this case the Torah (part of which figures in Christianity as the first five books of the Old Testament) and the voluminous exegetical writings known as the Talmud. From these are derived a number of moral guidelines with environmental significance. There is an emphasis on action and good deeds in this life, no concept of original sin but a desire for holiness, and no virtue in poverty. The rejection of the world and its ways seen in Christian veneration of the desert fathers and in the foundation of monasteries and convents finds no echo in Judaism. The structure of the world is, however, quite explicit. The account in Genesis 1: 28 where Adam is enjoined to fill the Earth and to subject it and to govern the fish of the sea and the fowl of the skies is not challenged. But this is no licence for unbridled exploitation: humans are to be stewards of what God created and to use it with tender mercy, not as an excuse for the imposition of slavery. Even further, humans are to share in God's creative work and carry it further to develop the Earth as a singing testimony (a psalm as it were) to God. Thus the planning of land and resource use, of population and the employment of technology are acceptable. In particular, wealth is not to be despised (poverty often inflicts environmental damage) and technology must be evaluated for its potential contributions to these processes and never despised as a form of general hubris. Proponents of the simple life can be regarded as misguided, although checks on consumption are foreshadowed in the concept of the Sabbatical Year, in which there was simply maintenance of production in the countryside and nothing new was planted.

So Judaism is environmentally tender but technology-friendly and so may well contain resources of relevance well beyond the boundaries of Israel or Stamford Hill. The rabbinical teaching against waste and destruction (*bal tashchit*: 'thou shalt not destroy') is part of the commandment to people to use their creative energies in imitation of their Maker, in the environment as elsewhere, since above everything else,

> . . . the land is Mine, because you are strangers and sojourners before me.
>
> (Leviticus, 25: 53)

Islam

The importance of the book and its exegesis is magnified in Islam, compared even with Judaism and Christianity. The Holy Qu'ran and its explications (*hāḍith*) are the basis for all proper human behaviour. The universe and its laws are the work of God and within them humanity has a special place. Our species is the only one on Earth which has the freedom of choice between right and wrong. Humans can therefore be at the pinnacle of creation but they can degrade themselves to be the lowest of the low. The proper place of a human ranks therefore as a *khalifah*: one who holds the Earth in trust. Some 500 verses of the Qu'ran have some applicability to the environment and they centre on humanity having access to the usufruct of the Earth (in modern terminology, the sustainable yield) but without damage or waste. This results from the human position as a tenant of the Earth but not the owner. Hence, corruption and pollution are the breaking of a covenant between God and humans, and always a matter of individual responsibility: it can never be shuffled off onto others.

One other important matter in environmental terms is the nature of Islamic science. Nowhere in Islam is there a distinction between the spiritual and the physical well-being of humanity and so science does not pretend to be value-free. It always takes place within Islamic law and is designed to show the will of God in all the regularities that are discovered. It may not necessarily, therefore, be aimed at providing more of most things, as is found in Western science and technology.

> Eat and drink, but waste not nor by indulging in excess; surely, God does not approve of the intemperate.
>
> (Holy Qu'ran 7:31)

Hinduism

Hinduism is a holistic religion which comprises a way of life rather than a set of beliefs. Indeed, 'Hindu' and 'religion' are both Western words: the Sanskrit phrase *sanatan dharma* is a closer description, meaning 'the true and eternal state'. Rooted in Indian culture and language, Hinduism suggests that everything from the rocks of the mountains to the whole cosmos is the home of God. So there has to be a sense of sacredness in the whole creation, especially but not only in life in all its forms; hence the need for vegetarianism.

Ethically, three principles follow from this world-view. The first is *vaina* or sacrifice. Ecologically, this is parallel to replenishing the Earth. If you take a tree, plant five more. Work to reduce your needs: if you can manage with five of anything, then do not take six. Secondly comes *dahna*, giving. Give to those who have given to you, whether of money, labour, intelligence or time. Since there is no separation in Hindu thinking between the natural and the social worlds, then giving to one is giving to all. Thirdly is *tapas*, self-control. In this practice, meditation may be very important, as may be abstinence of various kinds, including that of refraining from possessions. If may also include deliberate inaction; work, for example, is no virtue in itself and production has no value beyond our real needs. A common mantra, chanted before prayer, is 'Om shanti, shanti, shanti', in which *shanti* ('peace') means successively peace with nature, with society and with oneself. What we separate in the West as environment would embrace all of these. A valuable metaphor for the Hindu is to regard the whole world as a forest. The forest provides many things but especially prosperity from its products, if managed on a renewable basis. As well, it gives shelter from the hot sun and the torrential rains. Lastly, it is space for contemplation both on a daily basis and for the later phases of life which for the good Hindu involve the rejection of material wealth of any kind. All humans are, though, part of that forest: we can re-order the forest but we cannot by-pass it.

A final ethical precept is the one particularly associated with Mahatma Ghandi (1869–1948), that of non-violence. He interpreted this as applying beyond the political sphere, for he advocated a way of life which we would today characterize as essentially a low-impact solar-based economy, not far from ideas of deep ecology, for example. Development, to him, meant only more possessions and hence violence between people and to the land. This would sunder the essential wholeness of the cosmos as expressed in the *Isa Upanishad*,

> Everything within this world is possessed by God. He pervades both the animate and the inanimate. Therefore one should only take one's fair share, and leave the rest to the Supreme.

Buddhism

Like Western religions, this Asian 'religion' has an important central prophet. The title Buddha is given to Siddhartha Gautama, an Indian prince who reacted against the sacrificial practices of Brahminical Hinduism and finally achieved enlightenment after prolonged outdoor meditation in 525 BC. Buddhism is founded upon the idea of mind as the first cause: there is no need for deist beliefs. Because the human mind is at the root of cause and effect (the law of karma), it is human desire that produces suffering. All existence involves some suffering, and liberation from it is a desirable end. The word for this freedom is *nirvāṇa*, sometimes misunderstood in the West to be the equivalent of an Islamic or Christian heaven.

Since mind is a first cause, it can follow that interpenetration of minds and all other phenomena is central to the cosmos: in one is all and in all is one. At an ecological level, this leads us to a deeper understanding not simply of the nature of its relationships, so that when we look at a chair, we see the wood but we fail to observe the tree, the forest, the carpenter, or our own mind. This goes further: we are all dependent upon causes outside ourselves and so there is no independent, individual self. The notion of self separates us from the world and causes delusion and hence suffering. The search for wholeness then begins in ourselves and the need to discover our own enlightened selves or buddha-natures. To be enlightened is to discover how to live in this world, no more.

Since Buddhism centres on interdependent relationships, it is not surprising that it sees nature as being in a constant state of dynamism in which everything is ever-changing. But the course of change is interactive with mind so human actions and morality can affect the direction of change. Greed, hatred and delusion produce pollution within and without. Hence a non-aggressive attitude towards nature is as essential as it is towards people. The principle of non-harm (*ahimsa*) is a parallel to that of *bal tashchit* in Judaism.

Like its Hindu seed-bed, therefore, Buddhism is in favour of frugality. We may gather resources, as the bee gathers nectar, but only if we produce honey. Buddhism is rich in metaphors drawn from the open air, recalling that the Buddha received his enlightenment while sitting under a tree:

> Clouds of mosquitoes
> It would be bare
> Without them.

The Buddhist emphasis on the interpenetration of all phenomena is beautifully portrayed in an image from the Hua-yen Buddhism of 8th century China. This is the Jewel Net of Indra, conveyed for us by the American scholar F. H. Cook and hinted at earlier (p. 231),

> Far away in the heavenly abode of the great god Indra, there is a wonderful net which has been hung by some cunning artificer in such a manner that it stretches out infinitely in all directions. In accordance with the extravagant taste of deities, the artificer has hung a single glittering jewel in each 'eye' of the net, and since the net itself is infinite in dimension, the jewels are infinite in number ... If we now arbitrarily select one of these jewels for inspection and look closely at it, we will discover that in its polished surface there are reflected all the other jewels in the net, infinite in number. Not only that, but each of the jewels reflected in this one jewel is also reflected in all the other jewels, so there is an infinite reflecting process occurring ... This relationship is said to be one of simultaneous *mutual identity* and *mutual inter-causality*.
>
> (Cook's italics)

Such a world-view makes the connections of the Gaia hypothesis or chaos theory look over-simple, for it posits total mutual identity and inter-causality and a total lack of hierarchy. Neither humankind nor God is at the centre. A human being, a tree, a stone, a river exists only in relation to and dependence on all other things. Humans have then to adopt an ethical stance towards all things and promote their destinies as well. This may involve being used (e.g. being eaten by a tiger) as well as using.

Religions: an overview

The attempt to combine features of all the major religions into one faith is based on the notion that the one god is interpreted in different ways culturally and temporally. It is usually labelled syncretism. Modern attempts often add an evolutionary element to the picture, in the sense of a creative unfolding at all of cosmic, life-form and cultural levels: all these are god-making. A recent (1993) attempt in an environmental context by Henryk Skolimowski tries to link Franciscan Christianity, Heraclitus (500 BC) and Buddhism. In it the world is seen as sacred and has to be treated as a sanctuary. A spiritual guide can be drawn up on this basis which includes attributes of responsibility, the presence of a web that includes all life, and becoming a realized, celebratory being. Apart from the language, perhaps, much of it might appeal to an atheist as well. It does however point, along with the kind of material discussed above, to a renewed interest in the separate and combined interest of the major religions in matters of the environment.

This chapter has come a long way: from the relatively inflexible black letters of the law to the mysteries of God, should she exist. There are some features in common: all have the desire to lead people to a good life, with minimization of suffering; all share the need to be adapted to a local or regional culture even if a few absolutes are retained in all of them. In the world at present both are showing signs of tension. At one level there is global law and global belief; at another there are regional fragmentations into ethno-cultural groups showing a fierce independence, and fundamentalist movements insisting that they alone know the truth. All exist now within a scientific–technological framework from which even the most dedicated hermit can scarcely hope to escape. So the cultural ecology of the environment embraces not only the realized geography of the face of the world but the manifold processes of its becoming as well. Representing this complex and its possible futures in writing and a few pictures is not at all easy but is the inevitable theme of the last chapter.

Further reading

The literature on environmental law is difficult of access. It tends to be strictly national in scope and often technical as well, so that general accounts of convergence in environmental law-making are hard to find. A useful sequence from philosophy through to politics and law is Sagoff

(1988): its subtitle is 'philosophy, law and the environment'. The trouble with material on the emergence of international agreements and law is that they are well out of date by the time anything appears in this kind of print, and new ground is being broken (and indeed brokered) by the perceived need to deal with the new set of truly global problems, especially those of the atmosphere. But Caldwell (1990) is useful, and the WRI annual *World Resources* usually includes an update on recent international agreements. The other WRI annual, *State of the World*, sometimes has a section on the way in which international law is applied. Justice is dealt with at book length by Wenz (1988) and more succinctly by Hooker (1992). Likewise with the fine arts: many individual crafts have some attention paid to them in the environmental context but general syntheses are hard to find: Kemal and Gaskell (1993) have made an interesting, if necessarily selective, collection. Religion, on the other hand, is replete with books and papers. These are mostly from a Christian standpoint, but there is a series from Cassel in 1992, with various editors, which deals with the attitudes of the major world religions. The Native Americans are treated in Bierhorst (1994). A more syncretic alternative is offered by Skolimowski (1993). Religion, philosophy and the role of women are caught up by Gottlieb (1995).

Bierhorst J 1994 *The Way of the Earth: Native America and the Environment*. Morrow, New York

Caldwell L K 1990 *International Environmental Policy, Emergence and Dimensions*, 2nd edition. Duke Press, Durham NC and London

Callicott J B, Ames R T (eds) 1989 *Nature in Asian Traditions of Thought: Essays in Environmental Philosophy*. State University of New York Press, Albany

Gottlieb R S 1995 *This Sacred Earth: Religion, Nature, Environment*. Routledge, New York and London

Hooker C A 1992 Responsibility, ethics and nature. In Cooper D E, Palmer J A (eds) *The Environment in Question*. Routledge, London and New York: 147–64

Kemal S, Gaskell I (eds) 1993 *Landscape, Natural Beauty and the Arts*. Cambridge University Press

Skolimowski H 1993 *A Sacred Place to Dwell. Living with Reverence on the Earth*. Element Books, Rockport, Mass

Wenz P S 1988 *Environmental Justice*. State University of New York Press, Albany

CHAPTER 7

No foul casme

We might adopt as our motto for this book the statement of the 17th century Czech educator Jan Amos Comenius that knowledge should be 'universal, disgrac'd with no foul casme'. Yet all along it has been necessary to divide and to classify: knowledge now has an extensive network of chasms, with few people willing to jump across them from one cell to another. Such is the nature of specialization. All the time, too, the position of the chasms and of the solid ground between them is moving. So what are our chances of building a more unitary knowledge of ourselves and our environments which can be relevant to more than one individual human? It can be argued that the very act of writing makes the whole project impossible but what way is there other than establishing a text of some kind? Students often point out the importance of pictures in these studies but have to acknowledge that visual images rarely explain, compared with more abstract symbols such as words and numbers.

Papering over the gaps

Is it possible to reverse the human-produced processes of fragmenting knowledge: to squeeze out the chasms and meld the plates into one? It seems a tall order: would it really produce a better 'truth' than we now have? And how would we know? It seems worth exploring on the grounds of prudence, if there is any chance that more stable environments, more human dignity and justice might be the outcomes. (On present trends these qualities seem not to be increasing.) But we have to beware of a single over-arching theoretical base (whether it be the ideas of Marx, of environmental determinism or of deep ecology) which explains everything: that way lies the danger of totalitarianism.

The book so far: imparting knowledge

The distinctive feature of this book so far has been the variety of types of knowledge imparted. We have looked at measurements taken with complicated pieces of technology such as the sensors mounted in orbiting satellites. We have accepted the measurements taken in different places as providing a true pattern of sea temperature or atmospheric sulphur content. Equally, we can accord validity to descriptions of how a child sees its local surroundings and to reconstructions of how medieval people attributed sacredness to the nature around them. One feature these have in common is that they are selective: scientific measurements rarely take readings of everything all the time: nobody can afford that. So they sample, providing a set of discontinuous numbers which refer to a continuous pattern of variation. If we talk of a child's surroundings, then we suggest that we can disentangle its processing of 'environmental' information from the rest of its life. What we know of medieval life is mostly what those who were literate have left and what has survived the winnowing of time. So although the quantity of knowledge seems to be overwhelming, in fact we have only a little and that is systematically incomplete.

The book so far: a critical knowledge

The natural sciences and their imitators have sometimes made the claim that they produce a kind of disembodied truth which floats like a pure consciousness above the world and the people who produce it. No such ideas are usually claimed for other kinds of knowledge: these are acknowledged to be products of socially mediated time and place. But can this view be applied to the natural sciences?

At the world-view scale, several commentators have tried to look at the sciences as a whole and the directions of their findings. They point to the importance of prediction in both the formulation of theory and in practical applications and suggest that the overall aim (a phrase often used is 'the scientific project') is dominance of nature by humans. The key event in science is the experiment, whose major characteristic is its control by the investigator. Others go further and argue that the aim of the project is dominance by male humans. They compose the vast majority of the practitioners of science and so, feminists propose, science is a mostly male activity like, for example, watching football or warfare. A feminist science could be different not only in what it investigated but in the type of knowledge generated. The notion of beneficence introduced in Chapter 1 might need to be re-evaluated, as indeed it may for the poorer people of the world.

At a more everyday scale, it is clear that a social science which tries to produce objective description and explanation of societies is likely to be used as an agent of social engineering and control, most often in the service of the established social order. However, it is in principle at the service of the revolutionary as well, which is why censorship is often practised by those in power at a given moment, even when they themselves are former revolutionaries. Even the arts can be employed to confirm the propriety of the current power structures. Oil paintings, for example, seem to have acquired an authority in the European Renaissance that could be deployed to show approval of those in authority surrounded by their trappings (e.g. house and horses) or in the act of receiving the endorsement of God via the Virgin Mary or, in the case of the more modest, a well-regarded saint. From the time of its invention, writing has conferred authority. In its early stages, it enabled power to be deployed even in the absence of the ritual which normally validated priestly or kingly authority. It retained this power through much of history (think, for instance, of the bloody struggles over whether the Bible might be translated into vernacular languages and hence accessible outside a very narrow class), and even in an age of images writing retains this property, for example, the tabloid newspapers and the university library, odd bedfellows though they are.

The conclusion to be drawn is that any item of knowledge (including those about human–environment relations) is likely to have been produced in a social milieu and certainly will be received by the socially constructed. A new idea in science is very likely to start out as heresy, with scorn being poured on its proposer. Twenty years later he (*sic*) gets a Nobel prize for the same piece of work. Either way, marks on paper will have been critical.

In the cells

For most people interested in the human environment, the type of discussion just outlined is rather strange. The application of the sociology of science or notions of literary theory to CO_2 levels or the number of spotted owls in the Pacific Northwest of the USA seems absurd. It is clearly time to look at such operations in more detail: there is the threat here of much of our previous knowledge being undermined, perhaps to the point of accepting no limits to any kind of environmental impact. But the findings of the sciences are always admitted to be provisional, so that the 'problems' identified in Chapter 1 could be ameliorated by new discoveries in science or the technological applications of such knowledge.

The natural sciences revisited

Nobody ought to deny the fact that the information acquired by systematic scientific investigation since the 16th century is probably one of the highest achievements of humanity. The data are usually reliable, with clear estimates of possible error, and are independent of personality. This mode of knowing makes possible our ideas of the ultimate structure of matter or the placing of men on the Moon or the use of paracetamol to alleviate the headaches caused by both the writing and reading of books like this one. Since scientific knowledge is cumulative, each discovery building upon the last, an imposing edifice has been built, the like of which has never been equalled in the history of this planet.

Humans are not omniscient, however, and so all the knowledge that we have is still subject to the constraints of the human brain. For example, compared with some other animals, our direct perception of our surroundings is limited. True, we have stereoscopic colour vision able to focus accurately over a wide range of distances (though count how many of your friends need a technological aid to do so) but we are not so good at smells (as any walk with a dog will confirm) and we can detect only a limited range of electromagnetic wavelengths: walking under a high kilovolt powerline produces no detectable effect. True, we can extend our senses with instruments, but they are in turn inventions of the human mind and are thus limited by our imagination as well as our technology. An innovative biologist of this century, J. B. S. Haldane, summed it up by saying that the world was not only queerer than we supposed but queerer than we *could* suppose.

In its activities of describing, classifying, explaining, modelling, theorizing and the making of laws, science is searching for order. Its foundation therefore demands that there is order in the universe rather than simply a tendency in our minds to want to impose it. The kind of order to be found has exercised many minds down the centuries: Aristotle was convinced that it had a mathematical basis; Islamic and Christian theologians strongly believe that it has been created out of nothing by God according to some divine template; Darwin complicated matters by suggesting that it was open-ended, and modern chaos theory to a great degree extends this evolutionary model. If indeed the whole universe is evolving in the wake of a 'big bang', argue the cosmologists, then the very 'laws of nature' may not have been the same for all time. (Indeed, time itself is not the simple, linear progression from 0001 hrs to 2359 hrs that our senses suggest: Einstein is famous for wanting to think of it as curved.) If the very nature of time and space are debatable, then perhaps our failure to predict the behaviour of an ecosystem affected by human activities over the next 20 years is more excusable.

The cumulative nature of science and the way in which humans think result in some instability in the communication of the findings of science. For example, progressive knowledge of an entity means that what was called an electron in the 1920s and was described in a particular fashion may now, even if keeps the same name, be described in a very different way. In the 1930s an ecosystem was a static catalogue of plants (and perhaps some animals); now it is more like a dynamic interaction of flows of energy and matter. Important though mathematics and graphics are in the communication of science, the language of words is fundamental and so words have to be found for phenomena. The commonest way is that of the metaphor: one book in the 1960s was indeed titled *Physics as Metaphor*. A study of concepts in biogeography, for example, questions the use of the words 'invasion' and 'explosion' to describe the rapid immigration and expansion of populations of animals and plants into new territories, as if they were conquering armies given the goal of the conquest of territory. Thus much of science is, in the end, understood in terms of the metaphors used by its practitioners.

Interesting though these intellectual matters are, in the world of action science is used for mastery: over other people and over the natural world, on the home planet and elsewhere in the known universe. Just like any other tool of hegemony, its use has to be subject to moral considerations: is it right to do something simply because it is possible? The usual route has been for scientists to say 'I simply discover things. It's up to others to decide what to do after that.' Society has then to erect mechanisms for control, which get more and more difficult to emplace as technology becomes miniaturized and global communication more ubiquitous and rapid (Longino, 1990). The time may well be coming when the moral autonomy of the scientist will be our only safeguard against letting the wrong sort of genie out of unlabelled bottles. But what constitutes an acceptable moral autonomy when the long-term consequences of a given action can never be known? We have to reread *The Bhagavad-Gītạ*.

Technology in a social context

The conventional view of technology has been that it is a servant of human culture: that we determine what is and is not to be employed. But its role is more complex than that. The Western world-view and its partial extensions into developing countries mean that we live in a world where technology is as much part of the framework of life as the soil or the air, and is a great deal more central to many lives than the natural environment. We accept most of it without much thought and assume that it can yield a solution to most problems. Contraceptives will deal with population growth, iron filings in the sea will allow phytoplankton blooms to mop up excess CO_2, and giant mirrors in space will provide pollution-free energy. Though there is some questioning at an intellectual level of a few technologies (e.g. the private motor vehicle), very few moves to any wholesale redirection or abandonment has ever taken place. The Montreal Protocol of 1987, which aims to eliminate CFCs altogther, is a pioneering agreement.

The attraction of technique

Technology is the application of science, usually involving some form of machinery to enhance and apply the energy to which the inventors have access. Its status in Westernized societies gives it an appeal which may translate into 'if we've got it then we might as well use it'. The use of the latest technology is therefore a badge of modernity and progress (Winner, 1977). In 1964, the French philosopher Jacques Ellul (d. 1994) called all such machine- and information-led processes, *la technique*. It can be more than that: in some senses it can provide a validating medium for a culture. This was true of print in the era of the book and the popular newspaper (in the 1930s, for example, people might say, 'it must be true – I saw it in the paper') and is now the case with the small screens of TV and the computer. Note the appearance of the computer screen as an accessory to persuasion in TV advertisements (a person in the advertisement on screen will often use a computer screen as a confirmation that they are right) and in documentaries. In the latter, the books may also appear and give their *imprimatur*, though the appearance of call-mark labels on the spines often suggests that we are in the library rather than facing the interviewee's own collection.

The attraction of *technique* is not new, though the volume of it available is entirely a feature of this century. The US historian Lynn White (1967) pointed out that in medieval iconography the elect were usually those who had use of the latest machinery; the damned were bereft of such aids. The 20th century version has technology as a reducer of costs: people are usually more expensive and so capital can produce more profit by using more machines. This means that what can be done by machines *is* done; what cannot, remains undone. The computer, for example, has produced the saying, 'what can be counted, counts'. The application of technology world-wide is producing wholesale shifts in people's occupations: the notion of a full-time job occupying a person's life until their retirement seems to be fading in developed countries, just as there is a massive population shift to the cities in developing countries, as there was in the 19th century when the West industrialized.

Control of technology

The idea of a human-created and controlled set of servants runs deep. Dr Frankenstein's creation has inspired many a poor movie. But behind the stories there is a powerful mythological element, since in all cases the robot develops its own life: it gets out of control. The archetypal myth here is that of Prometheus, who stole fire from the gods and was condemned to have his liver pecked out by an eagle every day although it was renewed at night. Only Hercules was able to rescue him.

The control of technology is very difficult (Smith and Marx, 1994). In the Renaissance Utopia of Sir Thomas More, the guardians supervised all new inventions. A section of today's society would like to do that but is prevented not only by an atmosphere which sanctions freedom of information and use, but by the impossibility of regulating technologies which are so small that they are difficult to detect (witness the lengths to which airlines must go to scan baggage for terrorist bombs) or so anastomosed that they cannot be intercepted for control, as with satellite TV, fax transmission and the Internet. Even expensive government listening posts cannot deal with every electronic transmission, for instance. Control thus passes more into the hands of the providers than any other group: they use them for profit by persuading a large number of customers of the utility of one aspect of their use. In the case of the motor car, for example, this is done by failing to address all the external costs of the shift to private rather than public urban transport. If enough people are persuaded of the virtues of a new technology, then the label 'democratic' can be applied.

Overview

Technology is not going to go away: there is a great hunger for it which will not tolerate 'de-industrialization' in the Western world. In the developing nations, 'technology transfer' is seen as the main way forward. If there is enough energy to power all the technology for which there is, or can be generated, a demand then the challenge seems to be one of how to direct more of the development and applications of it towards all the people rather than simply the profit-takers.

The recurring theme of the chapter to this point has been that of critical knowledge. The researcher is always aware of the need to acknowledge the limits of the validity of any information acquired. The scientist is acutely sensitive to the fact that constraints of time and space may affect the applicability of generalizations emerging from the process of testing hypotheses and constructing theories. The social scientist and the humanist are always alive to the diversity of human behaviour and to the sudden tangential diversions of the pathways of history because of, for example, a charismatic individual. The consumers of knowledge, though, may approach less critically and assume that the scientists' predictions will be true for all times and all places or that history will indeed repeat itself. Since the cultural ecology of the environment rests upon so many kinds of knowledge, the difficulties are multiplied. This does not allow us to turn away in a despairing ignorance but it does make it imperative to think about the production of the knowledge we use and the language in which it is expressed.

Social theory and environment

The recorded history of humanity contains no lack of explanations about how humans are affected by their relationships with the non-human world about them. Astrology, environmental determinism and technophobia all in some way make statements about these relations. In the 19th century, the discipline of sociology crystallized out to formalize the study of society in imitation of the natural sciences. In other words, its aim was to help with control; phrases such as 'social physics' and 'social engineering' were heard. More recently, most of the other social sciences were also limited to responding to the findings of the natural sciences, which provided the 'facts' to which individuals and society had to respond. Now, there is a more reflexive attitude in, for example, sociology, political science and anthropology, in which the mediation of all cognitions through a social framework is held to be primary (Redclift and Benton, 1994). Some of this change may well have come about as a response to societies in the West that are no longer so certain about the virtues of the 'conquest' of nature as they were in the later 19th and early to mid 20th centuries.

Greener social thought

Many branches of social theory today have engaged with environmental concerns. Some, it is true, have dismissed the apprehensions as being those of the rich who want to pull up the ladder behind them, but the majority have espoused ideas of the mutual and necessary interconnectivity of humans and environment: the integrity of nature now depends upon a similar quality in human behaviour. The following paragraphs outline just a few of these areas of social thinking.

Marxism

In response to post-1960s thinking elsewhere, Marxians have turned to the master's analyses of capitalism as producing many more environmental pathologies than would happen under true socialism (Pepper, 1993). In particular, competitive capitalism in which large companies (especially MNCs) compete against each other to drive down costs are held responsible for degrading both environments and people where both lack the ability to resist (Schnaiberg, 1980). The appeal of Marxist ideas is at present in the penumbra of the discovery of just how degraded the environment of formerly communist states like Poland, the Czech Republic and Russia had become. But if state communism has now collapsed, more minds can be concentrated on the problems caused by capitalism.

Feminism

A number of branches of feminist thought have sought to qualify for the label of 'eco-feminist'. By far the most influential have been those who suggest that women are essentially different from men in their relations with nature. They take their icons from ancient history in the form of the 'Earth Mother' in her numerous avatars, and postulate a close and intuitive connection between the planet's rhythms and those of human reproduction, for example. This leads them to hold that the dualism of 'man and nature' is a male construction which allows the formation of a hierarchy which can be dominated by the men. Development away from this 'essentialist' movement leads to a cultural feminism which acknowledges that biology is important but not the total picture. Here, society constructs gender in the sense of what women are allowed to be, which is dominated by male hierarchies and activities such as technology of an aggressive kind. Both women and nature need to be validated as equals to men and indeed celebrated as such. There has been a vigorous discussion about whether a distinctive feminist science might exist which expressed a distinctively female temperament (showing for example co-operative and interactive social attitudes rather than competitive, controlling and individualist ones) or whether this would simply confuse the feminine in a biological sense with the feminist in a social setting (Plumwood, 1993; Warren, 1994).

There are many complications in the evolving strands of thought that make up eco-feminism, well described by Carolyn Merchant in her book

Radical Ecology (1992). What has often come out in practical terms is the direct action by women in defence of the natural environment and against environmental contamination. The Chipko movement of India, where women tried direct action to frustrate the logging of forests, is the best known but there are many other examples, which include the industrialized nations as well. In many instances, environmental concerns have allowed women to take leadership roles which hitherto had been reserved for men. It seems likely that women will increase their share of power in many societies through ecological and environmental action: it will be interesting to see how in the end the results will be different from movements led by men. Nevertheless, the freeing of women from a purely reproductive role seems to foreshadow many kinds of social change.

Post-modernism

Deriving essentially from French literary criticism of the 1970s, post-modernism is best know for its espousal of 'deconstruction'. Deconstruction thinks of the world as being constructed of language and hence there can be slippage between the object and its signification in words. So the foundations of rationality established in the Enlightenment of the 18th century no longer exist: there are only local and contingent truths (Rosenau, 1992). For environmental concern, this has a frightening aspect, for if the natural sciences indeed have no epistemological foundation, then there is no truthful information about the world, and humans might as well behave totally in their own immediate interests.

By contrast, there is also the less well-known reconstructive post-modernism, which would rejoice in the disappearance of the all-embracing aspects of the modern (such as Marxism and global capitalism) and their replacement by more locally self-sufficient communities. It often fuses with radical religious ideas in rejecting the notion of an off-the-planet male God in favour of a god of emergence and process, showing the influence of the process philosophy of Alfred North Whitehead (1861–1947), in which each stage in the evolutionary process represents an increase in divine goodness. All things have value in part of this process: the rocks and the sea anemone as well as Ludwig van Beethoven. All can lead to the realization of an intrinsic value for non-human nature.

Non-Western thought

The rational basis of thought and action which has characterized the West, especially since René Descartes, has clearly produced material abundance for many, but there has been a price. One aspect of the downside has been the virtual elimination of less complex cultures whose contact with nature was close and immediately dependent. It is often argued that their relationships were those of taking only what was strictly necessary: there was no waste. Even the disposal of unwanted remains was carried out according to approved ritual. One outcome, it is proposed, was that populations of animals were never hunted to extinction: that rituals and totemism ensured that over-hunting did not occur, at least until contact with Western hunters of meat and furs. It is also fair to say that others have argued that some native hunting practices were very wasteful and that only low population densities of humans saved some of the larger animals. Doubtless both have been true of some times and some places but the seed-bed for a different attitude to nature is still there, though assigned to a rather neglected part of the garden.

Deep ecology

In the search for radical alternatives to present ways of thinking and action, one movement that has gathered a coherent set of ideals is that of deep ecology. Though Western in origin (the father-figure is a Norwegian philosopher, Arne Næss) it draws upon a number of other traditions of thought. In it, humans are placed in a relationship of equality with all other biospheric life: no one species has an intrinsic value above another (Næss, 1989). At the personal level, a total intermingling of person and planet is required: the skin is a place of contact, not of differentiation. Industrial society is rejected as the normative paradigm for all societies and a bioregional settlement pattern based on non-industrial carrying capacity is favoured. Design with nature takes the place of the imposition of form on nature. The deep ecology movement is one of the few to try to think of the type of human communities which might best live in accordance with a regional ecology. At a time of many globalizing trends they can be regarded as a highly necessary adjunct to the global at a human and controllable scale or dismissed by opponents as irrelevant throwbacks.

Although deep ecology is regarded as an inconsequential cult by its detractors (especially those on the Left who are impatient with its overt lack of attention to social justice), the themes it puts forward find an echo in many wider circles; those concerned with environmental contamination and with resource scarcity, for example, as well as those worried that essential biodiversity is being lost by current practices. While humans have many exceptional qualities, they remain interdependent with the rest of the biosphere and thus part of a very intricate web in which any action may have long-term and unintended consequences. Although the inventiveness and power of humans seem to extend the capacity of the Earth for them, they are still not exempt from what seem to be ineluctable limits imposed by the biophysical composition of this planet. Escape to other planets seems an unlikely route for more than a handful of people and the accessible planets look somewhat hostile to human society. So, the influence of deep ecology is more than its role as an ethical guide for radical thinkers, activists and communities: it is singing a tune echoes of which are increasingly being hummed even if very quietly by politicians of the more growth-promising sort (Sessions, 1995).

Globalization

It is increasingly apparent that many social and economic changes are happening on a world-wide scale. Some of these developments, however, belong to systems which interact and coalesce over all the globe. This phenomenon is called globalization. It is made possible by instantaneous communication and two examples of relevance here are the transfers of capital by electronic means and satellite television. The former makes it possible for a company or nation state to borrow large sums of money to finance developments which may well have environmental consequences. The latter can penetrate into any corner of the world: a small dish aerial, a car battery and a TV set will suffice to receive the latest Australian soap, the richest American televangelist or the oldest BBC sitcom. It is too early to foretell what the outcomes might be in terms of attitudes towards the environment. The obvious result would be a raising of the material expectations of billions of people as they see ever more clearly how the rich 20 per cent of the world lives. There can be only a relatively small number of people who have no inkling of those ways of life. Even if they have not had TV they have rarely been that isolated.

Collapsing space and time may well mean increased opportunities for knowledge: it will be possible for the world in all its diversity to coalesce on any individual who has access to the technologies of communication. Immediately, this may lead to information overload. One consequence of this will be indifference to all except the most attention-grabbing events; another is confusion brought on by a very rapid pace of innovation as an individual or society tries to 'keep up' with all the feasible changes. The opposite may also be possible, no doubt. Some groups will want consciously to disengage themselves from the world, as millenarian communities (and pious Hindus) have done through the centuries. Almost certainly gaps between the rich and poor will become more obvious, though with what responses we cannot tell. The opportunities that will exist for media imperialism are already apparent, with the emergence of satellite barons in the mould of press lords. Such people and their corporate satraps will of course act as gatekeepers: they will try to determine who knows what in a way hitherto reserved for national governments. In this fashion, the imposition of a global culture is a possibility, with satellite TV the key piece of technology. Given the unpredictable nature of human affairs however, it may not work out in the obvious ways: we have not yet seen how countervailing trends may work (Brunn and Leinsbach, 1991).

The electronic networks do nevertheless comprise something close to how the Russian thinker Vernadsky (1863–1945) forecast a global implosion of consciousness, which he called the **noosphere**, as a complement to the biosphere and lithosphere. Just what the emerging thoughts will be like and whether they will differ in type and quality from their uncoalesced predecessors remains to be discovered.

The overwhelming conclusions of these discussions is that all knowing is mediated through a socially constructed framework: the world is indeed different from anything that our limited senses and minds can organize. The key point, perhaps, is that data of any kind (whether of raw perception or highly filtered through language) do not fall on a new surface in the brain. They enter a system which is already strongly structured both by its neurophysiology and its learning experiences. But none of these features of the process of environmental cognition preclude the origin of the signals in a 'real' world out there, whose reality is beyond human recognition and, almost certainly, control. This does not prevent us (and indeed never has) from constructing models both of ourselves and of a world out there' and of the relations between them. One problem is, simply, that the models may contain falsehoods about any of these components. An error like 'the Earth is flat' did not lead to any irretrievable disasters. If however, 'the world can support three times its present population at a Western standard of living' turns out to be an error then we may not find out until it is too late for both humans and the planet's systems. So all decision-making whether conscious or accidental is an act of uncertainty, with associated risks. In the field of the environment there is not the same consensus about how to behave in the face of risk as there is in, say, taking part in rock-climbing.

Attempts at integration

There are no prizes for realizing that there exists a huge diversity of types of information about the environment. Such a variety has in turn generated a great quantity of data, of which some is numerical and some graphical, but most is verbal; words are also the medium of most of the communication between humans about this topic, as indeed with other matters. Yet this intelligence is incomplete. The scientifically collected information is subject to constraints of time and money, of accessibility, of the possibilities of instrumentation, and of knowing what to look for. Further, it operates in a changing world, which is a condition even more true of the human sphere. There is the further complication that all knowledge is culturally filtered and that action passes through a further screen of culture, (including technology) so that it is doubly reflexive.

This last realization has added to the problems of the status of science. Whereas for much of this century it was a guarantor of truth, its capabilities and relativities are now being recognized. They include the following features:

- Uncertainty: in planning of developments, for example, which proposal will lead to less environmental damage? It is often impossible to tell (Beck, 1992).
- Phenomena may be on the margins of observability due to their location (e.g. in the deep oceans) or difficulties of measurement (e.g. in the early days of organochlorine pesticide use).
- The low level of development of theory of ecosystem behaviour may lead to poor predictability. For relatively stable systems, one physicist has posited that the percentage of the time up to which the behaviour of the system can be predicted with say better than about 66 per cent accuracy amounts to somewhere between 5 and 10 per cent of the length of time the system has been continuously observed in the near past up to the present time when the prediction is made. But many ecosystems are neither simple nor stable and so do not meet even these requirements.
- The complexity of large-scale phenomena, even below the global level. Natural scientists such as ecologists and physical geographers have not built the systems they are talking about, so there is a residuum of ignorance in every case. Further, other groups (corporations, governments) have a large degree of control or influence over changes in these systems.

Knowledge of the empirical shortcomings of these sciences has been extended by their use in public forums such as debates over nature conservation, waste dumping and, on a global scale, climatic change. One consequence, therefore, is that scientific evidence produced in any forum is not necessarily any stronger than that presented by other groups making claims on the environment based on social problems. This situation is exacerbated when disputants call to mind more fundamental epistemological characteristics of science.

Epistemologically, classic reductionism requires the metaphysical assumption that the smallest functional units are real and that cause can be understood in terms of interaction among these smallest units. But even facts derived from such units do not enjoy an unassailable status: uncertainty is a feature of science. Observation statements can, of course, withstand a variety of tests of intersubjectivity but they are practical achievements and cannot always be guaranteed. Furthermore, non-experts can have a view of events and processes only if instructed by the scientists who have, hence, monopoly ownership of this kind of knowledge, with full control over the way phenomena are construed, processed, interpreted and narrated. Challenges to this kind of authority are now more acceptable than they were 30 years ago. An outcome of these trends is a diversity of trajectory in the natural sciences, so that there is no longer a single strand, no shared methods, no common preoccupations, no values which all its branches share. Indeed it has become a problematical category.

So why bother?

All this questioning of the nature of our ignorance might well raise the question, is it worth bothering with any of this? Taoism, fatalism in the face of personal powerlessness and free-market ideology all alike would counsel that we go with the flow and then *que sera, sera*. Two counter-arguments stand out of the many. The first might be called responsibility, and centres on the idea that we all gain benefit from the environment, for it is life-supporting and therefore we have a reciprocal responsibility for ensuring its (and hence our) well-being. The second is simply that of prudence: if no care is taken, then we may soon diminish the Earth's capacity to support life, including that of our own species. So the ethical imperative to be concerned does not have to be very complicated and can appeal to mystics and free-market capitalists alike.

Why try to integrate knowledge?

A first reason will not surprise anybody reading a book written by an academic. It is that of intellectual adventure: can it be done? Can a human get his or her mind round these various pieces of knowing and by changing their relationships make them into a better understanding of the world we have to live in and our place in it? A second reason might focus more on the practical, for those who have to have such a motivation. Since the world currently supports more humans than it has ever done, it is clear that no outer limits of its capacity to sustain our species have yet been reached. But inner limits are binding in many places: where there is malnutrition, conflict over resources, poverty resulting from unjust access to resources, endemic diseases that result from under- and malnutrition, polluted places in rich countries and in nations recently released from undemocratic regimes, are all examples, though by no means an exhaustive list. Does anybody want to live in, for example, a food–people monoculture? So within such outer limits, striving for diversity and dignity, of people, cultures and the rest of the biosphere's inhabitants, seems a prudent and a worthy aim. But diversity has to be stressed: there are to be no Utopias imposed on everybody and everything simply 'for their own good'. Here is a world of fudge and compromise, with steps forward and backward, and crabwise as well.

Echoes of the resonance model

One way of looking at the human–nature relationship assumes that nature, or the environment, does not communicate directly with humans. This shows its origins in modern Western thinking, for in some cultural situations people believe that nature can communicate directly with them. Examples might be the essentialist type of eco-feminism, some 'New Age' mystics and other religious mystics. Those non-Western philosophies and religions which reject any dualism of humanity and nature must also accept a direct apprehension of the non-human world.

Most of us, though, talk to ourselves about it: in words, in pictures, by means of numbers; we analyse, discuss and exhort (Luhmann, 1989). In relatively simple societies, the basic human activities of production, reproduction and social control are not widely separated from each other: the same myth for example may deal with all three. But in advanced industrial societies the picture is different: all the activities and their control are split up into a form of specialization. If we use the image of a pipe organ, then there are different pipes for each of the activities of a 'society', bearing in mind that such a term may refer to any scale from a small town community to the emerging 'global village' made possible by electronic communication.

In this analogy, the 'pipes' are the familiar sectors of activity of a modern society; consider however how separated they often are: if we wanted to know about any of them, we would very likely have to consult a specialist who would disclaim any authority in any of the other channels. A non-exhaustive list would include the following:

- economy
- education
- material production
- environmental impact and change
- religion
- legal system
- ecological science
- healing
- caring

Let us re-emphasise the separation of most of these channels: there are remnants of former practices of course: the central role of the temples and their priests in the irrigation water distribution in Bali is an example. Then we need further to accentuate the fact that Western habits of dualism (in particular the mind–body, emotion–fact separation implanted so firmly by Descartes) have often provided binary opposites in each of these:

- economic/uneconomic
- rich/poor
- educated/ignorant
- faith/scepticism
- legal/illegal
- degradation/conservation
- science/'soft' knowledge
- ill/well
- caring/indifferent

Beyond this even, we know that one of each of these pairs is deemed to be 'good' and one to be 'evil' and that the one must overcome the other, preferably within our lifetime.

Nobody need be surprised, therefore, that there are some squawks from some of the pipes, and that when they sound together there is something less like harmony than discord. But the only way to control that is to have a score. In this context, having a pre-printed score is like having a Utopia, the drawbacks of which we have already discussed. (Note as well how musicians will quarrel over the interpretation of even a printed score.) Perhaps improvization is the better way ahead and if some of the chords are a bit peculiar, we can recall that Mozart and Beethoven were both criticized in their time for discordant sounds.

Emergent qualities

The lesson of historical study seems to be that complexity follows all forms of life and its webs. A world without humans is complex enough; with us there is a continual traffic in the creation of new genotypes and new ecosystems. Some of this results in greater biodiversity, as with the domestications of the Neolithic and the new applications of genetic engineering. Some results in biodiversity loss, as with the Pleistocene 'overkill' and loss of tropical forests today. Most have resulted in additional cultural complexity and indeed some societies became so complex in their organization that they seem to have collapsed under the weight of organization rather than from famine or climatic change. The difference today is in the number of humans and the speed with which change can be effected: yet another set of possibilities is opened up before the last group has been absorbed and understood.

This cultural complexity occurs under the umbrella of globalization: most major world trends of that kind initiate countervailing forces. The meanings which are drawn out of the phenomena of globalization are opposed by the concepts of post-modernism, in which the local negotiation of meaning becomes paramount: truth in effect becomes a local phenomenon which is not in any sense absolute. This should also apply, it would seem, to human–environment considerations as well. It is at once a frightening prospect (since it might mean that the findings of the natural sciences are not especially privileged above other forms of knowledge) and liberating, since it might also signify a true democracy resulting from people's empowerment. We might hope (but cannot be certain) that such units would be more environmentally tender. They are bound to have their problems with, for example, wildlife conservation (ask Indian herdsmen about tigers) but they may well act with prudence (i.e. in a mode which anticipates and avoids problems) since smaller units lack the perceived power to conquer that larger units, especially MNCs and states, are prone to possess.

One of the forms of power is that of prediction. A reason why TNCs and states harness the natural sciences and the positivist social sciences is to provide them with the best possible predictions of the outcomes of their actions. Will the development of this product make a medium-term profit for the company? Will this environmental policy produce a revolution or get the government voted out of power? One trouble is that prediction is not all that good. Limited powers are derived from relatively stable systems. In those undergoing rapid change and where there is emergence of novelty, they dwindle to almost nothing (Fig. 7.1). So any action based on prediction needs to be aware of its limitations and to build in wide margins of error at the very least. Even better, nothing irreversible should be done.

So why are we concerned: why are some people obviously highly anxious? If thinking about the future produces problems but the future is more or less unknowable, why worry? Why not go for business as usual: let the good times roll for the lucky ones at least. There are perhaps two scales of response to this type of *laissez-faire*. The first is basically metaphysical and it revolves around an evolutionary view. In biological terms, the palaeontological record sees species come and go, with often nothing but chance determining the survivors. But we see ourselves culturally as well as (often, rather than) biologically. In that light, we cannot subscribe to the notion that our evolutionary trajectory is to be a short but fiery one. We have not been around very long, the thinking runs, and it is inconceivable that *Homo sapiens* should not inhabit the Earth as long as life on that planet persists. So fundamental to much thinking about our environmental relations is the scare that the wrong actions may bring about the degree of environmental instability that will to all intents and purposes result in

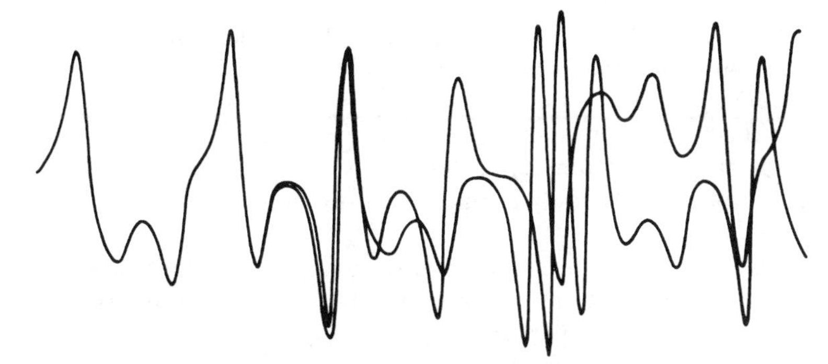

Figure 7.1 The two curves represent initial conditions which differ by only 0.0001. At first they coincide but their dynamics under conditions of chaos lead to independent and very different trajectories. This leads to the butterfly effect, where a single butterfly flapping its wings produces a tiny change in the conditions of the atmosphere that eventually leads to a storm somewhere else. The main point, however, is that the system's future is virtually unpredictable (from Stewart, 1990)

our extinction, as if the dinosaurs had brought about their own demise.

Inside this over-arching concern, there is nested, so to speak, another set of considerations. These involve the idea of equity between generations and between cultures. Surprisingly enough, it is more or less universally accepted (other than by the very conservative) that the rich owe a duty of assistance to the poor, and that one generation should not foreclose the opportunities available to its children and grandchildren. Just how we go about that is debatable, for one school holds that we might as well use up all the resources we feel we need, since the richer and more diverse we are the better the inheritance for later generations, and some of this will trickle down to the poor. It can be said that counter-arguments certainly exist.

Uses of the past

Most writers of history, science and poetry alike agree on the universality of historical change. John Donne sums them up with his river motif:

> Nor are, (although the river keep the name)
> Yesterday's waters, and today's the same.

Though we might add that humans are now much better at diverting and damming rivers than they were in the 17th century. Yet the human mind may also change.

There appears to be something like a triple helix in which nature affects human life and is in turn materially changed by it. Out of these two strands comes the third, which is that of ideas, so that the human mind collectively has altered what it apprehends, comprehends and communicates. For example, in the Western literate tradition there have been a number of phases of dominant world-view:

- In medieval times, the world was seen as a book. Not just any book, but a sacred book in which could be read the mind of God. It followed that every particle could be regarded as divine, even if it was incomplete and needed some human action to finish it to near-perfection.
- The Renaissance revived Classical ideas and one of them was that of Protagoras (481–411 BC) that humans are the measure of all things. A transition to the idea of the Earth as a storehouse of goods for that species is not therefore difficult.
- Following the discovery of planetary orbits, the world became realizable as a machine and the prospect of controlling it with the same predictable outcomes as an orrery was in view. The advent of steam power, even at low efficiencies, added credibility to such views.
- Since the 19th century, the dominant Western picture has seen the Earth become a cornucopia of material goods, with the deployment of energy (in the form of power of many kinds) as the chief means of technological access. Even those who are poor are promised that eventually a version of this world-view (usually labelled capitalism) will deliver the goods, rather in the manner of a cargo cult.

Models of a future

There is no shortage of ideas for exemplary worlds in which there is plenty, justice and fulfilment for all living things. Let us look at two that are relatively radical and one that looks more like the world we experience today.

- In his book on religious ecology-related philosophy, Henryk Skolimowski (1993) argues that the world must become a sanctuary. Having suggested that all humans have in common the need to behave responsibly, he adds the need to celebrate ourselves and the Earth. To add the mythic quality that makes ideas stick in people's minds and initiate action, he proposes the model of the world as a sanctuary, in which everything must be treated with care, and in which we move about as it were quietly. Apart from regarding everything in the sanctuary as having awe-inspiring qualities (and a sense of wonder at the world is a great help to us all), our role is to disturb it as little as possible.

- If our secular-minded world prefers models based on science rather than religion, then ecology gives the notion of a web whose connections are virtually infinite. The further developments of concepts in chaos theory and complexity suggest that the connections in the web may not always be obvious; some writings based on quantum theory allow for the interaction of subatomic particles at cosmic-scale distances. So if there is one or another kind of web which binds all the living and the non-living into one network of interactions, then we are inescapably part of it and rely on its integrity for our futures. The model seems little different from the Jewel Net of Indra except that it adds dynamic qualities to an otherwise static model.
- What strikes us most about the world as we experience it, however, is its messiness. Since I started to write this chapter there has on the one hand been the decision to create an Antarctic whale sanctuary; and on the other, there have been unprecedented ground-level ozone concentrations in British cities, causing an epidemic of asthma. There are simultaneously signs of hope and manifestations of tribulation. Can this be turned to advantage? It means abandoning all ideas of a predetermined goal, of a teleology. Instead there is a shorter-term view in which all that seems to be positive is seized upon and worked into the next stage. But fudge, compromise and uneasiness are part of the deal, and the future is open-ended. There are no grand plans: dignity consists in having any choice at all.

Another metaphor

If we believe any of the ideas in the last paragraph above, then perhaps a single metaphor like the medieval book and the 19th century machine is of no value. Yet in a globalizing world, the chances of large numbers of people hearing about such an idea are greater than ever. So let us try out the thought of a metaphor for the twenty-first century, to succeed those of the book, the man (*sic*), the machine and the cornucopia. Let us try the vision of the garden.

Unless, after all, the numbers of humans are drastically reduced by nuclear warfare or disease (neither are impossible, still), it is no use sighing for some kind of Arcadia in which most of the world is a wilderness and the imprint of humans virtually absent. But a heavily imprinted ecosystem which seems biased to creative rather than destructive existence is exactly what a garden can be. Gardens have the characteristics of being

- useful for production: of fruit, vegetables, herbs and medicinal plants, ponds with fish; basic subsistence may not be possible but a lot of variety and flavour certainly is;

- beautiful: there are places for flowering trees, flowers, fountains and shade; there may well also be wilderness areas where wild flowers grow and birds nest unseen by the human inhabitants; there will be quiet, hidden corners for meditation and for lovers;
- in need of a lot of attention: so there are co-operative tasks for a lot of people; further, it is next to where we live and so it is not detached from us: we are keenly aware of what is happening;
- for ourselves locally and so can reflect whatever we want: the balance of rigid-edged beds and untidy corners is up to us but a garden is never too large to be altered in the light of changing ideas and circumstances;
- primarily for ourselves: an obvious extension of our personalities in co-operation with the nature of the place, but not in isolation for we can swap produce, cuttings, seeds and ideas with the neighbours;
- a good place for music even if some kind of structure is necessary in many climates. The world could do with some more enchantment.

How do societies change?

The Spanish poet Antonio Muchado (1875–1939) in his *Proverbios y Cantares* gives us a memorable image of a journey whose outcome is in no way predetermined:

> Caminante, son tus huellas
> el camino, y nada más:
>
> (Traveller: your footsteps are the road, nothing else).

We have to consider how societies change even if we have no end in view. If we adopt the contingent rather than the predetermined, then knowing how novelty becomes the commonplace is useful. There is, first of all, lots of resistance and backlash. The rich and poor alike feel threatened and turn on innovators, who are mostly comfortable in their incomes and secure in their ideas. It has probably been ever thus and no amount of exhortation from either priest or professor is going to change the fact that (a) it happens and (b) it never in the end prevents metamorphosis. A major route of change is mutual coercion mutually agreed upon. This is the way of governments and of regulations, laws and penalties for not conforming. We have seen its complexity earlier and its seems agonizingly slow in the face of some of the problems. Yet many recent international and national agreements and arrangements would have seemed to have been impossible aims perhaps 15 years ago.

The other great avenue of modification is by people tuning intuitively to the resonances around them and by 'greening' themselves. They resist destruction as and when it is planned, they adopt low-energy life-styles and plan their family size. This road is the great unknown in the face of the communications explosion about to happen: will this result in 'greening' or in an immense onslaught on resources and environment to try to provide material benefits for those currently denied them?

We do not know. If we believe in the models of emergent, self-organizing systems which are open to the future then we cannot know. But we can work within all those changes to seize chances to enhance the garden, a place in which the chief promise as always is its diversity of all things, living and non-living alike.

Further reading

The intention of this chapter is at once to bring together matters normally kept separate and at the same time to encourage readers to make new connections for themselves. Thus the reference to *The Bhagavad-Gita* is to Krishna's instruction to 'let your rewards be in the actions themselves, not in their fruits' since nobody of course can tell what the eventual outcome might be. Frijthof Capra's evocation of himself as the dancing Shiva is in *The Tao of Physics* (1976). In the Western analytical mode, which tries to deal with people as individuals through to the global scene, an outstanding though little known book is *The World at the Crossroads. Towards a Sustainable, Equitable and Liveable World* (P. B. Smith *et al.*, 1994). Of all the religious books, the best in the Christian tradition seems to be *Ecology and Religion: Towards a New Christian Ecology of Nature* (Cobb, 1983). Those interested in the language that informs an emotional commitment to belief and action could try *Caring for Creation. An Ecumenical Approach to the Environmental Crisis* (Oelschlager, 1994). However, the radical views of Anna Primavesi (1991), *From Apocalypse to Genesis,* should raise a few hopes as well as hackles. Is there a non-material approach which is not tied down in the hierarchies, dogmas and orthodoxies of the major religions? Skolimowski (1993) has already been mentioned; *The Arrogance of Humanism* (Ehrenfeld, 1978) to some extent fills the bill. In the end, though, my advice is to read all the science that you can understand, let it foment for a while and then read some of the poets who have tried to get into the Wyndham Lewis-type inner-ness of the human–nature relationship. The early work of R. S. Thomas, for example, John Clare, Thomas Hardy, W. H. Auden and Gary Snyder might replicate the relations of malt and Milton referred to by A. E. Housman.

Capra F 1976 *The Tao of Physics*. Fontana Books, London: 9

Cobb M 1983 *Ecology and Religion: Towards a New Christian Theology of Nature*. Paulist Press, Ramsey, NJ

Ehrenfeld D 1978 *The Arrogance of Humanism*. OUP, New York

Oelschlager M 1994 *Caring for Creation. An Ecumenical Approach to the Environmental Crisis.* Yale University Press, New Haven, Conn. and London

Primavesi A 1991 *From Apocalypse to Genesis.* Burns and Oates, London

Skolimowski H 1993 *A Sacred Place to Dwell. Living with Reverence on the Earth.* Element Books, Rockport, Mass.

Smith M R, Marx L (eds) 1994 *Does Technology Drive History? The Dilemma of Technological Determinism.* MIT Press, Cambridge, Mass.

Glossary

Biodiversity A broad description of the variety of soil–vegetation–animal communities (*biomes*) at world or continental scale, or the diversity of ecosystems within those biomes, or the number of species within each ecosystem. Look carefully at the scale at which it is being used.

Biogeochemical cycles Descriptions of the flow of a chemical element or compound between inorganic (non-living) pools or states and organic (living) states. Carbon, for example, flows between pools in the atmosphere (non-living) and biota (living organisms), as well as between other reservoirs.

Biome The major climate-driven zones (latitudinal and altitudinal) of vegetation, soils and animal communities of the Earth, together with non-zonal formations such as the seas of the continental shelves, the open oceans, and estuaries.

Consensual By general agreement, sometimes of a majority rather than absolutely everybody.

Deciduous Applied to plants that drop their leaves in any season unfavourable for growth, such as the winter of the temperate zone.

Diversivore A term used of animals which eat a wide range of foods: most humans, for example, eat both animal and plant tissue. It is better than its synonym 'omnivore', which, strictly speaking, means an animal that eats everything.

Empirical Based upon observation or experiment rather than hypothesis or theory.

Epiphytes Plants which use other plants as their substrate rather than the soil. But they simply use it for physical support, unlike parasites, which derive nutrition from the plants upon which they grow. Orchids growing in the crooks of forest trees and the long dangling lianas of tropical forests (as used by Tarzan) are epiphytes.

Ethical systems A basis for human behaviour which has a systematic foundation; not based therefore on a fresh reaction to every situation. There may be obligatory rules such as 'do not kill'; the main purpose may be to maximize human welfare (utilitarianism); or there may be a flexible system which promotes the flourishing of each individual in a specific setting.

Euphotic The shallow layer of water in the sea which is penetrated by sunlight at an intensity high enough to allow plants to carry out photosynthesis.

Eutrophication The process of adding nutrients to an ecosystem, especially nitrogen and phosphorus. It happens in natural systems, as when a lake's vegetation adds mineral-rich detritus to its water and sediments but is commonly associated with human activity as when sewage is led into rivers or fertilizer runoff reaches lakes.

Gaia hypothesis J. E. Lovelock's model of this planet in which life is not simply adapting to the physical conditions of a cooling body but in which life is 'managing' the physical conditions (such as climate, ocean salinity levels, gaseous composition of the atmosphere) so as to maximize the opportunities for living organisms. The implications of 'purpose' inherent in such a model mean that it appears to be incompatible with most interpretations of Darwin's theory of evolution.

Hermeneutics Originally, this word was applied to the interpretation of scriptural texts, such as the Bible. Now concerned with the interpretation and significance of human actions and institutions. Goes beyond *empirical* knowledge to seek an understanding of human events and processes.

Holism The thesis that wholes are more than the sum of their parts. It suggests that wholes have characteristics that cannot be explained in terms of their constituent parts: *emergent qualities*. The freezing point of water, for example, cannot be predicted from our knowledge of hydrogen and oxygen. The adjective is *holistic*.

Justice Fair treatment for all who are deemed to come within its ambit. This has rarely included non-human entities in the Western world, so that radical environmentalists and environmental philosophers have proposed its extension to plants and animals.

Lifeworld The total 'envelope' in which individuals live, ranging over both their interactions with other entities and their emotions. Its formation and maintenance is thus critical to any understanding of environmental knowledge, for instance.

Neo-classical economics The basis of 'market economics' in which consumers possessed of perfect knowledge strike freely arrived-at bargains with suppliers at a price acceptable to both. It is not always suitable for environmental 'goods' since they may be unpriceable, though right-wing governments try hard to put a monetary number on almost everything.

Net primary productivity A measurement of the rate (quantity per unit area per unit time) of production of plant tissue. Thus data are available for the NPP of a suburban lawn or for the whole Pacific Ocean: it can be used at any scale. It is a smaller quantity than *gross primary productivity*, which is the amount of energy fixed in photosynthesis minus that used by the plant for its own metabolism and thus not appearing as added tissue.

Noosphere The concept of a layer of human thought analogous to the lithosphere and biosphere. Elaborated first in 1945, it appears to be coming to fruition in the form of digital electronics.

Normative The idea that statements can be made which are not simply descriptive or matters of fact but are concerned with rules, recommendations or prescriptions. There is also the implication that the norms (standards or values) involved are those of a social group rather than simply of an individual.

Objective Something which belongs not to the perceiving mind only, but which is presented to that mind from outside. The acceptance of the detachment of the observer from the observed is fundamental to the natural sciences of the West.

Permafrost Areas of land which contain a layer that is always frozen and never completely melts, even at the height of summer. It is especially characteristic of the *tundra* and also of parts of the Boreal Forest biome.

Phytotoxic Plants that exude or contain in their tissues chemicals which are poisonous to animals (as a defence against being eaten) or to other plants. Toxins secreted by root systems, for example, may lessen the competition for water, light or nutrients from other species or individuals of the same species.

Pollution In common parlance, the occurrence of a substance or other phenomenon (e.g. noise or smell) somewhere it is not wanted. It usually means the addition of something human-made (often a waste) to an existing environment, not necessarily a natural one. The term 'contamination' is probably preferable since pollution has acquired moral overtones.

Positivism Assumes that objectivity is possible, that observation is a necessary starting-point, and then goes on to make law-like statements that describe and explain regularities in the relations between phenomena.

Reductionist The opposite of *holistic*. In science, it argues that a system can be understood in terms of the isolated operation of its component parts or its subsystems. The idea of emergent qualities is not therefore accepted.

Rights Something that can be morally or legally claimed. Traditionally a Western concept and applied only to humans ('legal rights', 'human rights'). Can the concept now be extended to non-human entities?

Some environmental philosophers argue that this is possible conceptually; the practicalities seem daunting.

Subjective A category which centres on the thinking or perceiving [human] subject as distinct from an external or 'real' thing. In philosophy, idealism rests upon the notion that everything we can know is filtered through human consciousness and thus the external world is unknowable except as a human construct.

Sustainable A term applied to economies and to development. It implies that the 'capital' of the system is maintained and that societies live off the 'income'. In ecological terms, it implies stability through time and an absence of environmental degradation, howsoever that may be measured. There are similarities with the ecological notion of 'carrying capacity'.

Synergism A concept related to that of emergent properties: the greater effect of two substances when used together rather than separately, for example. Many drugs when combined with alcohol are more depressant, for example, than either separately.

Technocentric A world-view common today that science and technology are the keys to material prosperity and that any problems thus created can be solved by the application of more science and technology: the 'technical fix'. Its opposite is 'ecocentric'.

Value systems The way in which we arrive at the possibility of using the words 'good' and 'ought'. Most often used in a normative way but can also be applied in, for example, aesthetics: 'That was a good film' implies that a set of values has been used to evaluate its worth. So with 'a beautiful view'.

References

Anon. 1969 *Tyne Landscape*. Joint Committee as to the Improvement of the River Tyne, Newcastle-on-Tyne

Anon. 1991 Knowledge's outer shape, inner life. *Times Higher Education Supplement* August 16, 12

Archbold O W 1995 *Ecology of World Vegetation*. Chapman and Hall, London

Barbier E B (ed.) 1993 *Economics and Ecology: New Frontiers and Sustainable Development*. Chapman and Hall, London

Barbier E B *et al.* 1994 *Paradise Lost? Ecological Economics of Biodiversity*. Earthscan, London

Barrow C J 1995 *Developing the Environment, Problems and Management*. Longman, London

Beck U 1992 *Risk Society: Towards a New Modernity*. Sage, London

Beiswaner W L 1984 The Temple Garden: Thomas Jefferson's vision of Monticello landscape. In Maccobin P and Martins P (eds) *British and American Gardens in the 18th Century*. The Colonial Williamsburg Foundation, Williamsburg, Va

Bell M, Walker M J C 1992 *Late Quaternary Environmental Change, Physical and Human Perspectives*. Longman, London

Bennett J W 1976 *The Ecological Transition. Cultural Anthropology and Human Adaptation*. Pergamon Press, Oxford

Bierhorst J 1994 *The Way of the Earth: Native America and the Environment*. Morrow, New York

Bilsborough A 1992 *Human Evolution*. Blackie, London

Blotkamp C 1994 *Mondrian. The Art of Destruction*. Reaktion Books, London

Blowers A 1993 *Planning for a Sustainable Environment*. Earthscan, London

Blunden J 1991 Mineral resources. In Blunden J, Reddish A (eds) *Energy, Resources and Environment*. Hodder and Stoughton/Open University, London: 43–78

Bookchin M (1982) *The Ecology of Freedom. The Emergence of Dissolution of Hierarchy*. Cheshire Books, Palo Alto, Calif.

Boserup E 1994 *The Conditions of Agricultural Growth.* 2nd edition. Earthscan, London

Boston P J, Thompson S L 1991 In Schneider S H, Boston P J (eds) *Scientists on Gaia*. MIT Press, Cambridge, Mass. and London

Boulding K E 1981 *Evolutionary Economics*. Sage, Beverley Hills, Calif. and London

Briggs D, Courtney F 1985 *Agriculture and Environment*. Longman, London

British Airways 1996 *Holidays*. Brochure

Brown L R 1994 Facing food insecurity. In Brown L R (ed.) *State of the World 1994*. Earthscan, London: 177–97

Brown L R *et al.* 1995 *Vital Signs 1995–1996*. Earthscan Publications, London

Brown S 1990 Humans and their environments: changing attitudes. In Silvertown J and Sarre P (eds) *Environment and Society*. Hodder and Stoughton, London: 238–71

Brunn S D, Leinsbach T R (eds) 1991 *Collapsing Space and Time*. Harper Collins Academic, London

Burgess W R (ed.) 1978 *Lead in the Environment*. National Science Foundation, Washington, DC

Butcher S S, Charlson G H, Orians G H, Wolfe G V 1992 *Global Biogeochemical Cycles*. International Geophysics Series vol. 50, Academic Press, London

Caldwell L K 1990 *International Environmental Policy, Emergence and Dimensions*, 2nd edition. Duke Press, Durham, NC and London

Callicott J B, Ames R T (eds) 1989 *Nature in Asian Traditions of Thought: Essays in Environmental Philosophy*. State University of New York Press, Albany

Capra F 1976 *The Tao of Physics*. Fontana Books, London

Carson R 1963 *Silent Spring*. Hamilton, London

Chalmers A F 1982 *What is this Thing called Science?* Open University Press, Milton Keynes

Chalmers A F 1990 *Science and its Fabrication*. Open University Press, Milton Keynes

Charlson R J *et al.* 1992 In Butcher S S (eds) *Global Biogeochemical Cycles*. Academic Press, San Diego

Chorley R J (ed.) 1969 *Water, Earth and Man. A Synthesis*. Methuen, Calif. and London

Clarke I M 1985 *The Environment, Politics and the Future*. Wiley, Chichester

Cleary T 1993 *The Flower Ornament Scripture. A Translation of the Avatamaska Sutra*. Shambhala Publications, Boston and London

Cobb M 1983 *Ecology and Religion: Towards a New Christian Theology of Nature*. Paulist Press, Ramsey, NJ

Cook F H 1977 *Hua-yen Buddhism*. Pennsylvania State University Press, University Park and London

Cotgrove S 1982 *Catastrophe or Cornucopia: the Environmental Politics of the Future*, Wiley, Chichester

Curtis V 1986 *Women and the Transport of Water*. Intermediate Technology Publications, London: 26–7

Cushing D H 1975 *The Fisheries Resources of the Sea and their Management*. OUP, Oxford

Cushing D H 1988 *The Provident Sea*. Cambridge University Press, Cambridge

Czinkota M R *et al.* 1992 *International Business*, 2nd edition. Dryden Press, Orlando, Fla.

Daly H 1991 *Steady State Economics*, 2nd edition. Freeman, San Francisco

Dasgupta P 1982 *The Control of Resources*. Basil Blackwell, Oxford

Desmond A, Moore J 1991 *Darwin*. Michael Joseph, London (published by Penguin in 1992)

Durham County Council 1984 *County Durham Waste Disposal Plan*, 1st edition. Durham County Council: 20–1

Eckersley R 1992 *Environment and Political Theory. Towards an Ecocentric Approach*. UCL Press, London

Ehrenfeld D 1978 *The Arrogance of Humanism*. OUP, New York

Ellul J 1964 *The Technological Society*. Vintage Books, New York (originally published in French in 1954)

Elvin M, Su Ninghu 1995 Man against the sea: natural and anthropogenic factors in the changing morphology of Harngzhou Bay, circa 1000–1800. *Environment and History* **1** 7–8

Emery K O, Aubrey D G 1991 *Sea Levels, Land Levels and Tide Gauges*. Springer-Verlag, New York

Findlay A 1991 Population and Environment: Reproduction and Production. In Sarre P (ed.) *Environment, Population and Development*. Hodder and Stoughton, London: 3–38

Flavin C, Lenssen N 1991 Designing a sustainable energy system. In Brown L R (ed.) *State of the World 1991*. W W Norton, New York: 21–38

Forrest D M 1967 *A Hundred Years of Ceylon Tea 1867–1967*. Chatto and Windus, London

Fox M 1981 *Original Blessing*. Bear, Santa Fe, NM

Georgescu-Roegen N 1971 *The Entropy Law and the Economic Process*. Harvard University Press, Cambridge, Mass.

Glacken C J 1967 *Traces on the Rhodian Shore*. University of California Press, Berkeley and Los Angeles

Gleick P H 1993 *Water in Crisis: a Guide to the World's Fresh Water Resources*. OUP, New York

Gottlieb R S 1995 *This Sacred Earth: Religion, Nature, Environment*. Routledge, New York and London

Goudie A S 1981 *The Human Impact*, 3rd edition. Blackwell, Oxford

Goudie A S 1992 *Environmental Change*, 3rd edition. Blackwell, Oxford

Goudie A S 1993 *The Human Impact on the Natural Environment*, 4th edition. Blackwell, Oxford

Gould S J 1989 *Wonderful Life. The Burgess Shale and the Nature of History*. Penguin Books, London

Gregory R L 1972 *Eye and Brain: The Psychology of Seeing*, 2nd edition. Weidenfeld and Nicholson, London

Groombridge B (ed.) 1992 *Global Biodiversity, Status of the Earth's Living Resources*. Chapman and Hall, London

Hagget P 1979 *Geography: A Modern Synthesis*, 3rd edition. Harper and Row, New York and London: 16–17

Hall A S *et al.* (eds) 1986 *Energy and Resource Quality: The Ecology of the Economic Process*. Wiley, New York

Harris D R 1989 In Harris D R, Hillman G C (eds) *Foraging and Farming: The Evolution of Plant Exploitation*. Unwin Hyman, London

Harris D R 1990 *Settling Down and Breaking Ground: Rethinking the Neolithic Revolution*. Netherlands Museum of Anthropology and Prehistory, Amsterdam

Harris D R, Hillman G C (eds) 1989 *Foraging and Farming*. Unwin Hyman, London

Hayden B 1981 Subsistence and ecological relations of modern hunter–gatherers. In Harding R S O, Teleki G (eds) *Omnivorous Primates*. Columbia University Press, New York: 344–421

Hayward T 1994 *Ecological Thought: An Introduction*. Polity Press, Cambridge

Hill R 1949 *The Symphony*. Pelican Books, London

Holl O 1994 *Environmental Cooperation In Europe: The Political Dimension*. Westview Press, Boulder, Colo.

Hooker C A 1992 Responsibility, ethics and nature. In Cooper D E, Palmer J A (eds) *The Environment in Question*. Routledge, London and New York: 147–64

Hughes J D 1983 *American Indian Ecology*. Texas Western Press, El Paso

Huxley A 1932 *Brave New World*. Chatto and Windus, London

Huxley A 1962 *Island*. Chatto and Windus, London

Ironbridge Gorge Museum Trust 1979 *Coalbrookdale*. Ironbridge, Shropshire

Jaffe D A 1992 In Butcher S S (eds) *Global Biogeochemical Cycles*. Academic Press, San Diego, Calif.

Janke P J 1992 In Butcher S S (eds) *Global Biogeochemical Cycles*. Academic Press, San Diego, Calif.

Jantsch E 1980 *The Self-Organising Universe*. Pergamon Books, Oxford

Johnson D L, Lewis L A 1995 *Land Degradation: Creation and Destruction*. Blackwell, Oxford

Kemal S, Gaskell I (eds) 1993 *Landscape, Natural Beauty and the Arts*. Cambridge University Press

Kemp D D 1994 *Global Environmental Problems. A Climatological Approach*, 2nd edition. Routledge, London and New York

Kidron M, Segal R 1991 *The New State of the World Atlas*, 4th edition. Simon and Schuster, London

Kramer L 1995 *EC Treaty and Environmental Law*. Sweet and Maxwell, London

Lavine M J 1984 Fossil fuel and sunlight: relationship of major sources for economic and ecological systems. In Jansson A-M (ed.) *Integration of Economy and Ecology – An Outlook for the Eighties*. University of Stockholm Askö Laboratory, Stockholm: 121–51

Livi-Bacci M 1992 *A Concise History of World Population*. Blackwell, Oxford

Longino H 1990 *Science as Social Knowledge*. Princeton University Press, Princeton, NJ

Lovelock J E 1979 *Gaia. A New Look at Life on Earth*. OUP, Oxford

Lovelock J 1989 *The Ages of Gaia*. OUP, Oxford

Luhmann N 1989 *Ecological Communication*. Polity Press, Cambridge

Mandood Elahi K, Rogge J L 1990 *Riverbank Erosion, Flood and Population Displacement in Bangladesh*. Jahangirnagar University Riverbank Erosion Impact Study, Dhaka: 13

Mannion A M 1995 *Agriculture and Environmental Change. Temporal and Spatial Change*. Wiley, Chichester

Mannion A M, Bowlby S R (eds) 1992 *Environmental Issues in the 1990s*. Wiley, Chichester

Martin C 1978 *Keepers of the Game: Indian–Animal Relationships and the Fur Trade*. University of Georgia Press, Athens, Ga

Maslow A 1968 *Towards a Psychology of Being*, 2nd edition. Van Nostrand Rheinhold, London and New York

Mather A S, Chapman K 1995 *Environmental Resources*. Longman, London

Mathieson A, Wall G 1982 *Tourism: Economic, Physical and Social Impacts*. Longman, New York and London

Melosi M V 1982 Energy transitions in the nineteenth-century economy. In Daniels G H, Rose M H (eds) *Energy and Transport: Historical Perspective on Resource Issues*. Sage, Beverley Hills and London: 55–69

Merchant C 1992 *Radical Ecology*. Routledge, London and New York

Meyer W B, Turner B L 1992 Human population growth and land-use/cover change. *Annual Review of Ecology and Systematics* **23**: 39–61

Middleton N 1995 *The Global Casino. An Introduction to Environmental Issues*. Edward Arnold, London

Misch A 1994 Assessing environmental health risks. In Brown L R (ed.) *State of the World 1994*. Earthscan, London: 117–36

Meyers N, Simon J L 1994 *Scarcity or Abundance? A Debate on the Environment*. W W Norton, New York and London

Næss A 1989 *Ecology, Community and Lifestyle*. Cambridge University Press, Cambridge

Nitschke G 1991 *The Architecture of the Japanese Garden: Right Angle and Natural Form*. Benedikt Taschen Verlag GmbH, Köln

Oelschlager M 1994 *Caring for Creation. An Ecumenical Approach to the Environmental Crisis*. Yale University Press, New Haven, Conn. and London

Orwell G 1949 *Nineteen Eight-Four*. Secker and Warburg, London

Park C C 1992 *Tropical Rainforests*. Routledge, London

Passmore J 1980 *Man's Responsibility for Nature: Ecological Problems and Western Traditions*. Duckworth, London

Pearce D W 1983 *Cost–Benefit Analysis*, 2nd edition. Macmillan, London

Pepper D 1993 *Eco-socialism. From Deep Ecology to Social Justice*. Routledge, London and New York

Pepper D, Colverson T 1984 *The Roots of Modern Environmentalism*. Croom Helm, London

Perrings C 1987 *Economy and Environment. A Theoretical Essay on the Interdependence of Economic and Environmental Systems*. Cambridge University Press, Cambridge

Peters R H 1991 *A Critique for Ecology*. Cambridge University Press, Cambridge

Pickering K T, Owen L A 1994 *An Introduction to Global Environmental Issues*. Routledge, London and New York

Philander S G 1990 *El Nino, La Nina and the Southern Oscillation*. Academic Press, San Diego, Calif.

Pierce J T 1990 *The Food Resource*. Longman, London

Pickering K T, Owen L A 1994 *An Introduction to Global Environmental Issues*. Routledge, London and New York

Pickett S T A, Kolsa J, Jones C G 1994 *Ecological Understanding*. Academic Press, San Diego, Calif.

Pitt D C (ed.) 1988 *The Future of the Environment: The Social Dimensions of Conservation and Ecological Alternatives*. Routledge, London and New York

Plumwood V 1993 *Feminism and the Mastery of Nature*. Routledge, London and New York

Polden M 1994 *The Environment and the Law: A Practical Guide*. Longman, London

Primavesi A 1991 *From Apocalypse to Genesis*. Burns and Oates, London

Redclift M 1987 *Sustainable Development: Exploiting the Contradictions*. Methuen, London and New York

Redclift M, Benton T (eds) 1994 *Social Theory and the Global Environment*. Routledge, London and New York

Rees J 1990 *Natural Resources. Allocation, Economics and Policy*, 2nd edition. Routledge, London and New York

Roberts N 1989 *The Holocene. An Environmental History*. Blackwell, Oxford

Roberts R D, Roberts T M 1984 *Planning and Ecology*. Chapman and Hall, London

Rosenau P M 1992 *Post-Modernism and the Social Sciences. Insights, Inroads and Intrusions*. Princeton University Press, Princeton, NJ

Ross S 1993 Gardens, earthworks, and environmental art. In Kemal S, Gaskell I (eds) *Landscape, Natural Beauty and the Arts*. Cambridge University Press, 158–82

Sagoff M 1988 *The Economy of the Earth*. Cambridge University Press

Sandlund O T, Hindar K, Brown A H D (eds) 1992 *Conservation of Biodiversity for Sustainable Development*. Scandinavian University Press, Oslo

Sands P (ed.) 1993 *Greening International Law*. Earthscan, London

Schama S 1995 *Landscape and Memory*. Harper Collins, London

Schmidt A 1971 *The Concept of Nature in Marx* (trans. B Fowkes). New Left Review, London (originally published in German, 1962)

Schnaiberg A 1980 *The Environment from Surplus to Scarcity*. Oxford University Press, New York and Oxford

Schneider S H, Boston P J (eds) 1991 *Scientists on Gaia*. MIT Press, Cambridge, Mass.

Schumacher E F 1973 *Small is Beautiful*. Blond and Briggs, London

Serio J N 1978 The ultimate music is abstract. Charles Ives and Wallace Stevens. In Garvin H R (ed.) *The Arts and their Interrelations*. Bucknell University Press, Lewisburg, Pa; Associated University Presses, London, 120–31

Sessions G E (ed.) 1995 *Deep Ecology for the Twenty-First Century*. Shambala Press, Boston, Mass.

Simmons I G 1979 *Biogeography. Natural and Cultural*. Edward Arnold, London

Simmons I G 1982 *Biogeographical Processes*. Edward Arnold, London

Simmons I G 1989 *Changing the Face of the Earth*, 1st edition. Blackwell, Oxford (2nd edition published 1996)

Simmons I G 1991 *Earth, Air and Water*. Edward Arnold, London

Simmons I G 1993 *Environmental History. A Concise Introduction*. Blackwell, Oxford

Simmons I G 1996 *Changing the Face of the Earth*, 2nd edition. Blackwell, Oxford

Simon J L, Kahn H (eds) 1981 *The Resourceful Earth*. Blackwell, Oxford

Skinner B F 1948 *Walden II*. Macmillan, New York

Skolimowski H 1993 *A Sacred Place to Dwell. Living with Reverence on the Earth*. Element Books, Rockport, Mass.

Smil V 1991 *General Energetics: Energy in the Biosphere and Civilization*. Wiley, New York

Smil V 1993 *Global Ecology, Environmental Change and Social Flexibility*. Routledge, London and New York

Smith M R, Marx L (eds) 1994 *Does Technology Drive History? The Dilemma of Technological Determinism*. MIT Press, Cambridge, Mass.

Spretnak C, Capra F 1986 *Green Politics: The Global Promise.* Paladin Books, London

Stewart I 1989 *Does God Play Dice?* Basil Blackwell, London

Stewart I 1990 *Does God Play Dice: The New Mathematics of Chaos.* Penguin, London

Sugden D, Hulton N 1994 Ice volumes and climatic change. In Roberts N (ed.) *The Changing Global Environment.* Blackwell, Oxford: 150–72

Summerfield M A 1991 *Global Geomorphology: An Introduction to the Study of Landforms.* Longman, Harlow

Summers C M 1970 Energy use, conversion, transportation and storage. In *Energy.* W H Freeman, San Francisco. Readings from *Scientific American*

Swingland I R 1993 Tropical forests and biodiversity conservation: a new ecological imperative. In Barbier E B (ed.) *Economics and Ecology: New Frontiers and Sustainable Development.* Chapman and Hall, London: 118–29

Thompson R D 1995 The impact of atmospheric aerosols on global climate: a review. *Progress in Physical Geography* **19**: 336–50

Tooley M J 1994 Sea-level response to climate. In Roberts N (ed.) *The Changing Global Environment.* Blackwell, Oxford: 172–89

Tooley M J, Jelgersma S (eds) 1992 *Impacts of Sea-Level Rise on European Coastal Lowlands.* Blackwell, Oxford

Tooley R, Tooley M J 1982 *The Gardens of Gertrude Jekyll in the North of England.* Michaelmass Books, Durham

Treumann R A 1991 Global problems, globalization and predictability. *World Futures* **31**: 47–53

Trzyna T (ed.) 1995 *A Sustainable World.* Earthscan, London

Turner B L *et al.* (eds) 1990 *The Earth as Transformed by Human Action.* Cambridge University Press

Turner R K, Pearce D, Bateman I 1994 *Environmental Economics. An Elementary Introduction.* Harvester Wheatsheaf, London

Tu We-ming 1989 The continuity of being: Chinese visions of nature. In Callicott J B, Ames R T (eds) *Nature in Asian Traditions of Thought.* State of University of New York Press, Albany: 67–78

UNEP 1993 Environmental pollution. *Environmental Data Report 1993–94,* Part I. Blackwell, Oxford: 6–106

Vanecek M 1994 *Mineral Deposits of the World.* Elsevier, Amsterdam and London

Wall D 1990 *Getting There. Steps Towards a Green Society.* Green Print, London

Ware G W 1983 *Pesticides* Freeman, San Francisco

Warren K J (ed.) 1994 *Ecological Feminism.* Routledge, London and New York

Wenz P S 1988 *Environmental Justice.* State University of New York Press, Albany

White L 1967 The historic roots of our ecologic crisis. *Science* **155**: 1203–07

Williams M 1990 In Turner B L (ed.) *The Earth as Transformed by Human Action*. Cambridge University Press

Williams M A J *et al.* 1993 *Quaternary Environments*. Edward Arnold, London

Wilson E O 1975 *Sociobiology: The New Synthesis*. Harvard University Press, Cambridge, Mass.

Winner L 1977 *Autonomous Technology*. MIT Press, Cambridge, Mass.

World Resources Institute 1995 *World Resources 1994–95*. OUP, Oxford

World Resources Institute 1994 *World Resources 1992–3*. OUP, Oxford

Worster D (ed.) 1988 *The Ends of the Earth. Perspectives on Modern Environmental History*. Cambridge University Press

Worster D 1993 *The Wealth of Nature*. OUP, New York and Oxford

Yearsley S 1991 *The Green Case. A Sociology of Environmental Issues, Arguments and Politics*. Harper Collins, London

Ziman J 1980 *Teaching and Learning about Science and Society*. Cambridge University Press

Ziman J 1994 *Prometheus Bound: Science in a Dynamic Steady State*. Cambridge University Press

Index